RING-CHAIN TAUTOMERISM

RING-CHAIN TAUTOMERISM

Raimonds E. Valters
Faculty of Chemistry
A. Pelshe Riga Polytechnical Institute
Riga, USSR

and

Wilhelm Flitsch
Organisch-Chemisches Institut
der Westfälischen Wilhelms-Universität
Münster, Federal Republic of Germany

Edited by
Alan R. Katritzky
University of Florida
Gainesville, Florida

PLENUM PRESS • NEW YORK AND LONDON

Library of Congress Cataloging in Publication Data

Valter, Raimond Eduardovich.
  Ring-chain tautomerism.

  Includes bibliographies and index.
  1. Tautomerism. 2. Ring formation (Chemistry) 3. Chemistry, Physical organic. I.
Flitsch, Wilhelm. II. Katritzky, Alan R. III. Title.
QD471.V26 1985   541.2′252   85-3447
ISBN-13: 978-1-4684-4885-6   e-ISBN-13: 978-1-4684-4883-2
DOI: 10.1007/978-1-4684-4883-2

© 1985 Plenum Press, New York
Softcover reprint of the hardcover 1st edition 1985
A Division of Plenum Publishing Corporation
233 Spring Street, New York, N.Y. 10013

Foreword

The understanding of known chemistry and the prediction of new reactions depends on the rationalization of reactivity in terms of molecular structure. Unless the correct molecular structure is appreciated and used, attempts at rationalization are bound to fail. Potentially tautomeric compounds offer many pitfalls in this respect, and the situation is not made easier by the fact that some otherwise completely sane and reasonable organic chemists still persist in the use of incorrect tautomeric structures!

There have been two major guides to tautomeric structure available: R. Valter's book on ring-chain tautomerism, and the book *The Tautomerism of Heterocycles* with which the writer was intimately concerned and which appeared as Supplement I to *Advances in Heterocyclic Chemistry*. These books complement each other perfectly, as the latter does not cover the field of ring-chain tautomerism. Unfortunately, the former has been available only in Russian; it was therefore a great pleasure to be able to catalyze the production of the completely revised and updated version of Dr. Valter's book which now lies before you and which has been written in full collaboration with Professor W. Flitsch.

I believe that the availability of this book in English will help considerably in the rationalization of the structure and reactivity of a large and important group of compounds. I have learned much from its detailed perusal and commend it to you. It has been a pleasure and an honor to act as an Editor.

A. R. Katritzky, November 1984

Preface

After studying ring-chain tautomerism of ketoamides and related derivatives of functionalized carboxylic acids for more than ten years, the authors consider it useful to summarize available results on these prototropic equilibria.

First attempts to systematize the material were published by Jones in 1963 (Chapter 1, ref. 11). Much, sometimes contradictory, experimental data were scattered about the literature at that time; spectroscopic methods, applied to this field during the last two decades, were needed to revise several previous concepts.

In the following years special aspects of ring-chain tautomerism have been discussed occasionally, but no attempt was previously made to cover the whole field.

This review is designed to provide a comprehensive compilation of ring-chain tautomerism with one exception: carbohydrates which have already been treated repeatedly, have been omitted.

The book is based on a monograph published in Russian: R. E. Valters, Ring-Chain Isomerism in Organic Chemistry. Zinatne. Riga, 1978. Therefore, the arrangement and development of the theme is due mainly to one of the authors (R.E.V.). In the present work the literature has been covered until the end of 1982.

Most of the processes reported here apply to prototropic tautomeric equilibria. In addition, reversible conversions of open structured compounds into cyclic isomers and *vice versa* have been considered. Contrary to simple prototropy such as keto-enol tautomerism, ring-chain prototropic equilibrations are more complicated processes: not only do migration of protons and changes in electron distributions along the bonds of molecules occur, but the formation or cleavage of C—X bonds (X = O, N, S) also takes place. Nonprototropic ring-chain equilibria will be considered only if they contribute to a discussion of related prototropic processes.

This book has been organized along logical lines. An introduction into the phenomenon of ring-chain tautomerism is given in the first chapter, together with a critical discussion of applications of physical, mainly spectroscopic, methods. Chapters 2-4 refer to particular tautomeric systems. These have been arranged according to the interacting functional groups to facilitate the location of individual cases. General rules concerning structural as well as external influences have been discussed in Chapter 5, which should help to appraise unknown ring-chain tautomeric equilibria and cyclization reactions.

Ring-chain tautomerism is of great significance both in the field of synthesis of heterocyclic compounds and for a discussion of reaction mechanisms, wherever ring closure, ring opening, recyclizations of heterocycles, and migration of acyl and other functional groups are involved. Therefore, this monograph is primarily addressed to chemists in academic and industrial laboratories. It should also be useful for biochemists and biologists since ring-chain tautomerism often plays a decisive role in biochemistry.

The authors would like to express their thanks to everybody who has rendered his help in writing the book. Professors O. Neilands, A. Strakovs, and J. Freimanis have read the whole of the manuscript published in Russian and made valuable comments. Critical comments of Professors E. Gudriniece, A. N. Kost, and Yu. N. Sheinker were taken into account. Great help was rendered by A. Bace for the design work of the book and by D. Murniece and L. Kundzina for the translation into English. The first English version was extensively polished by one of the authors (W.F.).

R. E. Valters (Riga)
W. Flitsch (Münster)

Contents

Introduction

1.1. RING-CHAIN ISOMERIC INTERCONVERSIONS. GENERAL CONSIDERATIONS

The problem of ring-chain isomeric interconversions of organic compounds had already aroused the interest of chemists at the end of the last century. However, the use of chemical methods done for the investigation of these interesting phenomena frequently gave erroneous results, even in those cases when the type of structure (ring or chain) was to be determined. Spectroscopic methods (electron and vibration spectroscopy, and nuclear magnetic resonance) are much more efficient for this purpose. Since these methods were developed only after the first active period of study of ring-chain tautomerism, a new and extensive reexamination of the phenomena was to be expected as Hammond had suggested in 1956 [1]. Indeed, this field of organic chemistry has had a real revival during the last three decades.

Ring-chain isomeric interconversions of organic compounds can be divided into two groups. The first group concerns interconversions where a linear system forms a ring system (or *vice versa*) by means of electron rearrangement without the migration of atoms or groups. Such interconversions are generally called [2, 3] valence isomerism. Woodward and Hoffmann [4] have proposed rules governing "electrocyclic reactions", i.e., the process of single bond formation (and its reverse) between the ends of a conjugated linear system containing n π-electrons:

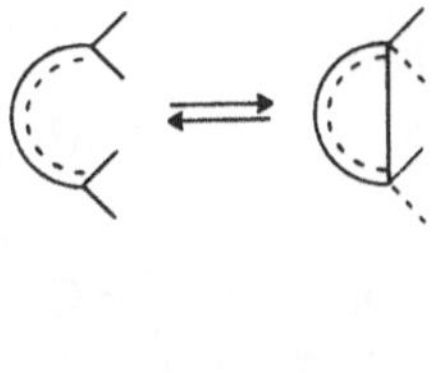

Ring systems which are formed in this way contain (n − 2) π-electrons. The equilibrium of benzofurazane oxide with o-dinitrosobenzene [2] and tautomerism of azide and tetrazole [5–8] are given as examples:

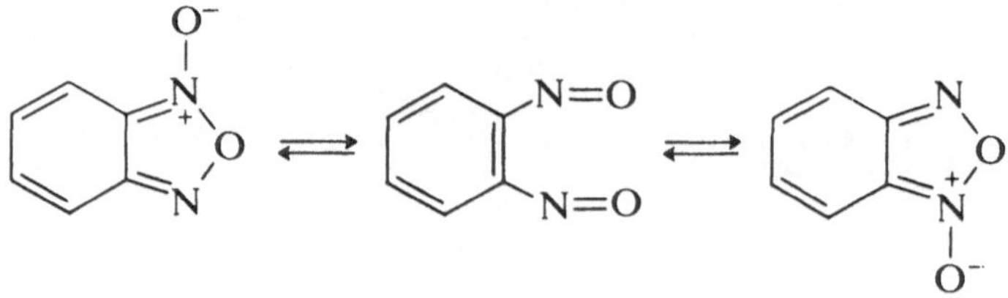

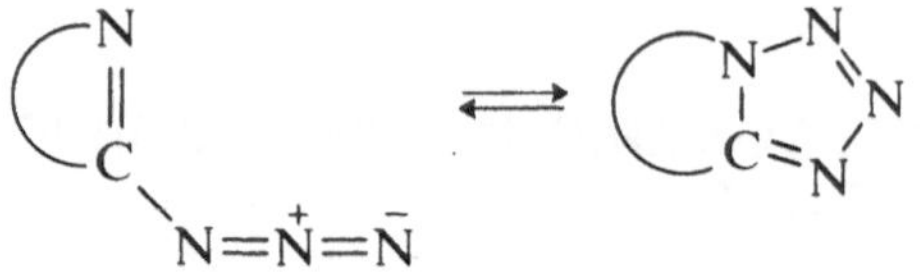

Many other examples of such valence isomerism are known [4, 9, 10].

In a second group of ring-chain isomeric interconversions, a linear system transforms into a cyclic one by intramolecular addition of the functional group to a polar multiple bond $(1\text{-}4A) \rightleftarrows (1\text{-}4B)$. Here and subsequently the letter A refers to open isomers and the letter B to cyclic ones. The molecular fragment connecting both mutually reacting functional groups is designed by a semicircle.

The process of the interconversion of a linear isomer into a ring one involves formally three operations: 1) Q–X bond cleavage, 2) Q–Y or Q–Z bond formation (ring closure), and 3) Z–X or Y–X bond formation.

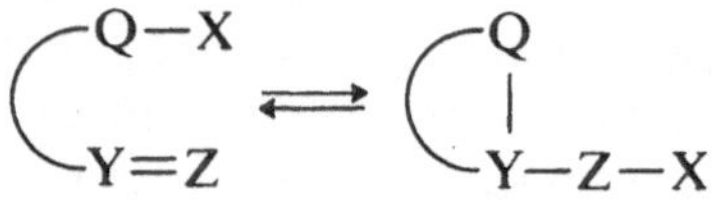

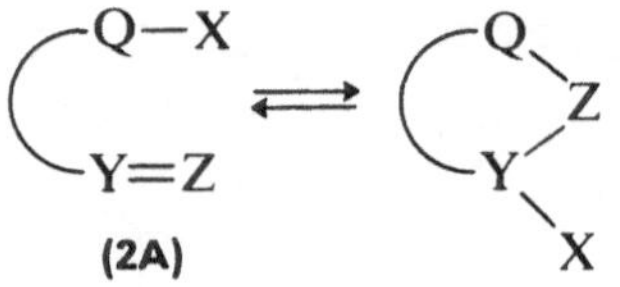

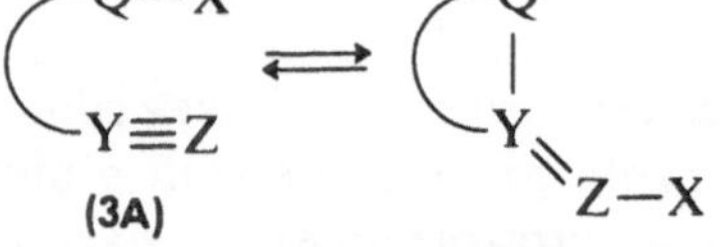

$$\begin{matrix} \text{Q—X} \\ \text{Y}\equiv\text{Z} \end{matrix} \rightleftharpoons \begin{matrix} \text{Q} \\ \text{Y} \end{matrix}$$

(4A)

(4B)

As in the case of electrocyclic reactions, a π-bond is transformed into a σ-bond. However, simultaneously with the ring closure (or opening), the migration of atom or group X occurs. Most commonly a proton is the migrating particle (prototropy). These isomeric interconversions may be precisely named "additive tautomerism" [10]. They are both widespread and diverse in organic chemistry.

There are numerous polar multiple bonds Y=Z or Y$\equiv$Z, such as C=O, C=N, N=C, C=C, C$\equiv$N, C$\equiv$C, a.o., and a great many functional groups Q-X, such as O-H, N-H, S-H, CO-Hal, which are capable of intramolecular addition to these bonds. The connecting fragment (designed by a semicircle) can also possess various structures: however, it must comply with one important condition, i.e., it must assure a favorable steric disposition of the interacting functional groups Q—X and Y—Z for the realization of reversible addition-cleavage reactions.

The present monograph deals with the study of the second type of ring-chain isomerism of organic compounds. The immense variety of these phenomena has made it impossible for the authors to present a thorough examination of all possible cases of ring-chain tautomerism. The main attention is paid to work published during the last three decades in which modern spectroscopic and other physical methods have been widely used. The results of earlier investigations will be discussed as necessary.

The first systematic study of the problem of ring-chain tautomerism of organic compounds was by Jones [11] in 1963. A significant part of the material covered in his review refers to the period when merely chemical methods were used.

Earlier reviews of the present authors [12–15] and other investigators [1, 16–19] were also useful in the writing of this monograph. Experimental studies carried out by the authors refer to the ring-chain isomerism of amides and other derivatives of keto-, imino-, and cyanocarboxylic acids [11–15, 20, 21].

Reporting the literature data published on ring-chain isomeric interconversions of organic compounds the main attention is paid to the following aspects:

1) Structural influences such as the polarity of Q—X and Y=Z bonds, electron and steric effects of the substituents at these bonds, the structure

of the linking fragment and other structural (internal) factors on the relative stability of isomers and the character of their interconversions (mobility of the system and equilibrium position);

2) The influence of external factors (temperature, physical state of the substance, properties of solvents, presence of catalysts) on the characteristics of equilibrium systems;

3) A search for selective preparations of open and ring isomers;

4) The estimation of the potential of spectroscopic and other physical methods for the identification of isomers and the determination of equilibrium and rate constants.

These regularities of ring-chain tautomerism are of great significance in the chemistry of heterocyclic compounds.

In many cases the estimation of the influence of structural factors on intramolecular cyclization reactions allows prediction of the ease of formation of a heterocycle and its stability [1, 14, 22, 23]. The reaction of intramolecular additions are often followed by elimination of water, hydrochloric acid, etc., thus resulting in the stable heterocycles.

The structural factors favoring or preventing the cyclization act similarly in many chemical reactions, including those which proceed through cyclic transition states or intermediates. They relate to chemical interconversions which proceed according to the following schemes:

$$chain \rightarrow [ring] \rightarrow chain$$

or

$$ring \rightarrow [chain] \rightarrow ring$$

Familiar examples of the first scheme comprise a) intramolecular catalysis with the neighboring group participation [24–28], b) intramolecular migration of acyl and other groups [24, 29–31] proceeded by the formation of cyclic tetrahedral intermediates [32], and c) hydroxyacyl and aminoacyl incorporation into peptides and depsipeptides [33, 34].

The second scheme includes the majority of ring transformation reactions of heterocycles [35–37].

Study of these reactions necessitates estimation of the influence of the molecular structure on the ability of the two atoms or functional groups to mutually approach thus enabling their intramolecular reaction. Information of that kind can be obtained from the ring-chain equilibrium constants in systems in which the structure of the linking fragments are sequentially modified.

1.2. TAUTOMERISM OR ISOMERISM? SYSTEM MOBILITY AND EQUILIBRIUM POSITION

Ring-chain isomeric interconversions are often referred to as ring-chain tautomerisms [1, 11, 38–40]. Strictly speaking this term should only be used for certain cases of the phenomena because ring-chain equilibria can be of very different mobility. Thus, solutions of 2-acylbenzoic acids in neutral solvents at room temperature display a tautomeric equilibrium between open and cyclic forms **(5A)** ⇌ **(5B)** [41]. Esters of these acids are usually stable under these conditions, and acid or alkaline catalysis is necessary [42] to achieve the equilibrium **(6A)** ⇌ **(6B)**.

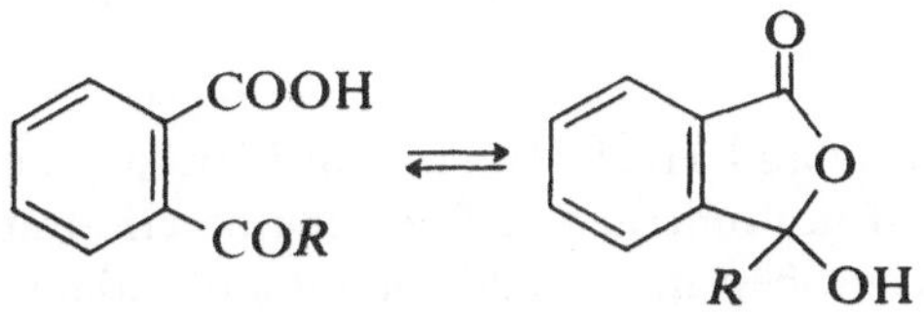

(5A)   **(5B)**

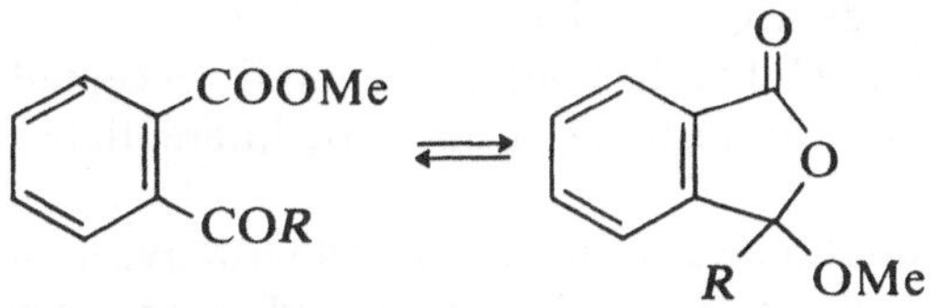

(6A)   **(6B)**

There are cases where the equilibrium, reached only under drastic conditions, is totally displaced toward one of the isomers. For example, *N*-monosubstituted 2-cyanobenzamides, on heating or with base catalysis, undergo isomerization into 3-iminoisoindolinones **(7A)** → **(7B)** [43].

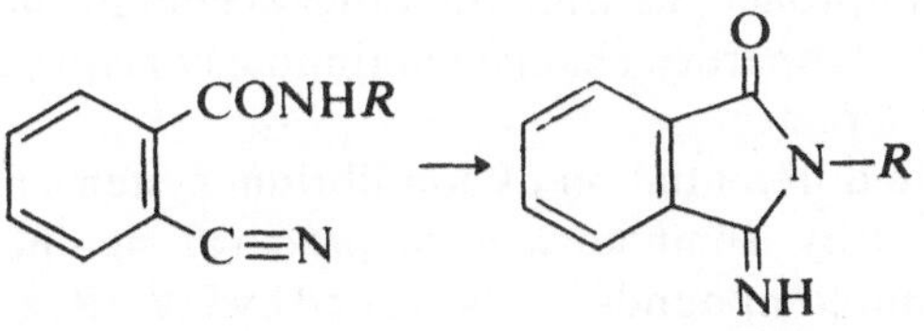

(7A)   **(7B)**

Small modifications of the structure of substituents at interacting functional groups can change the position of the equilibrium completely. For instance, 2-(2,4-dimethylbenzoyl)benzoyl chloride is obtained only as the cyclic 3-aryl-3-chlorophthalide **(8)**, whereas 2-(2,4,6-trimethylbenzoyl)ben-zoylchloride exists as such **(9)** [44].

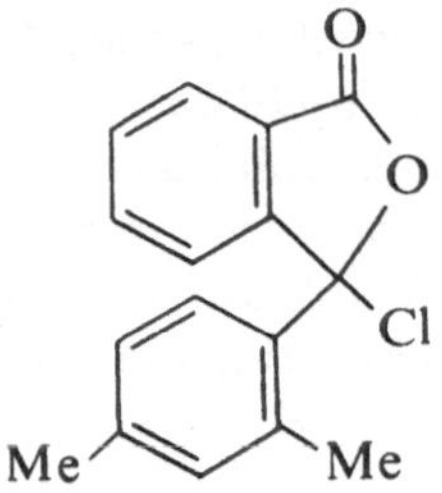

(8)

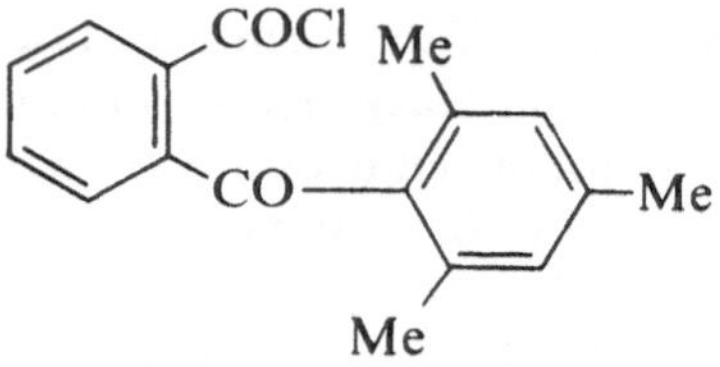

(9)

Recently Zefirov and Trach [10], in considering general problems and the classification of tautomerism, defined tautomeric transformation as: a reversible dynamic process in a formally monomolecular system, proceeding by the cleavage or formation of bonds between atoms (taking into account individual atoms of the same type) on condition that both tautomeric forms can be separately identified analytically.

The definition, while reflecting the generally accepted meaning of the term "tautomerism"†, unfortunately does not differentiate clearly tautomerism from isomerism.

Such differentiation was achieved in an alternative though less accepted approach by Temnikova [17]. Following the ideas of Favorskii and Knorr, Temnikova divides equilibration phenomena into tautomerism and isomerism. Equilibrium interconversions are referred to as tautomerism if the mobility of the system is high, i.e., when equilibration is observed in solutions at room temperature. When the mobility of the system is very small and isomeric transformation occurs only at elevated temperature or in the presence of active catalysts, then such interconversions are examples of isomerism. This approach, as with any other scientific classification, is not unequivocal since properties change continuously giving any classification an intermittent character.

However, the differentiation of equilibrium systems according to their mobility proposed by Temnikova is of practical significance. Today the structure of organic compounds is established by UV, IR, and NMR spectroscopic methods. It is indispensible for an interpretation of these spectra to draw a strict borderline between the two cases of equilibrating systems: 1) the equilibrium is reached at room temperature in a neutral solvent; 2)

---

† A. Ihde, *J. Chem. Educ.* **1959**, *36*, 330.

adjustment of the equilibrium affords more drastic conditions such as heating and/or acidic or basic catalysis.

Minkin and coworkers [31, 45], using thermodynamic and kinetic criteria, postulated a more rigorous definition of the term "tautomerism" in the following way.

Let us consider the tautomeric system

$$A \underset{k_2}{\overset{k_1}{\underset{K}{\rightleftarrows}}} B$$

and its energy profile (Fig. 1).

The quantitative characteristic of the equilibrium (K or $\Delta G°$) and of the rate (k or $\Delta G^{\neq}$) must be considered in relation to experimental method by which the equilibrium information was obtained. Hence, standard conditions for the investigation of equilibrium systems and in particular the sensitivity of the experimental methods must be considered. The choice of the conditions is a matter of chemical experience: atmospheric pressure and room temperature are taken as normal. The sensitivity of the most precise methods of determination of small equilibrium constants establishes constants with a lower limit 10^{-6}-10^{-8} which corresponds to a free energy difference $\Delta G° = 8$-9 kcal/mole.

As already mentioned above, it is rather difficult to draw a line between a dynamic, rapidly attainable tautomeric equilibrium and a slow isomerization. The barrier which, according to the general opinion of chemists, separates a mobile system from a stable one, is the possibility of preparative

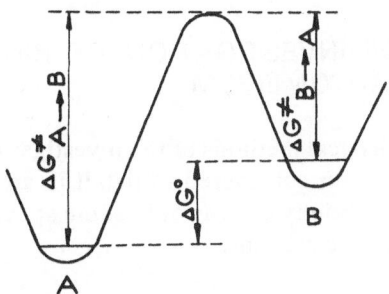

Figure 1. Energy profile of equilibrium $A \rightleftarrows B$. The value of the free energy of the reaction $\Delta G°$ determines the equilibrium constant K, the values of the free energy of activation $\Delta G^{\neq}$ determine the rates of the reactions $A \rightarrow B$ and $B \rightarrow A$ according to Eyring's equation.

isolation under normal conditions. In terms of the rate constant and activation energy this corresponds approximately to $k_{25} \sim 10^{-5}\,\mathrm{s}^{-1}$ and $\Delta G_{25}^{\neq} = 20\text{-}25\,\mathrm{kcal/mole}$.

Thus the approximate thermodynamic and kinetic line to be drawn between tautomeric and isomeric systems is (under normal conditions): $\Delta G_{25}^{\circ} < 8\,\mathrm{kcal/mole}$ and $\Delta G_{25}^{\neq} < 25\,\mathrm{kcal/mole}$.

In discussing ring-chain tautomeric equilibria we shall try to give an accurate characterization of the conditions under which the equilibrium is observed to avoid difficulties caused by the uncertainty of the term "tautomerism." If the equilibrium position within the possible limits of the spectroscopic determination is totally displaced toward one of the isomers, e.g., (B), we shall deal with the isomerization $(A) \rightarrow (B)$.

It is not possible to consider all the investigations in which the isomers have not been isolated, or in which the equilibrium has not been strictly proven but postulated for the explanation of a mechanism of the formation of reaction products. The intramolecular rearrangement $(10) \rightarrow (12)$ is an example which proceeds by the intramolecular migration of the acyl group between two nucleophilic groups (OH, NH, SH) via the intermediate ring isomers (11). Due to their limited stability these compounds have only rarely been isolated or proven to be present in solutions [29, 32, 46-48].

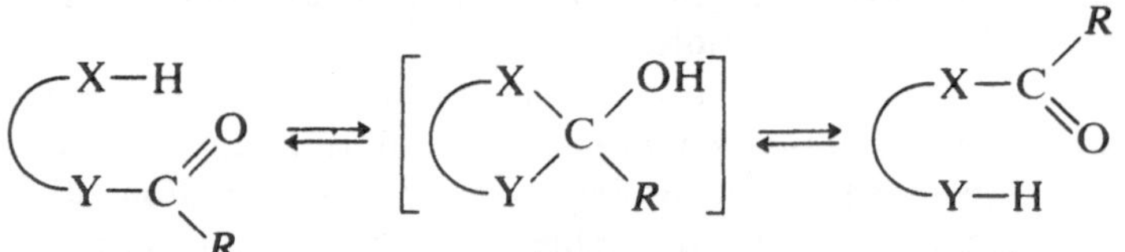

X, Y = O, NR, S

1.3. METHODS OF INVESTIGATION OF RING-CHAIN ADDITION TAUTOMERISM

Chemical and physical methods of the investigation of the tautomerism of heterocycles have been considered in detail in an excellent monograph [3]. We shall therefore briefly discuss only some specific aspects typical for the ring-chain tautomeric systems.

1.3.1. Chemical Methods

As a rule the formation of derivatives is not admissible evidence for the determination of the structure of ring-chain tautomeric compounds

[1, 3, 11, 49]. A solely chemical approach for this purpose is unreliable and frequently leads to erroneous results.

Since such molecules of either open or ring structure contain at least two reactive centers, ring isomers can form derivatives not only from cyclic but also from the acyclic structure, depending on the nature of reagents. On the other hand, cyclic derivatives may be obtained from open isomers. This is demonstrated using the reactions of amines with 3-chloro-3-phenylphthalide (13), as an example. The nucleophile according to its nucleophilicity may attack $C_{(1)}$ or $C_{(3)}$ of the phthalide yielding derivatives of either open and/or ring structure. Reaction of (13) with t-butylamine gave the open-chain derivative N-(t-butyl)-2-benzoylbenzamide (14), [50] while the less nucleophilic diphenylamine yielded the cyclic 3-(N,N-diphenylamino)-3-phenylphthalide (15), [51].

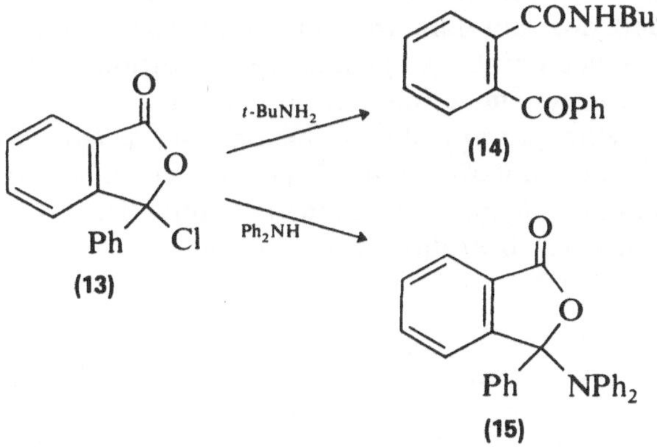

When aniline is used as a nucleophile, the attack proceeds at both reactive centers ($C_{(1)}$ and $C_{(3)}$ of phthalide (13) leading to a mixture of the products of type (14) and (15) [51]. Sometimes, it is essential to take into account possible isomerization of the starting material or products depending on the reaction conditions (acid or alkaline catalysis, heating). The proportion of the ester isomers which result from Fischer–Speier esterification of ketocarboxylic acids depends not only on the structure of the initial acid and alcohol but also on the time of the reaction (kinetic or thermodynamic control) [42, 52]. It is easily seen from this example that the chemical method of synthesis of the compounds capable of the isomeric interconversions cannot serve as a proof for the structure of the reaction product.

A spectroscopic determination of ring-chain (on any other) equilibrium has the advantage in not affecting the equilibrium position or causing isomerization during the course of the investigation.

1.3.2. Visible and UV Spectroscopy

During the interconversion of a cyclic compound into the open isomer (the isomerization **(B)** →**(A)**) a new multiple bond (i.e., a chromophore) is formed which causes a corresponding change in the electronic spectrum. If this chromophore (C=O, C=N) in the open isomer is a conjugated with a π-electron system, then by interconversion **(B)** → **(A)** and an intensive band appears in the experimentally accessible region of the UV spectrum. However the ultraviolet or visible spectra of equilibrium mixtures cannot be interpreted in such detail as the corresponding IR spectra which allows unambiguous determination of the presence of a particular multiple bond.

Usually the spectra of model compounds with fixed open or ring structure are used for a reliable interpretation of the electronic spectra of the compounds capable of ring-chain tautomerism. Provided the absorption of the model compound, in which the migrating hydrogen is substituted by a methyl group, does not differ from that of the corresponding H-analogue, it is possible to determine the quantitative proportion of the forms **(A)** and **(B)** in tautomeric systems or mixtures of isomers.

Thus, 2-methoxycarbonylbenzophenone anils **(17)** were used as model compounds in a quantitative study [53] of ring-chain tautomeric equilibria of 3-arylamino-3-phenylphthalides **(16A)** ⇌ **(16B)**. The UV spectra of these anils have longwave absorption at 322–348 nm.

(16A)   **(16B)**   **(17)**

The second model compound, i.e., 3-(N,N-diphenylamino)-3-phenylphthalide **(15)** shows no absorption in the above-mentioned region. The proportion of the open form **(16A)** of the equilibrium mixture was determined supposing the intensities of the longwave absorption bands in the spectra of the isomer **(16A)** and of the model compound **(17)** to be equal.

The UV spectroscopic method was used [54] to establish rate constants for both the base and acid-catalyzed rearrangements of a series of ring isomers of 8-benzoyl-1-naphthoates and 2-benzoylbenzoates to form the chain isomers.

The simplest way is to use visible spectra for the identification of isomers. Thus, all the open isomers or benzil-o-carboxylic acid derivatives are yellow while ring isomers are colorless [55, 56].

1.3.3. IR Spectroscopy

IR spectroscopy provides valuable information concerning the existence or absence of double or triple bonds in organic molecules. The information obtained by an examination of the region of multiple bond valence vibrations is the most important [57]. Additionally, absorption bands of OH, NH, and SH bonds as well the complex bands (e.g., amide-II) may be used to identify open and ring isomers.

Identification of the ring isomers of the derivatives of γ-ketocarboxylic acids utilizes the fact that the ν_{CO} band is shifted by 30–50 cm^{-1} to higher frequencies on cyclization, which is due to C=O group bond angle strain in the five-membered ring [58]. This band usually appears in a region in which the open chain isomers do not absorb. The band is therefore used for quantitative measurements.

The identification of open and ring isomers by the IR method is sometimes hampered by the formation of strong intermolecular (more rarely intramolecular) hydrogen bonds. For example, in IR spectra of the crystalline ring isomers of ketocarboxylic acid amides (hydroxylactams), the lactam C=O band is greatly shifted to lower frequencies in comparison with solution spectra and sometimes even split into doublets (see Table 7). Such a shifting and splitting is caused by hydrogen bonds OH···O=C in the solid state. In these cases the spectra in the double bond region should be recorded for solutions in proton acceptor solvents (e.g., in dioxan). Because of the competitive formation of the intermolecular hydrogen bonds OH···solvent, it is possible to observe the absorption of free C=O groups in such solutions.

The interpretation of IR spectra can also be hampered by the overlap of bands, as in the spectra of some ketocarboxylic acids where the C=O bands of carboxyl and ketone groups obscure each other, and in the spectra of some amides of ketocarboxylic acids where the band of ketogroup is overlapped by the amide-I band. In such cases, it is wise to measure the integral intensity of the total band and to compare it with the sum of intensities of both groups obtained from the spectra of the corresponding model compounds [50, 59, 60].

A more detailed discussion on the problem of the application of IR spectroscopy is given in the chapter concerning acylcarboxylic acids and their amides (see Sections 2.1.1 and 2.1.3).

1.3.4. Nuclear Magnetic Resonance

In recent years ^{1}H-NMR spectroscopy has been used frequently for the structure determination of open and cyclic isomers and for the quantitative determination of equilibrium mixtures. Commonly this method deals with the differences in the chemical shifts of protons located at the carbon atoms in α-positions to a polar multiple bond or directly attached to the carbon atoms forming a double bond (aldehydes, aldimines).

Using the ^{1}H-NMR method for the analysis of ring-chain equilibrium systems requires consideration of the rate of tautomeric interconversions: two cases may be distinguished [61, 62]:

1) If the tautomeric interconversion is slow on the ^{1}H-NMR time scale the signals of protons of both tautomeric forms are observed separately and the tautomeric composition is determined according to their relative intensities;

2) In the case of a rapid tautomeric interconversion the signals of protons of both forms coalesce and only averaged chemical shifts are observed. For quantitative measurements, the ratio of the forms (A) and (B) is assumed to be inversely proportional to the distances of this signal from the corresponding signals in the spectra of model compounds of fixed structure [41, 63].

The replacement of a less polar solvent by a more polar one or the addition of trace amounts of acids or bases can accelerate tautomeric interconversions and substantially change the appearance of the ^{1}H-NMR spectrum [41, 64, 65].

Thus, for example, in the ^{1}H-NMR spectrum of 2-acetylbenzoic acid in DMSO-d$_6$ two signals for methyl groups of open (MeCO) and cyclic (MeC(OH)O) tautomeric forms were observed. After the addition of a trace of hydrochloric acid these signals coalesced into an averaged one [66].

In studies of the kinetics of ring-chain tautomeric interconversions rate constants were determined by full line-shape analysis of ^{1}H-NMR spectrum [67, 68] employing special programs for computers [69]. Only in the case of slow tautomerizations ($k \leqslant 10^{-4}$ s^{-1}) is it possible to determine the rate constants following the intensity changes of the signals of both forms during the approach to the equilibrium [70].

^{13}C-NMR spectra were used for the identification of open and cyclic isomers of 2-acylbenzoic acids [71], phthaloyldichloride and 2-acetoxybenzoylchloride [72].

Activation parameters ($\Delta H^{\neq}, \Delta S^{\neq}$) for the tautomerization of 3-oxo-*trans*-bicyclo[4.4.0]decane-1-carboxylic acid (18A) → (18B) were estimated by full line shape analysis of the ^{13}C-NMR signals of the carboxyl and lactone carbonyls at different temperatures. However, these results have low reproducibility due to the presence of variable trace amounts of mineral acid [73].

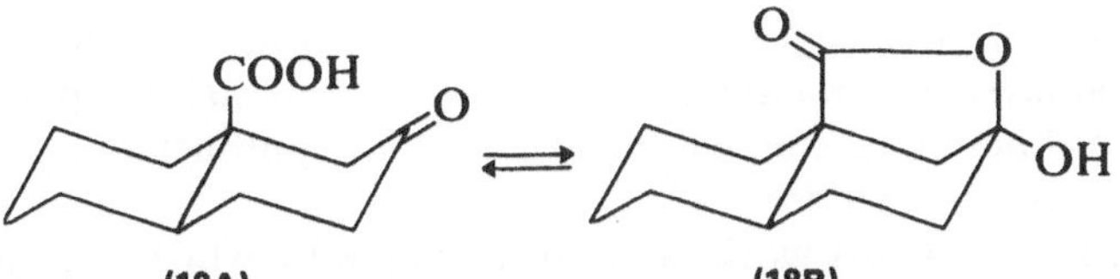

(18A)   **(18B)**

In the estimation of ring-chain equilibrium constants of 2-hydroxy-phenyliminosphoranes **(19A) ⇌ (19B)** by two independent methods, i.e., ^{1}H- and ^{31}P-NMR spectroscopy, precisely coincident results were obtained [74].

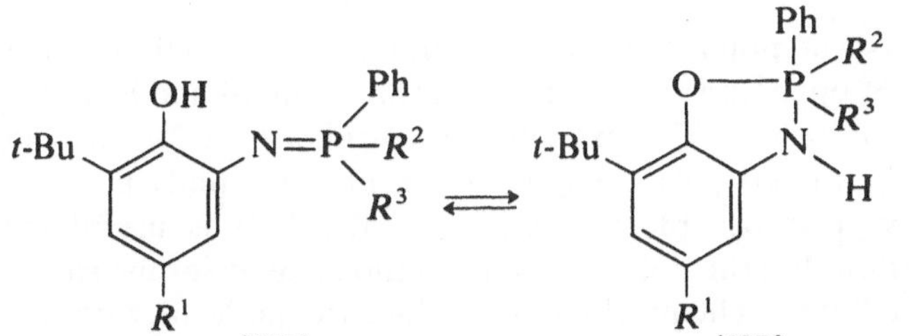

(19A)   **(19B)**

1.3.5. Mass Spectrometry

In a mass spectrometer the substance is in the vapor phase at high vacuum and collisions between molecules practically do not occur. These conditions essentially differ from those in the condensed phase where UV, IR, and NMR spectra are investigated. Therefore the results obtained by a mass spectrometric investigation of ring-chain equilibria [75] can supplement the information about this tautomerism acquired by other methods.

A few studies [76–79] of mass spectra of ring-chain equilibrium systems show that the two isomers possess quite different fragmentation schemes.

Investigation of ring-chain tautomeric compounds in the gase phase should allow for a separation of structural influences from solvation effects. Differences of solvation of both isomers cannot be ruled out but the few investigations which have been carried out so far do not point in this direction (see Chapter 2, Ref. 297).

1.3.6. X-Ray Diffraction Analysis

X-Ray diffraction analysis may be used to establish the structure of ring-chain isomers in the crystalline state. More important results, were obtained by Chadwick and Dunitz [73] who studied the equilibrium of keto acids such as **(18A) ⇌ (18B)**. Their work was based on ideas of Bürgi and

Dunitz [80–82] who suggested that structural information, mainly obtained by x-ray analysis of intentionally selected bifunctional compounds, could be used for finding out the optimal geometry of approach of the interacting groups during the reaction.

As a result of this work the optimal angle by which the nucleophilic nitrogen approaches the carbonyl group was found to be approximately 107° and not 90° as was assumed previously. For oxygen, as a nucleophile, the corresponding angle is in the limits of 100–110° [83].

1.3.7. Polarography

The use of polarography as an independent method for a structure detection of open and ring isomers is rather limited and has sometimes led to erroneous results. For example, the structure of 2-benzoylbenzanilide was misinterpreted [84] using polarographic methods and later defined exactly by spectroscopic methods [20, 51, 85]. It is useful to take into consideration that the polarographic method possesses the same shortcomings as the "pure" chemical methods since the molecular structure may be modified at the electrode layer before reduction. Therefore polarographic data on tautomerism require a careful explanation and the results should be compared with those of spectroscopic investigations [86]. Combining polarographic with spectroscopic methods an important problem can be solved, i.e., the determination of rate constants of isomeric interconversions [87].

1.3.8. Other Physical Methods

Optical activity has a direct relationship to ring-chain tautomerism since in most instances a new asymmetric carbon atom is created during the formation of a ring isomer. However, with the exception of studies of the mutarotation of carbohydrates, optical activity measurements have rarely been used for the investigation of ring-chain tautomerism [11].

Open and ring isomers differ in molar refractivity. This can be used for quantitative investigations of ring-chain equilibria of pure liquids [88–90]. However, the determination of the proportion of isomers in equilibrium mixtures is hampered by the necessity to estimate accurately the molar refractivity of both isomers which sometimes are difficult to isolate as pure compounds. The results of molar refractivity measurements have been compared with those obtained by NMR [88, 91, 92].

The use of chromatographic methods for an analysis, or for separation of mixtures of open and ring isomers, not capable of equilibrium interconversions in a neutral solvent, does not always yield reliable results. Initially, it should be shown that the adsorbent used (Al_2O_3, SiO_2) does not act as a catalyst for the isomerization process. A case is known [75] where the ring isomer of levulinic acid anilide was converted into the open isomer by passing a methanolic solution through a column filled with silica gel.

Evidently for the same reason, gas-liquid chromatography is rarely used for the analysis of mixtures of otherwise stable isomers. Nevertheless, quantitative results concerning methyl 2-acylbenzoates, obtained by this method, agree with the data obtained by ^{1}H-NMR spectroscopy [42].

Potentiometric titration and other methods of establishing protolytic constants have been successfully used for a determination of the ring-chain equilibrium constants of acylcarboxylic acids [93–97].

Studies of compounds capable of ring-chain isomeric interconversions at or below the melting point should be regarded with caution. For example, the real melting point of the ring isomers of some ketoamides cannot be established [98, 99]. If the temperature of thermal isomerization of the cyclic isomer into the open one **(B)** →**(A)** is lower than the melting point of the isomer **(B)** and if the isomerization process proceeds sufficiently rapidly, the melting point of the open isomer **(A)** instead of the isomer **(B)** is observed. In other words, the ring and the open isomers of these ketoamides show identical melting points and they cannot be investigated by a mixed melting points. If the temperature of the isomerization **(B)** →**(A)** is above the melting point of the ring isomer **(B)**, double melting points can sometimes be observed: the compound melts, the rising temperature results in crystallization and subsequently again melting [98, 99].

REFERENCES

1. G. S. Hammond, In: *Steric Effects in Organic Chemistry*, Ed. M. S. Newman, John Wiley and Sons, Inc., New York, 1956, Chapter 9.
2. T. V. Domareva-Mandel'shtam and I. A. D'yakonov, *Uspekhi Khimii* **1966**, *35*, 1324.
3. J. Elguero, C. Marzin, A. R. Katritzky, and P. Linda, *The Tautomerism of Heterocycles*, Ed. A. R. Katritzky and A. J. Boulton, Academic Press, New York, 1976.
4. R. B. Woodward and R. Hoffmann, *The Conservation of Orbital Symmetry*, Weinheim, Verlag Chemie, 1970, Chapter 5.
5. M. Tisler, *Synthesis* **1973**, 123.
6. V. Ya. Pochinok, L. F. Avramenko, T. F. Grigorenko, and V. N. Skopenko, *Uspekhi Khimii* **1975**, *44*, 1028.
7. R. N. Butler, *Chem. Ind.* **1973**, 371; R. N. Butler, *Adv. Heterocycl. Chem.* **1977**, *21*, 323.
8. L. A. Burke, J. Elguero, G. Leroy, and M. Sana, *J. Am. Chem. Soc.* **1976**, *98*, 1685.
9. T. L. Gilchrist and R. C. Storr, *Organic Reaction and Orbital Symmetry*, Cambridge, Cambridge University Press, 1972, Chapter 3; G. Maier, *Valenzisomerisierungen*, Weinheim, Verlag Chemie, 1972.

10. N. S. Zefirov and S. S. Trach, *Zh. Org. Khim.* **1976**, *12*, 697; N. S. Zefirov and S. Trach, *Chem. Scripta* **1980**, *15*, 1, 4.
11. P. R. Jones, *Chem. Rev.* **1963**, *63*, 461.
12. R. E. Valters, *Uspekhi Khimii* **1973**, *42*, 1060.
13. R. E. Valters, *Uspekhi Khimii* **1974**, *43*, 1417.
14. R. E. Valters, *Uspekhi Khimii* **1982**, *51*, 1374.
15. W. Flitsch, R. Heidhues, H. Peters, E. Gerstmann, V. v. Weissenborn, H.-D. Bartfeld, B. Müter, and K. Gurke, *Forschungsber. des Landes Nordrhein-Westfalen*, N 2220, Westdeutscher Verlag, Opladen, 1972.
16. V. M. Andreev, G. P. Kugatova-Shemyakina, and S. A. Kazaryan, *Uspekhi Khimii* **1968**, *37*, 559.
17. T. I. Temnikova, *The Course of Theoretical Principles of Organic Chemistry*, Khimiya, Leningrad, 1968 (In Russian).
18. V. Balasubramaniyan, *Chem. Rev.* **1966**, *66*, 567.
19. R. Escale and J. Verducci, *Bull. Soc. Chim. Fr.* **1974**, 1203.
20. W. Flitsch, *Chem. Ber.* **1970**, *103*, 3205.
21. R. E. Valters, *Ring-Chain Isomerism of Keto, Imino and Cyano Carboxamides*, Synopsis of Thesis of Doctoral Dissertation, Riga, 1975 (In Russian).
22. E. L. Eliel, *Stereochemistry of Carbon Compounds*, McGraw-Hill Book Company, Inc., New York, 1962, Chapter 7.
23. G. Illuminati and L. Mandolini, *Acc. Chem. Res.* **1981**, *14*, 95.
24. T. C. Bruice and S. J. Benkovic, *Biorganic Mechanisms*, Vol. 1, W. A. Benjamin, Inc., New York, 1966.
25. W. P. Jencks, *Catalysis in Chemistry and Enzymology*, McGraw-Hill Book Company, New York, 1969.
26. M. I. Page, *Chem. Soc. Rev.* **1973**, *2*, 295.
27. B. Capon and S. P. McManus, *Neighbouring Group Participation*, Vol. 1, Plenum Press, New York and London, 1976.
28. A. J. Kirby, *Adv. Phys. Org. Chem.* 1980, *17*, 183.
29. L. V. Pavlova and F. Yu. Rachinskii, *Uspekhi Khimii* **1968**, *37*, 1369.
30. R. M. Acheson, *Acc. Chem. Res.* **1971**, *4*, 177.
31. V. I. Minkin, L. P. Olekhnovich, and Yu. A. Zhdanov, *Molecular Design of Tautomeric Systems*, Rostov on Don University Publishing House, Rostov on Don, 1977 (In Russian); V. I. Minkin, L. P. Olekhnovich, and Yu. A. Zhdanov, *Acc. Chem. Res.* **1981**, *14*, 210.
32. B. Capon, A. K. Ghosh, and D. Mc L. A. Grieve, *Acc. Chem. Res.* **1981**, *14*, 306.
33. M. M. Shemyakin, V. K. Antonov, A. M. Shkrob, V. I. Shchelokov, and Z. E. Agadzhanyan, *Tetrahedron* **1965**, *21*, 3537.
34. G. Lucente, F. Pinnen, G. Zanotti, S. Cerrini, W. Fedeli, and F. Mazza, *J. Chem. Soc., Perkin Trans. I* **1980**, 1499 (see references cited therein).
35. O. P. Shvaika and V. N. Artemov, *Uspekhi Khimii* **1972**, *41*, 1788.
36. A. N. Kost, S. P. Gromov, and R. S. Sagitullin, *Tetrahedron* **1981**, *37*, 3423.
37. H. C. van der Plas, *Ring Transformations of Heterocycles*, Vol., 1 and 2, Academic Press, London, 1973.
38. C. K. Ingold, *Structure and Mechanism in Organic Chemistry*, 2nd edition, Cornell University Press, Ithaca, 1969, Chapter 11.
39. K. F. Reid, *Properties and Reactions of Bonds in Organic Molecules*, Longmans, Green and Co. Ltd., London, 1968, Chapter 21.
40. J. Mathieu and R. Panico, *Mecanismes reactionnels en chimie organique*, Hermann, Paris, 1972, Chapter 5.
41. K. Bowden and G. R. Taylor, *J. Chem. Soc., Sect. B* **1971**, 1390.

42. K. Bowden and G. R. Taylor, *J. Chem. Soc., Sect. B* **1971**, 1395.
43. R. E. Valters, A. E. Bace, and S. P. Valtere, *Latv. PSR Zinat. Akad. Vestis, Kim Ser.* **1972**, 726.
44. R. E. Valters and A. E. Bace, *Latv. PSR Zinat. Akad. Vestis, Kim. Ser.* **1971**, 335.
45. V. I. Minkin, L. P. Olekhnovich, and Yu. A. Zhdanov, *Zhurn. Vses. Khim. Ova.* **1977**, *22*, 274.
46. B. Helferich and H. Liesen, *Chem. Ber.* **1950**, *83*, 567.
47. J. Hine, D. Ricard, and R. Perz, *J. Org. Chem.* **1973**, *38*, 110.
48. G. A. Rogers and T. C. Bruice, *J. Am. Chem. Soc.* **1973**, *95*, 4452; *Ibid.* **1974**, *96*, 2481.
49. A. N. Nesmeyanov and M. I. Kabachnik, *Zh. Obshch. Khim.* **1955**, *25*, 41.
50. R. E. Valters and S. P. Valtere, *Latv. PSR Zinat. Akad. Vestis, Kim. Ser.* **1969**, 704.
51. R. E. Valters and V. P. Ciekure, *Khim. Geterotsikl. Soedin.* **1972**, 502.
52. M. S. Newman and C. Courduvelis, *J. Am. Chem. Soc.* **1964**, *86*, 1893.
53. R. E. Valters and V. P. Ciekure, *Khim. Geterotsikl. Soedin.* **1975**, 1476.
54. K. Bowden and F. A. El-Kaissi, *J. Chem. Soc., Perkin Trans 2* **1977**, 1927.
55. R. E. Valters and S. P. Valtere, *Latv. PSR Zinat. Akad. Vestis, Kim. Ser.* **1971**, 213.
56. A. Hantzsch and A. Schwiete, *Ber. Dtsch. Chem. Ges.* **1916**, *49*, 213.
57. A. R. Cole, In: *Elucidation of Structures by Physical and Chemical Methods*, Part 1, Ed. K. W. Bentley, Intersci Publishers, New York, 1963, Chapter 3.
58. L. J. Bellami, *Advances in Infrared Group Frequencies*, Methuen and Co, Ltd., Bungay, 1968, Chapter 5.
59. R. E. Valters, A. E. Bace, and S. P. Valtere, *Latv. PSR Zinat. Akad. Vestis, Kim. Ser.* **1972**, 61.
60. R. E. Valters, S. P. Valtere, and A. E. Kipina, *Zh. Org. Khim.* **1968**, *4*, 445.
61. L. M. Jackmann and S. Sternhell, *Application of Nuclear Magnetic Resonance Spectroscopy in Organic Chemistry*, 2nd edition, Pergamon Press, Oxford, 1969, Chapter 5-3.
62. A. I. Kol'tsov and G. M. Kheifets, *Uspekhi Khimii* **1971**, *40*, 1646.
63. P. R. Jones and P. I. Desio, *J. Org. Chem.* **1965**, *30*, 4293.
64. B. Paul and W. Korytnyk, *Chem. Ind.* **1967**, 230.
65. B. Paul and W. Korytnyk, *J. Heterocycl. Chem.* **1976**, *13*, 701.
66. J. Finkelstein, T. Williams, V. Toome, and S. Traiman, *J. Org. Chem.* **1967**, *32*, 3229.
67. L. Yu. Yuzefovich, B. M. Sheiman, T. M. Filippova, and V. G. Mayranovskii, *Khim. Geterotsikl. Soedin.* **1978**, 758.
68. H. B. Stegmann, R. Haller, A. Burmester, and K. Scheffler, *Chem. Ber.* **1981**, *114*, 14.
69. N. I. Borisenko, L. P. Olekhnovich, and V. I. Minkin, *Teor. Eksp. Khim.* **1976**, *12*, 825.
70. R. B. Kampare, R. E. Valters, E. E. Liepins, and G. A. Karlivans, *Latv. PSR Zinat. Akad. Vestis, Kim. Ser.* **1981**, 244.
71. J. H. P. Tyman and A. A. Najam, *Spectrochim. Acta* **1977**, *33A*, 479.
72. W. J. Elliot and J. Fried, *J. Org. Chem.* **1978**, *43*, 2708.
73. D. J. Chadwick and J. D. Dunitz, *J. Chem. Soc., Perkin Trans. 2*, **1979**, 276.
74. H. B. Stegmann, R. Haller, and K. Scheffler, *Chem. Ber.* **1977**, *110*, 3817.
75. O. S. Anisimova, Yu. N. Sheinker, E. M. Peresleni, P. M. Kochergin, and A. N. Krasovskii, *Khim. Geterotsikl. Soedin.* **1976**, 676.
76. O. Buchardt, A. M. Duffield, and C. A. Djerassi, *Acta Chem. Scand.* **1968**, *22*, 2329.
77. O. Keller and V. Prelog, *Helv. Chim. Acta* **1971**, *54*, 2572.
78. M. F. Rennekampf, J. V. Paukstelis, and R. G. Cooks, *Tetrahedron* **1971**, *27*, 4407.
79. O. S. Anisimova, Yu. N. Sheinker, and R. E. Valters, *Khim. Geterotsikl. Soedin.* **1982**, 666.
80. H. B. Bürgi, J. D. Dunitz, and E. Shefter, *J. Am. Chem. Soc.* **1973**, *95*, 5065.
81. H. B. Bürgi, J. D. Dunitz, J. M. Lehn, and G. Wipf, *Tetrahedron* **1974**, *30*, 1563.
82. H. B. Bürgi, *Angew. Chem.* **1975**, *87*, 461; see also H. B. Bürgi and J. D. Dunitz, *Acc. Chem. Res.* **1983**, *16*, 153.

83. H. B. Bürgi, J. D. Dunitz, and E. Shefter, *Acta Crystallogr.* **1974**, *B30*, 1517.
84. S. Wawzonek, H. A. Laitinen, and S. J. Kwiatkowski, *J. Am. Chem. Soc.* **1944**, *66*, 830.
85. R. E. Valters and S. P. Valtere, *Latv. PSR Zinat. Akad. Vestis. Kim. Ser.* **1969**, 753.
86. J. Stradins, *The Polarography of Organic Nitro Compounds*, Riga, 1961 (In Russian).
87. H. P. Rettig and H. Berg, *Z. phys. Chem.* (*Leipzig*) **1963**, *222*, 193.
88. A. A. Potekhin, *Zh. Org. Khim.* **1971**, *7*, 16.
89. K. v. Auwers and A. Heinze, *Ber. Dtsch. Chem. Ges.* **1919**, *52*, 584.
90. E. D. Bergmann, E. Gil-Av, and S. Pinchas, *J. Am. Chem. Soc.* **1953**, *75*, 358.
91. A. A. Potekhin and T. F. Barkova, *Zh. Org. Khim.* **1973**, *9*, 1180.
92. A. A. Potekhin and B. D. Zaitsev, *Khim. Geterotsikl. Soedin.* **1971**, 301.
93. R. P. Bell, *The Proton in Chemistry*, 2nd edition, Chapman and Hall, London, 1973, Chapter 10.
94. C. Pascual, D. Wegmann, U. Graf, R. Scheffold, P. F. Sommer, and V. Simon, *Helv. Chim. Acta* **1964**, *47*, 213.
95. R. Scheffold and P. Dubs, *Helv. Chim. Acta* **1967**, *50*, 798.
96. R. P. Bell, B. G. Cox, and B. A. Timini, *J. Chem. Soc., Sect. B* **1971**, 2247.
97. R. P. Bell and A. D. Covington, *J. Chem. Soc., Perkin Trans. 2* **1975**, 1343.
98. R. E. Valters, *Latv. PSR Zinat. Akad. Vestis, Kim. Ser.* **1970**, 223.
99. R. E. Valters, S. P. Valtere, and A. E. Kipina, *Biological Active Compounds*, Nauka, Leningrad, 1968, p. 213 (In Russian).

Intramolecular Reversible Addition Reactions to the C=O Group

2.1. ALDEHYDO- AND KETO-CARBOXYLIC ACIDS AND THEIR DERIVATIVES MODIFIED AT THE CARBOXYLIC GROUP

Ring-chain isomeric interconversions proceeding by intramolecular reversible addition reactions to the double bond of the C=O group have been well studied, particularly as regards γ- and δ-aldehydo- and keto-carboxylic acids (3- and 4-acylcarboxylic acids) and their derivatives (see Refs. [1–3]).

Compounds possessing an open chain structure were originally termed normal derivatives, ring (cyclic) isomers were designated pseudo or ψ-derivatives (pseudoacids, pseudoacid chlorides, pseudoesters, etc.).

Such terminology is not convenient, and now hardly ever is used, the more current terms being ring and chain isomers or tautomers. Closed (cyclic) and open structures will be applied in this book.

The structures of the most important derivatives **(1)**–**(5)** of the aldehydo (R=H) or keto (R= alkyl, aryl) carboxylic acid isomers are depicted below.

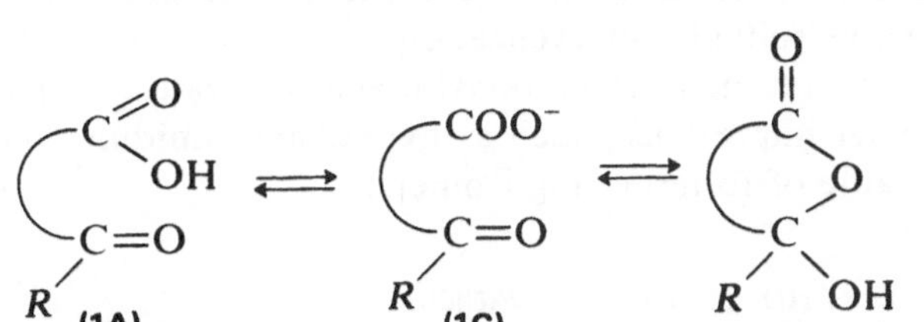

| (1A) Acyl carboxylic acid | (1C) Open anion | (1B) Hydroxylactone (lactol) | (1D) Cyclic (closed) anion |

| (2A) Acyl chloride | (2B) Chlorolactone | (3A) Amide | (3B) Hydroxylactam |

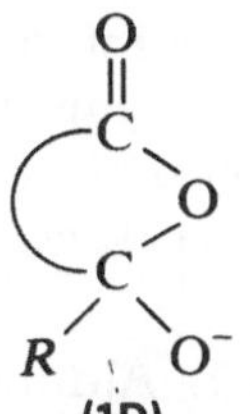

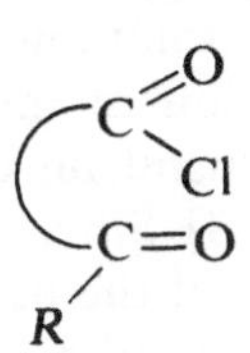

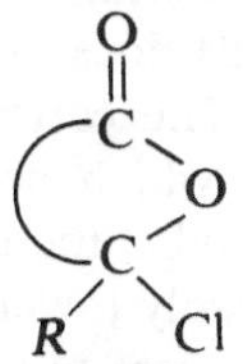

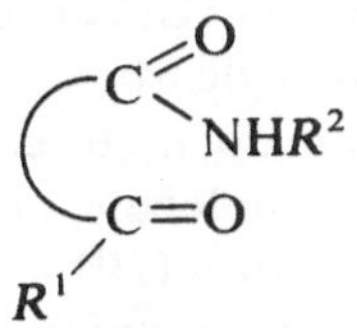

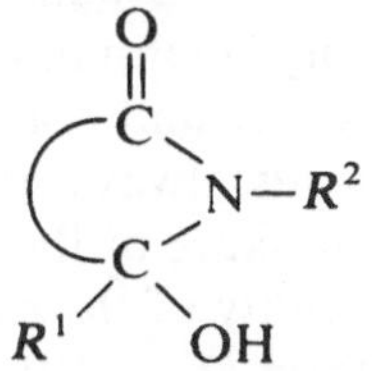

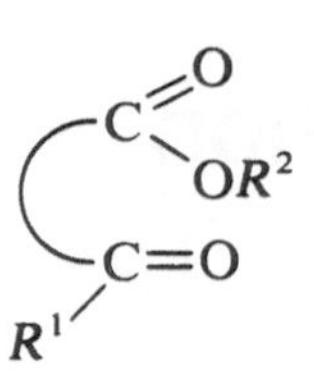

(4A)
Ester

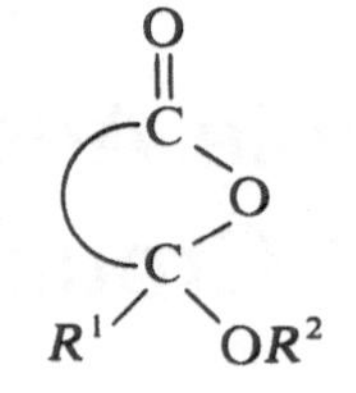

(4B)
Alkoxylactone

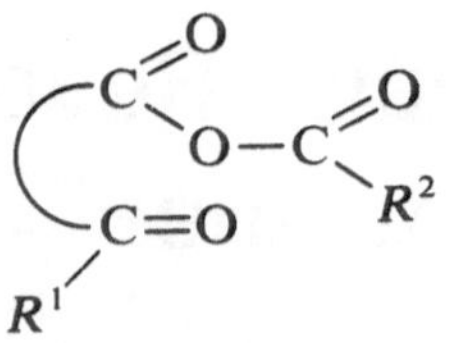

(5A)
Mixed
anhydride

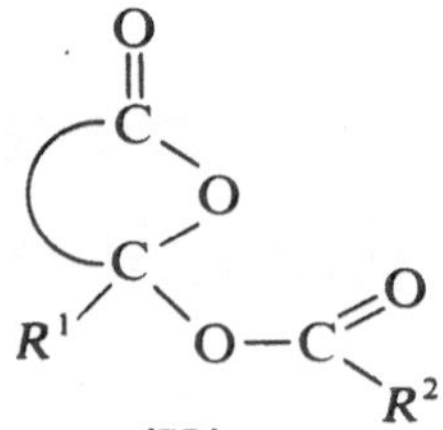

(5B)
Acyloxy-
lactone

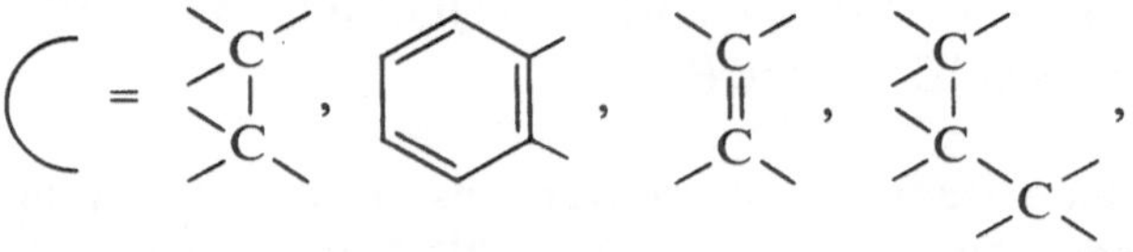

a   b   c   d

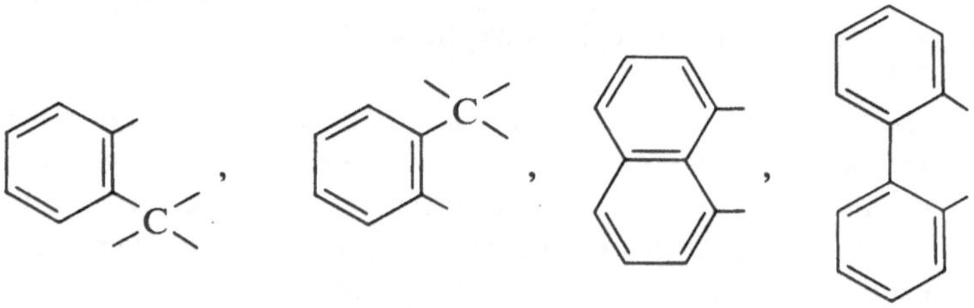

e   f   g   h

Some of the possible connections (a)–(h) of two mutually interacting groups which enable the formation of cyclic isomers are given; they certainly do not reflect all the cases suitable for cyclization.

Besides the above mentioned acylcarboxylic acid derivatives **(1)–(5)**, other derivatives of lesser importance, such as hydrazides, amidines, amidrazones, etc., are capable of forming ring isomers.

2.1.1. Aldehydo- and Keto-carboxylic Acids

The intramolecular nucleophilic addition of the carboxylic group to the C=O bond is a comparatively rapid reversible reaction [4–6] and the ring-chain tautomerism **(1A)** $\rightleftarrows$ **(1C)** $\rightleftarrows$ **(1B)** does occur in the solutions of 3-acyl and some 4-acylcarboxylic acids. The carboxylate anion of these acids always has an open structure **(1C)** which does not depend on the structure of the acid molecule and on the substituent R [5, 7–18]. Since the acidity of the cyclic form **(1B)** is considerably lower than that of the open form **(1A)** [19] the cyclic anion **(1D)** does not appear in the equilibrium mixture in any detectable concentration.

Solutions of the methylsubstituted levulinic acids in the solvents of low polarity show [19] the equilibrium depicted below:

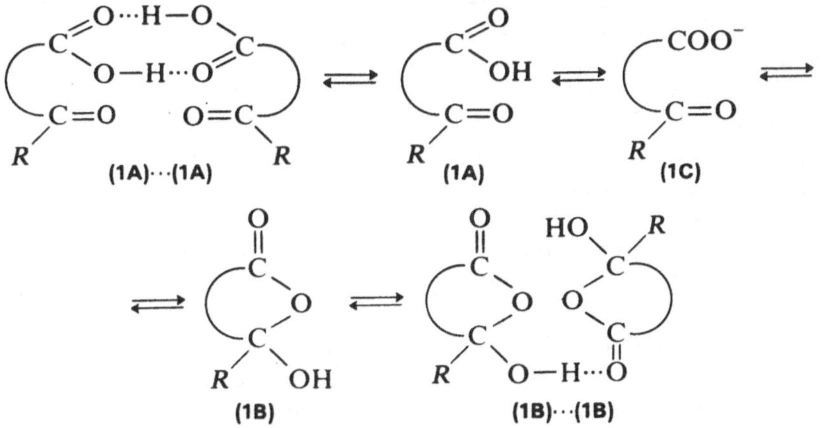

The authors [19] suggested that the ring-chain equilibrium constant†K_T should be derived from the equation:

$$K_a = \frac{K'_a}{1 + K_T}$$

where K_a is the dissociation constant of the acylcarboxylic acid directly measured by means of a potentiometric method and K'_a means the "true" (hypothetical) acid dissociation constant of the acyl carboxylic acid under the condition that the formation of the ring isomer does not take place. The necessity to estimate the K'_a constant, which can be done only indirectly, limits the applicability of this method [5, 19, 20, 21]. To estimate K'_a one usually measures the dissociation constants for model compounds of similar structure but incapable of ring isomer formation.

Using this method, Bowden and Henry [22] obtained ring-chain equilibrium constants K_T of a series of Z-3-acylacrylic acids. They estimated the "true" K'_a values comparing K_a values of Z- and E-3-acylacrylic acids. For structural reasons the E-acids are incapable of cyclization.

Bell and co-workers derived the "true" constant K'_a of levulinic acid and its chain methylsubstituted derivatives [23], of 2-formylbenzoic acid [5], of a series of 2-acylbenzoic, 3-acetyl-2-naphthoic, and 8-acetyl-1-naphthoic acids [24] measuring the catalytic effect of the respective anions on the decomposition of nitramide and on the mutarotation of glucose [5, 25]. For all of these well-established relations are known between the catalytic

† Here and elsewhere $K_T = [B]/[A]$.

constants of carboxylate anions and the dissociation constants of the corresponding acids. Using these K'_a constants the authors have calculated K_T constants of the acylcarboxylic acids examined in the aqueous solution (Table 3). Without doubt this method is more precise than that mentioned above.

Bowden and co-workers [26, 27] solved the reverse task: by using the K_T constants, determined by NMR method, and the K_a constants, determined by potentiometric method, they have calculated the "true" K'_a constants for a series of 2-acylbenzoic and Z-3-acylacrylic acids in 80% methoxyethanol-water solution.

There are many examples where the ring-chain equilibrium constants of the acylcarboxylic acids have been measured using UV, IR, and NMR methods (see Table 3).

The determination of K_T constants by means of electronic spectroscopy is supported on the assumption that position and extinction coefficients ε of the analytical bands in the spectra of the model compounds having a fixed open structure (usually open ester (4A)) and of the tautomeric forms (1A) are equal [28–30]. Prior to these measurements the UV spectra of model compounds having a closed structure (usually alkoxylactones (4B)) had to be studied to make sure that the ring form (1B) did not absorb in the region of the analytical band.

The use of infrared spectroscopy in the determination of ring-chain equilibrium constants is more useful for the 3-acylcarboxylic acids forming a five-membered ring. Here, due to C=O group angle strain the C=O band of five-membered hydroxylactones (1B) is shifted to higher frequencies compared with the carboxylic or ketonic C=O band of the open form (1A) and it usually appears in solution spectra at $1800-1770 \text{ cm}^{-1}$ (Table 1). It may also serve as analytical band for quantitative measurements of K_T constants.

The determination of K_T constants for 2-aroyl [31, 32], 2-acylbenzoic [26] and Z-3-aroylacrylic acids [27] is based on the assumption that the integrated intensities of C=O bands in the spectra of open and closed form are equal to the intensities of the corresponding bands of fixed model compounds, i.e., esters ((4A), ν_{COOR} $1740-1720 \text{ cm}^{-1}$, $\nu_{C=O}$ $1700-1670 \text{ cm}^{-1}$) and alkoxylactones ((4B), $\nu_{C=O}$ $1785-1770 \text{ cm}^{-1}$) [26, 27, 36, 37]. The reproducibility of determinations of tautomeric equilibria in solution is within 2-3% [26, 32].

Lactone C=O bands in the spectra of 4-acylcarboxylic acids, which form six-membered hydroxylactones (1Bd-g), are but little shifted to high frequencies as compared with the carboxyl C=O band. These bands thus overlap, which hampers analytical measurements. Nevertheless K_T constants have been determined for 8-aroyl-1-naphthoic acids by means of the

Table 1. C=O Band Frequencies in IR Spectra of 3-Acylcarboxylic Acids (1a–c) in Solution

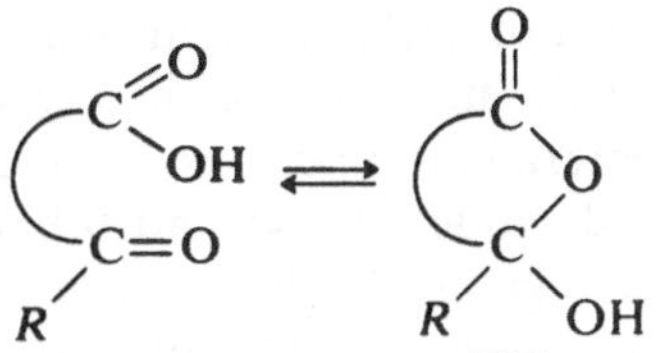

(1A)   (1B)

|  | R | Solvent | C=O in (1A) | | C=O in (1B) | Refs. |
			Ketonic or aldehydic	Carboxylic	Lactonic	
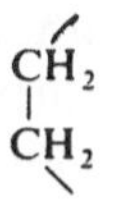 CH₂ / CH₂ group	H	CCl₄	1749	1711	1795	30
	Me	CCl₄	1726	1758 (free) 1714 (bonded)	—	19
	Ph	dioxan	1690	1735	—	33
	CCl₃	CHCl₃	1767	1724	1808	34
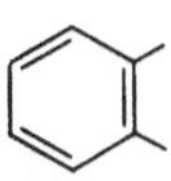 benzene ring	H	dioxan	1699	1723	1778	26
	Me	dioxan	1706	1724	1776	26
	Ph	dioxan	1681	1726	1781	26
	CCl₃	CHCl₃	1754sh	1721	1789	34
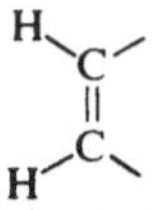 CH=CH group	H	(KBr)	—	—	1795, 1761	35
	Me	dioxan	—	—	1763	22
	Ph	dioxan	1676	1713	1769	27
	CCl₃	CHCl₃	1786sh	1715	1812	34

IR method [38, 39]: the ketonic band at 1670–1660 cm^{-1} [39] was used as the analytical one.

The qualitative and quantitative determination of the composition of acylcarboxylic acid capable of ring-chain tautomerism by ¹H-NMR spectroscopy is most frequently accomplished using signals of alkyl (mostly methyl) protons, located at the ketone group (AlkC=O in (1A) and AlkC(OH)O in (1B)) [15, 19, 20, 26, 29, 40–42]. The signals of protons located at the carbon atoms in the chain between ketone and carboxyl groups are only rarely used [27, 43]. The signals of aldehyde and methine protons of aldehydrocarboxylic acids (HC=O in (1A) and HC(OH)O in (1B)) are also suitable for analytical measurements [11, 12, 14, 16–18, 21, 30].

The ¹H-NMR spectrum of 2-acetylbenzoic acid in DMSO-d₆ solution shows signals at 2.45 ppm (CH₃CO) and at 1.78 ppm (CH₃C(OH)O), the relative intensities of which indicated $K_T = 2$. Addition of a trace of hydrochloric acid caused fast interconversion of the tautomers and the collapse of these signals into one averaged signal at 2.02 ppm [42].

Most ^{1}H-NMR spectra of acylcarboxylic acids show a single averaged signal, indicating tautomeric interconversions fast on the NMR time-scale [19, 20, 26, 29, 41]. K_T determination can be carried out in such cases using ^{1}H-NMR spectra of fixed models ((4A) and (4B)) for calculation of the positions of the averaged signals [26, 29].

The K_T constant for 3-formylpropionic acid was obtained from the ratio of the aldehyde (9.54–9.66 ppm) and methine (5.66–5.83 ppm) proton signal intensities [30].

Rate constants for 2-formylbenzoic [5] and 2,3,3-trimethyl-levulinic [6] acid tautomeric interconversions in aqueous solution were determined by Bell and Cox by relaxation time measurements. Using the observed dissociation constants K_a, "true" dissociation constants K'_a and pH relaxation times, measured by means of the temperature-jump method, the authors calculated the rate constants for the equilibria (1) and (2), postulating them as rate-determining proton transfer reactions (Table 2). The last equation describes the ring-chain interconversion ($K_T = k_4/k_3$).

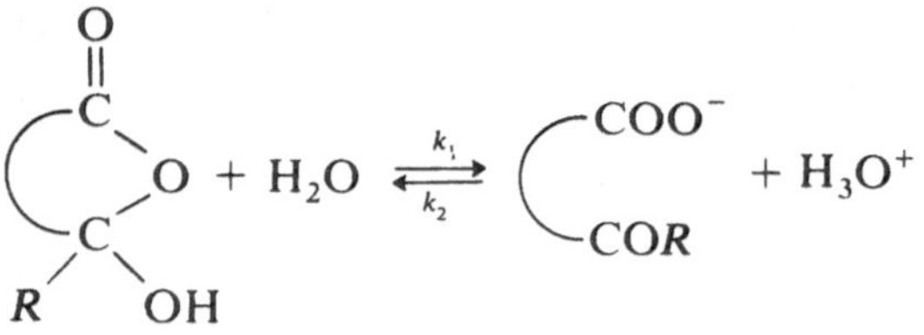

$$\text{(1)}$$

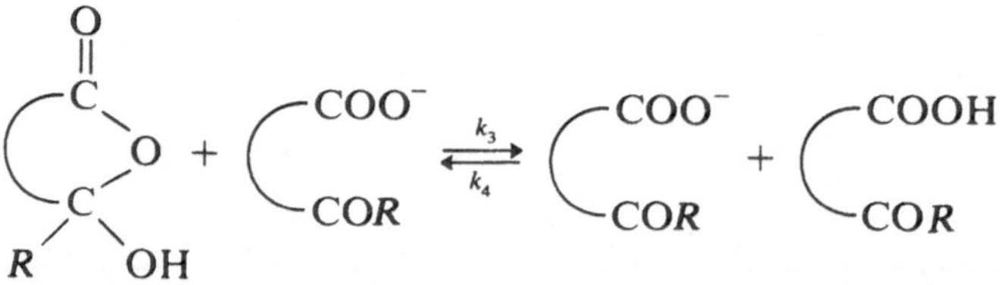

$$\text{(2)}$$

We now discuss the influence of the structure of acylcarboxylic acids (1a–h) on the ring-chain equilibrium constants given in Table 3. Cases are considered in which either the open or the closed form predominates in the solid state.

Table 2. Protolytic and Tautomeric Rate Constants of 3-Acylcarboxylic Acids ($1 \cdot mol^{-1} \cdot s^{-1}$) in Aqueous Solution [6]

Acid	k_1	k_2	k_3	k_4	K_T
2,2,3-Trimethyllevulinic	9	1.15×10^8	9.3×10^4	6.5×10^5	6.9
2-Formylbenzoic	23	2.5×10^7	2.2×10^5	3.2×10^6	14.8

Table 3. Ring-Chain Equilibrium Constants ($K_T = [B]/[A]$) of Acylcarboxylic Acids

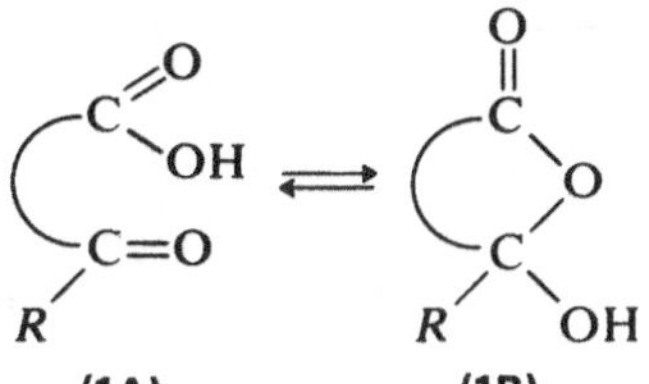

		R	K_T	Refs.

 R^1	R^2	R^3	R^4	R	K_T	Refs.
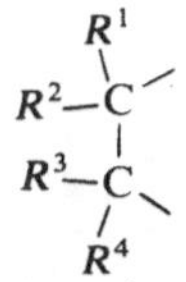 H	H	H	H	H	0.82^a	30
H	H	H	H	Me	0^b	23
Me	H	H	H	Me	0^b	23
Me	Me	H	H	Me	1^b	23
H	H	Me	Me	Me	0.2^b	23
Me	Me	Me	H	Me	6.9^b	23
H	H	H	H	t-Bu	0.2^b	23
H	H	H	H	Ph	0^b	23

R	K_T	Refs.
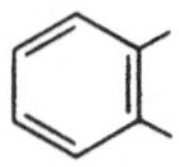 H	14.8^b	5
	$6.7^c; 4.6^d$	26
Me	$4.7^c; 4.3^d$	26
	3.2^b	24
Et	$4.3^c; 3.5^d$	26
i-Pr	$4.8^c; 2.2^d$	26
	5.6^b	24
t-Bu	$5.3^c; 5.7^d$	26
PhCH$_2$	$5.5^c; 3.0^d$	26
Ph	0.07^c	26
	0.033^c	32
4-MeC$_6$H$_4$	0.018^c	32
4-ClC$_6$H$_4$	0.034^c	32
4-BrC$_6$H$_4$	0.043^c	32
4-CNC$_6$H$_4$	0.078^c	32
4-NO$_2$C$_6$H$_4$	0.092^c	32
3-NO$_2$C$_6$H$_4$	0.067^c	32
3-NO$_2$-4-MeC$_6$H$_3$	0.057^c	32
3-NO$_2$-4-ClC$_6$H$_3$	0.102^c	32

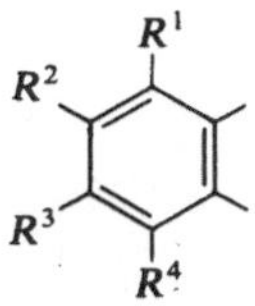

R^1	R^2	R^3	R^4	R	K_T	Refs.
Me	H	H	H	Me	16^b	24
H	H	H	Me	Me	0.52^b	24
H	Me	Me	H	Me	2.9^b	24
H	MeO	MeO	H	Me	5.1^b	24
Cl	Cl	Cl	Cl	Me	12^b	24

continued

Table 3. (*cont.*)

	R	K_T	Refs.
	Me	6.3^b	24
	t-Bu	94^e	22
	1-Adamantyl	70^e	22
	Ph	0.3^e	22
	Me	34^e	22
	4-MeOC$_6$H$_4$	1.0^c	27
	4-MeC$_6$H$_4$	3.0^c	27
	4-BrC$_6$H$_4$	4.3^c	27
	4-ClC$_6$H$_4$	4.3^c	27
	Ph	10.6^c; 8^d	27
	4-MeOC$_6$H$_4$	1.7^c; 1.1^d	27
	4-MeC$_6$H$_4$	4.5^c; 2.5^d	27
	4-BrC$_6$H$_4$	7.6^c	27
	4-ClC$_6$H$_4$	6.6^c	27
	4-NO$_2$C$_6$H$_4$	26^c	27
	H	1200^e	38
	Me	130^e;	38
		100^b	24
	Et	210^e	38
	i-Pr	420^e	38
	t-Bu	1300^e	38
	Ph	1.41^c;	38
		2.34^c; 1.56^f	39
	4-MeOC$_6$H$_4$	0.14^c;	38
		0.64^c; 0.12^f	39
	4-MeC$_6$H$_4$	0.43^c	38
	4-BrC$_6$H$_4$	1.60^c	38
	4-ClC$_6$H$_4$	1.66^c	38
	3-MeC$_6$H$_4$	0.84^c	38
	3-ClC$_6$H$_4$	2.63^c	38
	3-CF$_3$C$_6$H$_4$	3.06^c	38
	H	3000^e	38

3-Acylpropionic Acids and Substituted Derivatives (1a). The electrophilicity of the carbon atom in the group RCO decreases in the series $R = H$, alkyl, aryl. As a consequence, the stability of the cyclic form of acylcarboxylic acids is lowered in the same order. In solutions of 3-formylpropionic acid, tautomeric equilibrium involves approximately equal amounts of both forms [30]. Substitution in the chain shifts the equilibrium towards the ring form [19, 21, 44–46].

In solutions of levulinic acid, the ring form is not observed [19, 23]. Substitution of two or three methyl groups at the carbon atoms in the chain forces the C=O and COOH groups into closer proximity and shifts the equilibrium towards the cyclic form (see Table 3). The largest equilibrium constant is found for 2,2,3-trimethyllevulinic acid (6.9 in aqueous solution [6], 7.7 in an 80% aqueous solution of methylcellosolve [19]). This has been interpreted as a result of the Thorpe–Ingold effect, sometimes called the gem-dimethyl or gem-dialkyl effect [47–51]. The most satisfactory explanation was given by Allinger and Zalkov [52] who showed that branching of the chain decreases the free energy of ring formation through more favorable $\Delta H°$ and $T\Delta S°$ terms and consequently shifts the equilibrium toward the ring form. This effect plays an important role in many ring-chain equilibrium systems, as will be shown later. It also increases the rates of cyclizations and other reactions proceeding through cyclic transition states in the rate-limiting step.

Recently it was shown experimentally [53], and corroborated by force field calculations [54], that the dominant role of rate enhancement of 3-(2-hydroxyphenyl)propionic acid lactonization by methyl substituents in the chain, and in the 3- and 6-positions of the benzene ring, is ground state strain, i.e., an enthalpy effect. This work [53, 54] disproves the former concept of stereopopulation control, suggested by Milstien and Cohen [55], according to which the entropy effect dominates.

According to IR investigations, 3-aroylpropionic acids [33, 56–58], and some of their more complicated derivatives [59], occur exclusively in the open form in the solid state as well as in solution. It has been shown by means of dissociation constant measurements [23] that an aqueous solution of 3-benzoylpropionic acid does not contain the cyclic form. 3-(3-Indolylcarbonyl)propionic acid also possesses the open structure as shown by IR and UV spectra [60].

Footnotes to Table 3

The K_T constants are determined by:
[a] UV method in dioxan;
[b] Dissociation constants measuring and true dissociation estimating in water;
[c] IR method in dioxan;
[d] ^{1}H-NMR method in dioxan:
[e] In the same way as *b*, but in 80% 2-methoxyethanol–water solution;
[f] IR method in DMSO.

A series of 3-trichloroacetylcarboxylic acids based on 3-trichloro-acetylpropionic acid shows a heightened tendency to form the cyclic tautomers [13, 34]. Obviously, the −I-effect of the trichloromethyl group enhances the electrophilicity of the carbonyl carbon atom. The relatively high stability of the ring form of bicyclic ketocarboxylic acids **(6)–(9)** is a result of the fixed mutually eclipsed conformation of the COOH and C=O groups providing the sterically most favorable condition for intramolecular addition.

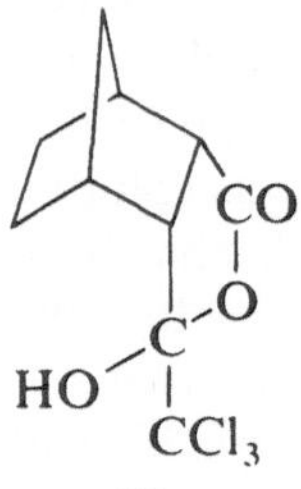

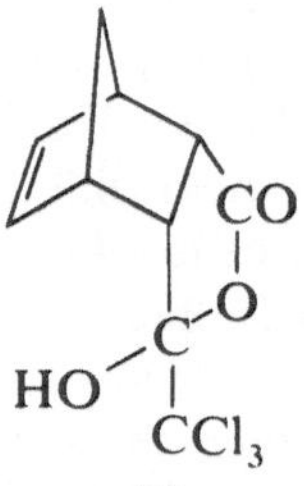

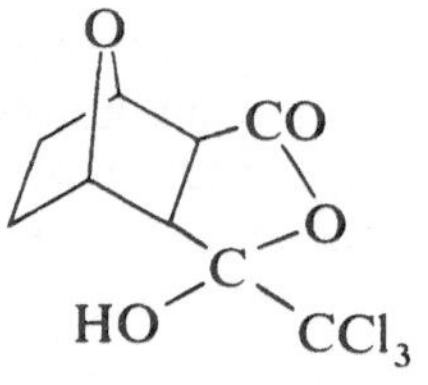

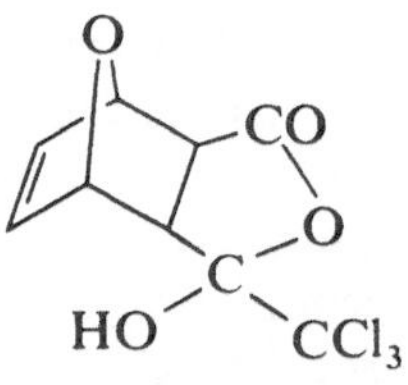

2-Acylbenzoic Acids (1b). The ring-chain tautomerism of 2-acylbenzoic acids has been thoroughly studied [1, 3]. Here the C=O and the COOH groups are nearer to each other than in 2-acylpropionic acids. Furthermore, rotation of the connecting C–C bond is impossible. Steric models show [61] that the open tautomers **(10A)** can exist only in a non-planar conformation, which inhibits the conjugation of C=O and COOH groups with the benzene ring.

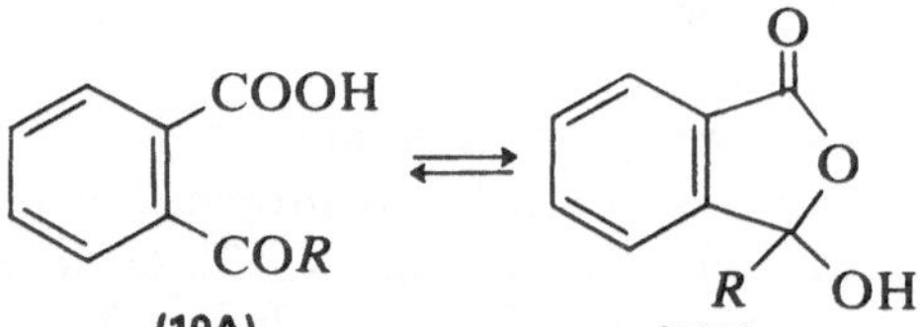

2-Formylbenzoic acid in the solid state exists as the cyclic hydroxyphthalide (**10B**, *R* = H) [62–64], and this form also predominates in the solutions (see Table 3) [5, 8, 12, 26, 65]. The same behavior applies to solutions of 2-formyl-5,6-dimethoxybenzoic (opianic) acid [16, 18], but for 2-formyl-3-hydroxy-2-methylpyridine-4-carboxylic acid the equilibrium is shifted toward the open form [14, 18].

Many papers [29, 41, 62, 64, 66] present spectral evidence for the existence of 2-acetylbenzoic acid in solid state as the 3-hydroxy-3-methyl-phthalide (**10B**, *R* = Me). However, the dependence of the solid state structure on the substituent R (10) is not straightforward. Some 2-acylbenzoic acids with R = alkyl or a more complicated substituent crystallize in the cyclic form [7, 13, 34, 67], but others solidify in the open chain [68] or as mixtures of both forms [68–73].

The isolation of one definite isomer as a solid substance does not necessarily imply its prevalence in solution [29, 68, 71], since the minor tautomer can possess the lower solubility [48]. The isomeric composition of the solid acylcarboxylic acids isolated from solution depends not only on the structural factors, but also on the method and conditions of the isolation, such as crystallization or precipitation by acidification from alkaline solutions, the temperature of the solution during crystallization or precipitation, polarity and proton acceptor properties of the solvent, pH value of aqueous solution after acidification, etc.

With variable or not clearly defined isolation conditions, little regularity exists between the tautomeric composition of the isolated solid and the structure of such acids (10).

Both isomers are observed [74, 75] for solid benzil-o-carboxylic acid: the yellow form (11A), $\nu_{C=O}$ 1698, 1683 cm^{-1}, ν_{OH} 2632–2513 cm^{-1}; the colorless form (11B), $\nu_{C=O}$ 1745, 1692 cm^{-1}, ν_{OH} 3268 cm^{-1} [62]. The isomerization (11B) → (11A) is accomplished by heating.

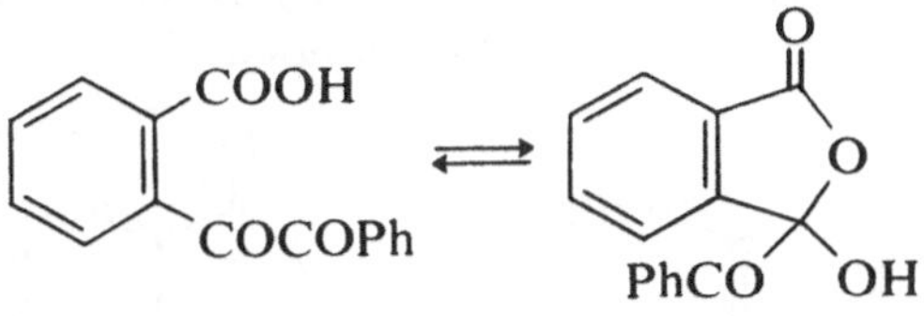

(11A)   (11B)

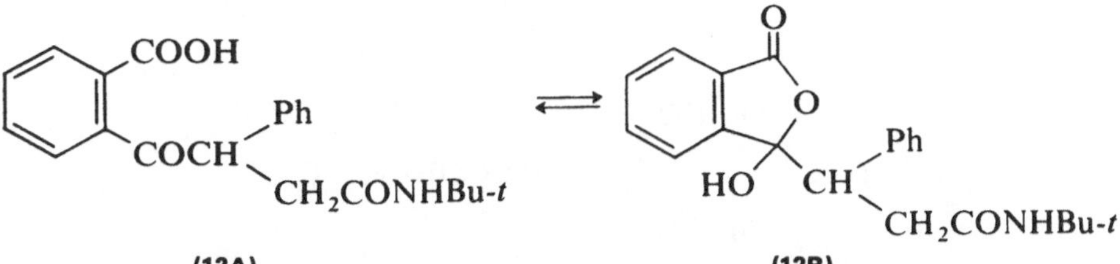

(12A)   (12B)

3-(2-Carboxybenzoyl)-3-phenylpropionic N-(t-butyl)amide (12) was isolated [73] in three crystalline modifications which differ from each other concerning relative amounts of ring and chain forms (12A) and (12B). This is because there are various possibilities for intermolecular hydrogen bonding in this molecule.

The acid (13) was obtained [76] in the lactone form (13B) from acid solution or as a betaine (13A) from alkaline solution.

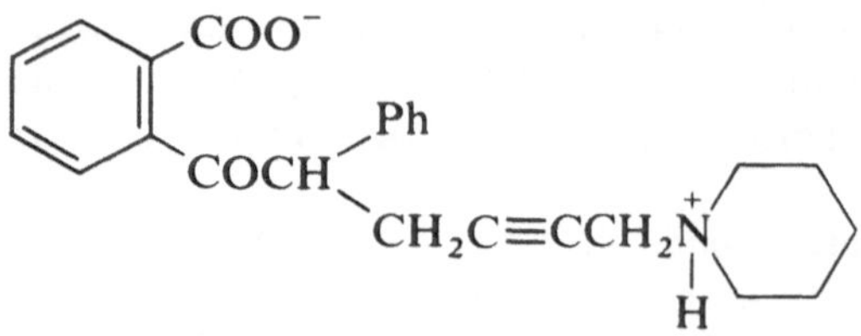

(13A)

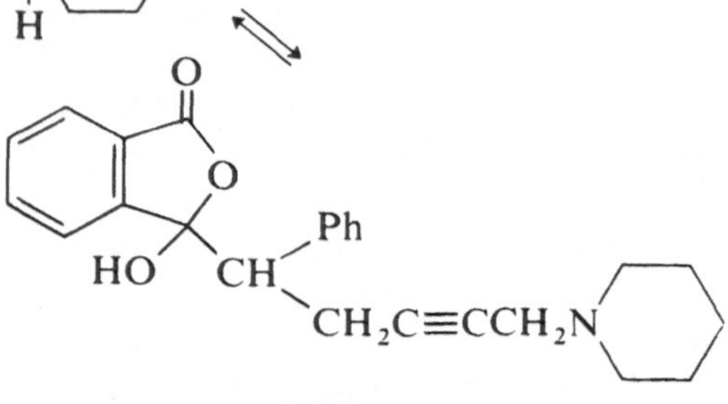

(13B)

Korshak and co-workers [77] have isolated both forms of 2,5-diben-zoylterephthalic acid **(14A)** and **(14B)**. The open form, obtained by acidification of a solution of the disodium salt, is less stable and it is easily transformed into the ring form **(14B)** by acids, by suspension in dioxan, benzene, ether, methanol or acetic acid, and by recrystallization. The authors [77] suggest that intermolecular hydrogen bonds

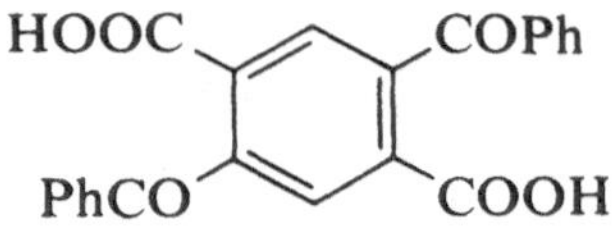

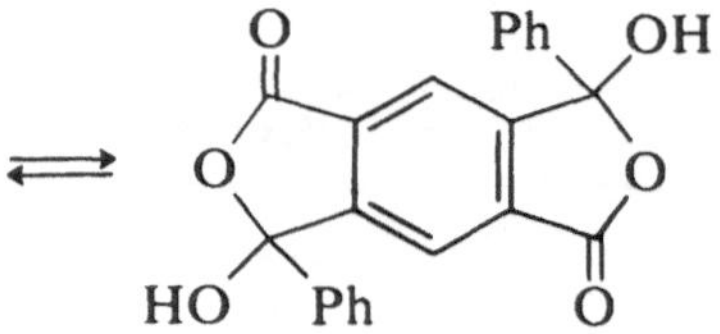

(14A)   **(14B)**

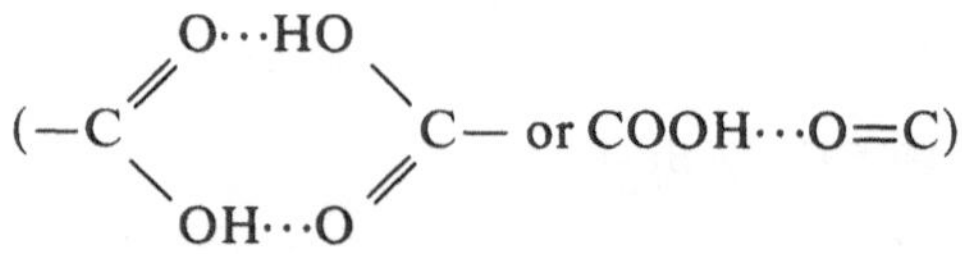

$$(-C\underset{OH\cdots O}{\overset{O\cdots HO}{\diagup}}C- \text{ or } COOH\cdots O=C)$$

stabilize the open form. Heating or the action of solvents breaks these bonds and a C=O group protonation takes place which is followed by cyclization.

IR spectroscopy has been used [29, 62, 64] to determine the structures of 2-acylbenzoic acids in the solid state. Unfortunately, this method is hampered by the presence of strong intramolecular hydrogen bonds OH···O=C (ring form), which cause splitting of the lactol νC=O band and shifting toward the lower frequencies in comparison with solution spectra (see Table 1, Fig. 2). Concentrated solutions of 2-acylbenzoic acids and other γ-ketocarboxylic acids [19, 41, 68] in solvents of low polarity show, in addition

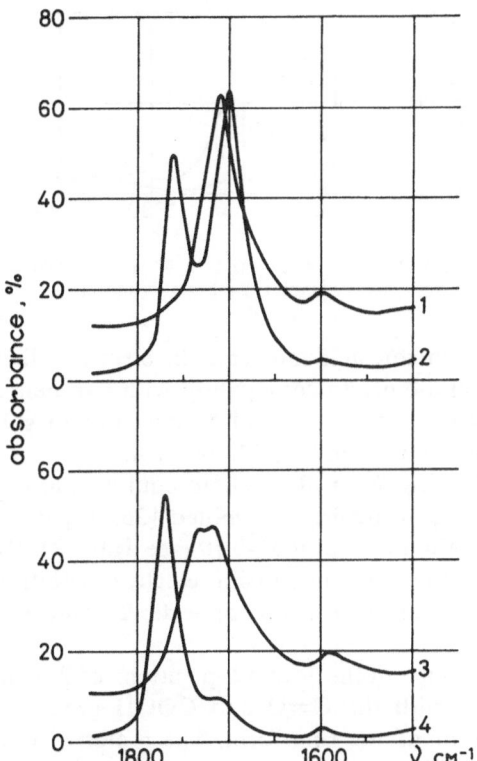

Figure 2. The IR spectra of 2-acylbenzoic acid ring isomers: 1—3-hydroxy-3-methylphthalide in the solid state; 2—its solution in dioxan **(A)** ⇄ **(B)**; 3—3-bromomethyl-3-hydroxyphthalide in solid state; 4—its solution in dioxan ($c = 5 \times 10^{-2}$ mol/l).

to the free lactol band C=O group at 1770 cm^{-1}, a band for the bonded C=O group at 1750 cm^{-1} (Fig. 3). The dilution of the solution decreases the intensity of this band (1750 cm^{-1}) and such a change in the spectrum makes it possible to distinguish the band of the bonded lactol C=O group from the absorption of the free COOH group of the open isomer that usually appears in the same range (~1750 cm^{-1}).

The basic proof of the cyclic structure of acylcarboxylic acids in the solid state is the presence of an intense broad band at 3350–3150 cm^{-1}, attributable to the bonded OH group (OH$\cdots$O=C) in their IR spectra. Ketocarboxylic acids of the open structure **(1A)** show the OH-absorption of dimeric COOH groups as a broad diffuse band at lower frequency (Fig. 4).

Tautomeric equilibria of 2-acylbenzoic acids **(10, R = alkyl)** [7, 24, 26, 29, 41, 64, 68–71, 73] in solution are displaced in favor of the ring forms **(10B)** (see Table 3).

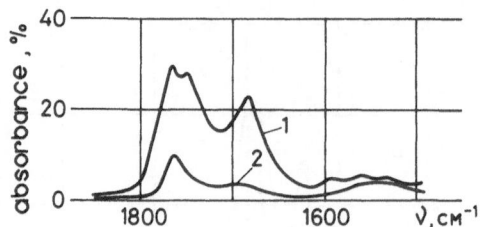

Figure 3. The IR spectra of 2-(2-phenylbutyryl)benzoic acid **(A)** ⇄ **(B)** in CCl_4: 1—c = 5×10^{-2} mol/l; 2—c = 10^{-2} mol/l.

In the 2-aroylbenzoic acid molecule the electrophilicity of a ketogroup is low owing to ketone group conjugation with two benzene rings. Hence, 2-aroylbenzoic acids both in the solid state and in solutions exist predominantly as the open form [26, 29, 64, 78–80], unless special structural factor stabilize the ring form. Ring-chain equilibrium constants of 2-(3-X or 4-X-benzoyl)benzoic acids, determined [26, 32] in dioxan by the IR method, lie within the range of 0.03–0.10 (see Table 3). Higher values have been observed for the acids possessing electron-withdrawing substituents X. A satisfactory linear correlation connects K_T and the σ-coefficients of the substituents X.

The methyl groups in the 3- and 6-positions of 2-benzoylbenzoic acids interfere sterically with the C=O and COOH groups, thus shifting the equilibrium in favor of the ring form. The 6-methyl group has a greater influence than the 3-methyl group [24]. Methyl groups in the 2'- and 6'-positions of the benzoyl substituent sterically hamper the cyclization [28,

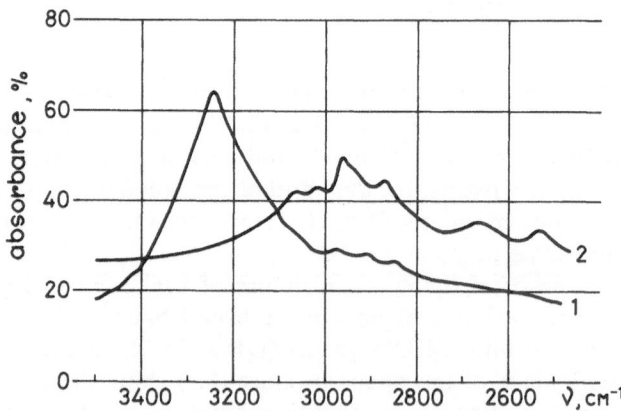

Figure 4. The IR spectra of ring and chain isomers of 2-acylbenzoic acids in solid state at 3500–2500 cm^{-1}: 1—3-hydroxy-3-methylphthalide; 2—2-(2-phenylbutyryl)benzoic acid.

81, 82]. 2-Benzoyl-3,4,5,6-tetrachlorobenzoic acid reportedly [24, 29, 32] exists preferentially in the cyclic form.

Reportedly [83], 2-acetyl-3-nitrobenzoic acid is cyclic, whereas 2-acetyl-3-hydroxy- and 2-acetyl-3-amino-benzoic acids exist in open forms. This is probably due to the ability of the last two acids to form intramolecular hydrogen bonds as shown (15). Additionally, the electron-donor groups OH and NH_2 decrease the electrophilicity of the keto group and thus stabilize the open structure.

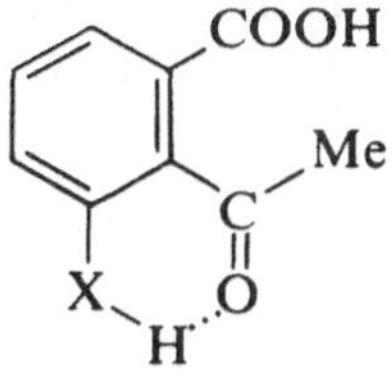

(15)

Increasing the electron donating effect ($X = NO_2 < H < OH < NH_2$) in a number of 2-acetyl-6-X-benzoic acids by contrast shifts the equilibrium in favor of the ring form [42]. In this case the conjugation of the carboxylic group with an electron-donating group increases the nucleophilicity of COOH group oxygen atoms, thus stabilizing the ring form.

The introduction of nitro groups into the phthaloyl ring irrespective of its position stabilizes the open structure [29, 32, 78], which is difficult to explain.

Some 2-heteroylbenzoic acids (10A, R = 2-pyrrolyl, 1-methyl- and 1-benzyl-2-pyrrolyl [84], 2-pyridyl [85], 3-indolyl [60]) also possess the open structure.

In contrast to 2-formylbenzoic acid (K_T = 14.8 in aqueous solution [5]), the ring form of its heterocyclic analogue 3-formylthiophene-2-carboxylic acid is not observed in solution [86], perhaps due to the unfavorable bond angles of the five-membered thiophene ring.

Anthraquinone-1- (16A) [87] and fluorenone-1-carboxylic (16B) [26] acids represent special cases among 2-aroylbenzoic acids. Their C=O group is planar to the aromatic system and cannot rotate around the CO—Ar bond (shown by arrows in formulae (16A) and (16B)). As a consequence, the approach of the carboxyl group to the carbonyl group, which requires strict stereoelectronic control†, cannot take place and the cyclic isomers of the acids (16A) and (16B) do not appear in solution.

---

† P. Deslonghamps, *Stereoelectronic Effects in Organic Chemistry*, Pergamon Press, Elmsford, N.Y., 1983.

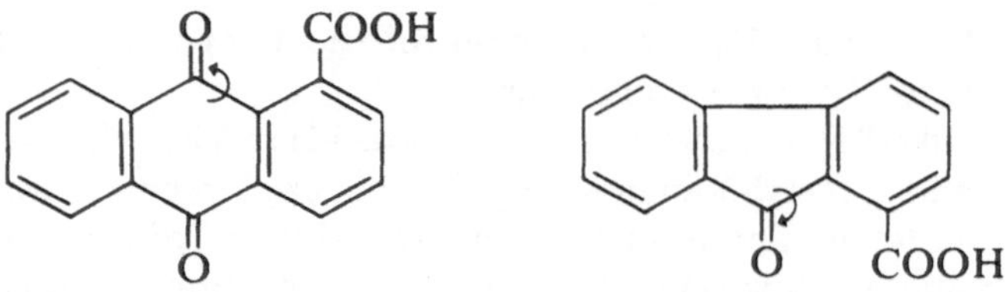

(16A)   (16B)

Interestingly, the COOH group of anthraquinone-1-ylacetic acid, the homologue of **(16A)**, is capable of an intramolecular addition to the C=O group. The cyclization is accompanied by 1,4-dehydration to the lactone **(17)** [88–90]:

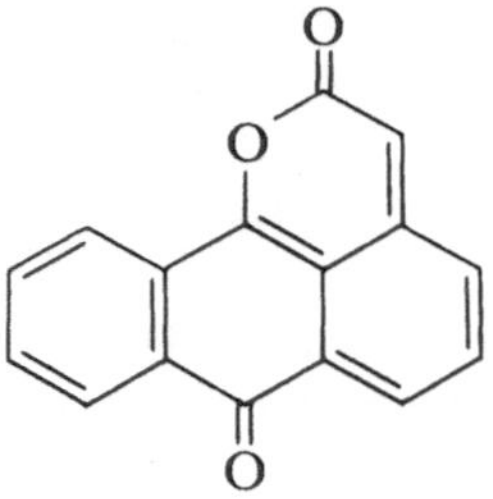

(17)

Z-3-Acylacrylic Acids (1c). Z-3-Formylacrylic acid [2, 35, 91, 92], its 2-methyl [93], 3-methyl [94], 2,3-halogeno [11, 17, 95–98], and other 2,3-substituted derivatives [99], and also Z-3-acetylacrylic acids [15, 20, 100–103], exist in the solid state, as well as in solution, entirely in the ring form **(18B)** as does penicillic (Z-3-methacryl-3-methoxyacrylic) acid [104, 105].

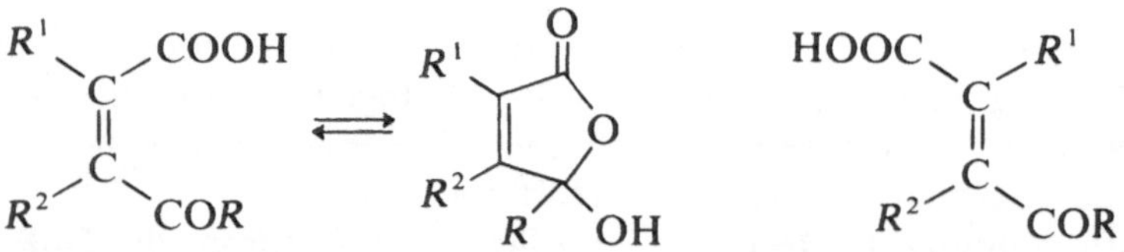

(18A)   (18B)   (18C)

By means of acid ionization constant measurements [19], high ring-chain equilibrium constants have been obtained [22] for Z-3-acylacrylic acids in 80% aqueous 2-methoxyethanol solution (see Table 3).

Z-3-Aroylacrylic acids have been isolated in solid state in cyclic or open form [9, 10, 106–109], but in solution the equilibrium is displaced in favor of the ring form. In a number of Z-3-(4-X-benzoyl)-3-methylacrylic acids, a good linear correlation has been obtained between K_T constants and inductive (σ_I) and resonance (σ_R^0) constants of the substituent X:

$$\log K_T/(K_T)_0 = \rho_I\sigma_I + \rho_R\sigma_R^0$$

Bowden and Henry [27] in a series of six compounds found: $\rho_I = 0.374$, $\rho_R = 1.885$ (standard deviation $s = 0.072$, correlation coefficient $r = 0.990$).

Polarography established a very low value of the ring-chain equilibrium constant [110] for aqueous solutions of Z-3-formylacrylic acid ($K_T = [(18B)]/[(18A)] = 1.82$; $R = R^1 = R^2 = H$). This differs greatly from the results obtained for other Z-3-acylacrylic acids by potentiometric or spectroscopic methods (see Table 3). Unfortunately Z-3-formylacrylic acid has not been investigated potentiometrically or spectroscopically.

In the study of 3-acylacrylic acid ring-chain tautomerism, it is sometimes necessary to differentiate not only ring and chain forms (18A) and (18B) but also Z- and E-isomers of the chain forms (18A) and (18C). This can be achieved by ^{1}H-NMR spectroscopy using spin-spin coupling constants of olefinic protons [15, 20, 102, 103], or by the IR technique utilizing the-out-of-plane deformation C—H vibrational mode at 980 cm^{-1} of the E-isomers [43, 111].

By the action of thionyl chloride on E-3-acetylacrylic acid, Scheffold and Dubs [20] obtained the ring isomer of the acyl chloride, i.e., 5-chloro-5-methyl-2,5-dihydro-2-furanone. Its careful hydrolysis led to the ring form of Z-3-acetylacrylic acid (18B, $R = Me$). Interconversion into the more stable E-isomer took place during sublimation. The authors [20] did not succeed in obtaining Z-3-acetylacrylic acid in the open form (18A).

Sugiyama and co-workers [103] carried out the isomerization (18C)→(18B) by irradiation of E-3-acetylacrylic acid with a mercury vapor lamp. Attempts to obtain the open isomer by isomerization of E-3-pivaloylacrylic acid (18C, $R = $ t-Bu) were unsuccessful. These experiments confirm the exceptionally high stability of the Z-3-acylacrylic acid ring form, which is probably due to the steric proximity of the interacting groups in the open structure.

Kuchar [111] has obtained a mixture of three isomers (18A), (18B), and (18C) by acidification of an alkaline solution of 3-(1-adamantylcarbonyl)acrylic acids; however, the presence of the open Z-form (18A) was not strictly proven. An equilibrium between the open and closed forms was observed in acidic and basic solutions of 3-(1-adamantylcarbonyl)-2-(or 3)-halogenoacrylic acids [43, 112]; however, the configuration of the open isomers was not established.

Methyl substituents at the olefinic carbon atoms of Z-3-acylacrylic acids stabilize the cyclic form (see Table 4), the methyl group at $C_{(3)}$ exerting a

Table 4. The influence of Methyl Substituents at Olefinic
Carbon Atoms on the Ring-Chain Equilibrium Constant of
Z-3-Acetylacrylic acids (**18**, R = Me)[†] [20]

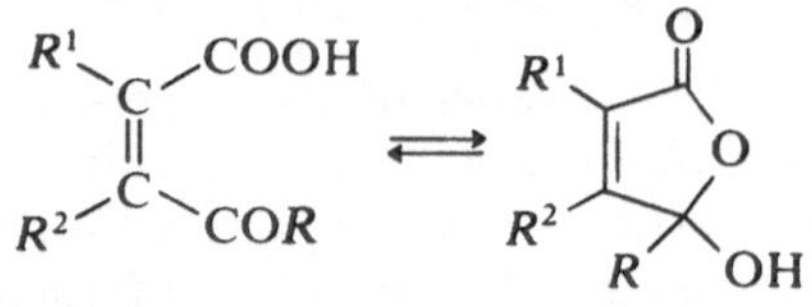

(18A)   (18B)

R^1	R^2	pK_a determined	pK_a' estimated	log K_T
H	H	6.7	5.5	1.2
Me	H	7.60	5.7	1.9
H	Me	9.14	5.5	3.6
Me	Me	10.9	5.7	5.2

[†] Determined by means of potentiometric method in 80% aqueous methylcellosolve.

greater influence than that at $C_{(2)}$ [20, 27]. An analogous phenomenon is
observed [43] for 3-(1-adamantylcarbonyl)-2-(or 3)-halogenoacrylic acids.

The increasing stability of the ring form by a halogen atom at position
3 can be explained as the result of its −I-effect, which operates through
two bonds and increases the electrophilicity of the keto group. Simultaneously the +C-effect operating through the double bond increases the
nucleophilicity of carboxylic group oxygen atoms.

In the case of E-3-acyl-2-halogenoacrylic acids, the halogen atom has
a weaker influence on the ring stability, because its −I-effect operates on
the keto group through three bonds, but the +C-effect operating through
the double bond decreases the electrophilicity of the keto group.

Ring-chain and keto-enol tautomerism appear simultaneously in
maleiylacetone **(19A)** ⇌ **(19B)** [113, 114].

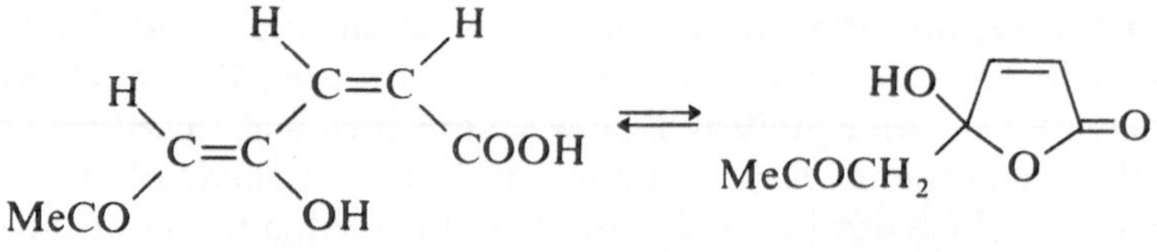

(19A)   (19B)

4-Acylbutyric Acids (1d). Pentin and co-workers [57, 115–117] investigated a number of 4-acylalkanecarboxylic acids. According to IR and UV

data these acids in the solid state and also in CHCl$_3$ and CCl$_4$ solutions possess an open structure. Evidently, the great conformational mobility of a chain having three sp^3-carbon atoms does not favor the formation of the six-membered hydroxylactone ring. Unfortunately, no data are available of a spectroscopic examination of 4-acylbutyric acids possessing gem-dialkyl substituents in the chain which could favor a ring closure by the Thorpe-Ingold effect discussed on page 27.

Meerwein [118] found that the ionization constant decreases on going from 4-formyl-4-methyl-3-phenylvaleric acid to its 2-methyl derivative. This and some other differences in the physical and chemical properties of both acids lead to a supposition that the introduction of a methyl or more generally an alkyl group at C$_{(2)}$ favors the hydroxylactone formation. In this case the chain between the interacting groups contains four substituents including a gem-dimethyl group.

2-Acylmethylbenzoic and o-Acylphenylacetic Acids (1e, 1f). In these acids the chain containing two aromatic and one sp^3-carbon atoms and possessing therefore less conformational mobility than in the foregoing acids (**1d**) leads to a greater stability of the ring form.

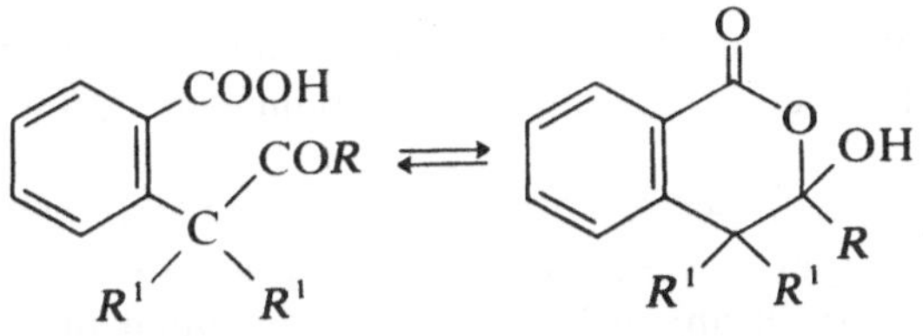

(20A)   (20B)

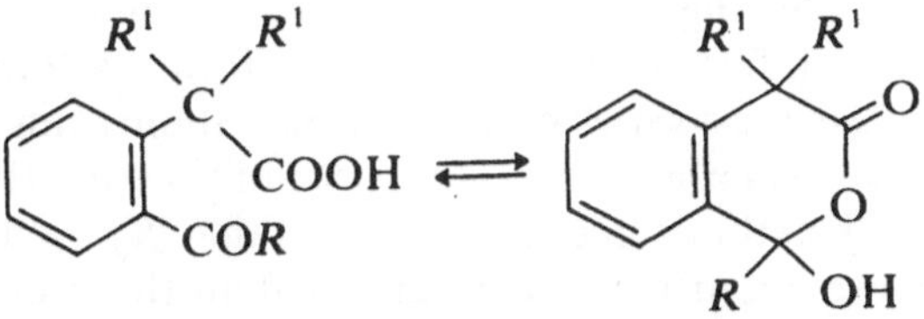

(21A)   (21B)

The structures of crystalline 2-acetonyl- and 2-phenacylbenzoic **(20a–c)**, and o-acylphenylacetic acids **(21a–d)** have been examined by IR (see Table 5) [119–122]. The position of the ring-chain equilibrium has only been approximately determined from chemical reactions. Acids **(20b, c)** and **(21c)** containing gem-dimethyl groups are cyclic in the solid state. The acid **(21d)**, containing a keto group conjugated to two benzene rings, possesses an open structure, regardless of the presence of a gem-dimethyl group. From these acids only **(20b)** gives the cyclic 3-methoxy-3,4,4-trimethyl-1-isochromanone

Table 5. The Structure of 2-Acylmethylbenzoic **(20)**
and *o*-Acylphenylacetic Acids **(21)** in the Solid State
[119, 120]

Compound	R	R^1	Isomeric structure obtained in solid state
(20a)	Ph	H	A
(20b)	Me	Me	B
(20c)	Ph	Me	B
(21a)	Me	H	A
(21b)	Ph	H	A
(21c)	Me	Me	B
(21d)	Ph	Me	A

in a reaction with diazomethane. The equilibrium of acid **(20b)** in solution is shifted entirely in favor of the ring form due to the absence of a keto group conjugation with the benzene ring and the presence of a gem-dimethyl group [120].

Solid 2-formylmethyl-5,6-dimethoxybenzoic acid possesses an open structure but a DMSO-d$_6$ solution shows a balanced equilibrium ($K_T = 1$) [18].

8-Acyl-1-naphthoic Acids (1g). Among 4-acylcarboxylic acids, these acids represent a special case. The interacting groups are rigidly fixed at a shorter distance from one another than in 2-acylbenzoic acids.

Lansbury and Bieron [40] were the first to show, using ^{1}H-NMR spectroscopy, that the ring-chain equilibrium in a solution of 8-acetyl-1-naphthoic acid is shifted toward the cyclic form. Additional indirect chemical proof exists for a series of 8-acyl-1-naphthoic acids [81, 123–126].

A quantitative examination of these equilibria by IR [38, 39] and potentiometric [38] methods has clearly shown that K_T values of 8-acyl-1-naphthoic acids are about two orders higher than those of 2-acylbenzoic acids having a corresponding substitution pattern (see Table 3). This difference is caused by a greater proximity of interacting groups in 8-acyl-1-naphthoic acids, which can be confirmed using molecular diagrams. Figure 5 shows that the nucleophilic oxygen atom of a planar carboxyl group is closer to the electrophilic carbonyl carbon atom in the naphthoic acid derivative.

Increasing size of the alkyl substituent at the keto group in a series of 8-acyl-1-naphthoic acids (Me < Et < *i*-Pr < t-Bu) favors the cyclic form [38]. A similar, but smaller effect has also been observed for the corresponding 2-acylbenzoic [26] and Z-3-acylacrylic [22] acids.

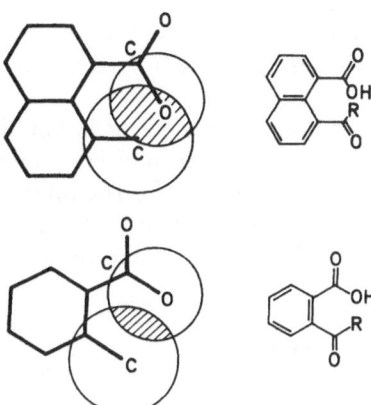

Figure 5. The molecular diagrams of 8-acylnaphthoic and 2-acylbenzoic acids demonstrating the hypothetical planar conformations. The overlapping areas of circles with the effective radii of a keto group carbon and of a carboxylic group oxygen atom are shaded.

Expanding the bulk of substituent R probably increases the steric strain of the open structure much more than that of the cyclic one. Moreover, sterically demanding substituents cause the keto group to turn out of the plane of the aromatic ring or of the ethylene bond, which also favors cyclization.

8-Benzoyl-1-naphthoic acid is certain to have at least two crystallinic modifications and both possess cyclic structure [127]. By contrast, 8-(4-methoxybenzoyl)-1-naphthoic acid exists as an open isomer in the solid state [128].

The presence of increasingly electron-withdrawing substituents X in a series of 8-(3-X- or 4-X-benzoyl)-1-naphthoic acids causes a displacement of the equilibrium in favor of the ring form, as demonstrated by a linear correlation [38]:

$$\log K_T/(K_T)_0 = \rho_I\sigma_I + \rho_R\sigma_R^0$$

A series of eight compounds shows $\rho_I = 1.286$; $\rho_R = 2.918$ ($s = 0.073$; $r = 0.970$).

Carboxylic Acids with Acyl Groups in Position 5. Unlike the 3- and 4-acylcarboxylic acids and their derivatives, often undergoing ring-chain tautomerism, 5-acylcarboxylic acids show interconversions leading to the seven-membered hydroxylactones very rarely.

Christiaens and Renson [81] did not succeed in an attempt to detect the ring form of 2-acyldiphenyl-2'-carboxylic acids by chemical methods, although the very type of structure might give the essential steric proximity of COOH and C=O groups.

A ring-chain equilibrium **(22A)** ⇌ **(22B)** has been reported [129] for solutions of *N*-pyruvoylanthranilic acid. However, later these data were rejected [130].

The only 5-acylcarboxylic acid present in equilibrium with the seven-membered hydroxylactone, which even predominates, is the 4-formyl-5-phenanthroic acid **(23A)** ⇌ **(23B)** [38, 81]. The interacting groups are rigidly fixed at a short distance, which is shown by the high equilibrium constant ($K_T = 3000$) determined by the potentiometric method [38].

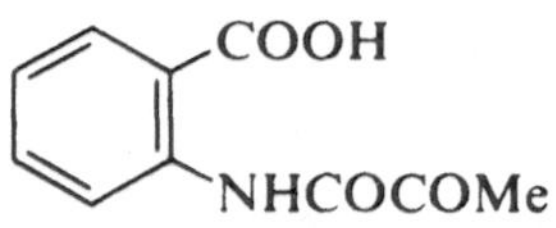

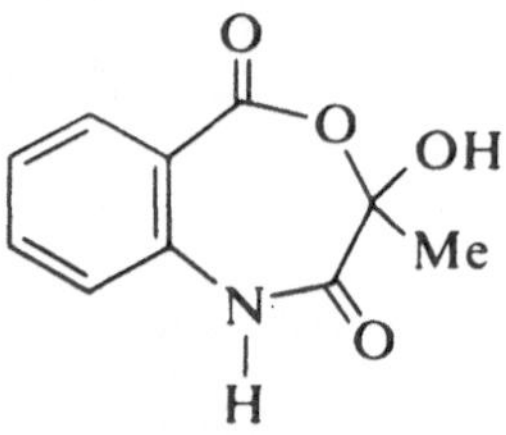

Bowden and Last [38] have suggested the following order of increasing stability of the ring form provided that the substitution pattern at the keto group is unchanged: 2-acylbenzoic < Z-3-acylacrylic < 8-acyl-1-naphthoic < 4-acyl-5-phenanthroic acids. Obviously, this order reflects the distances between the interacting groups.

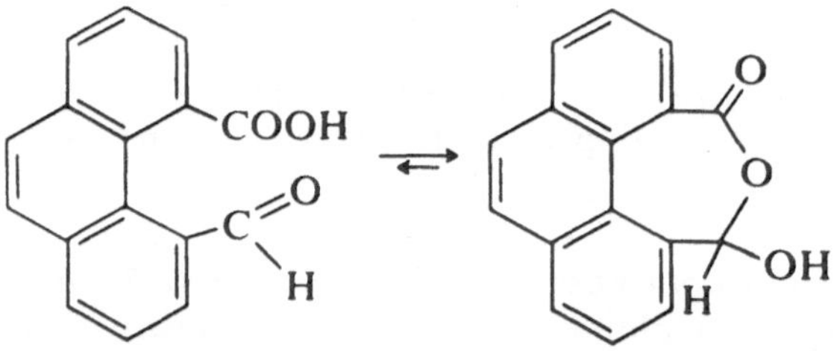

The Acylcarboxylic Acid Tautomeric Equilibrium Dependence on Solvents, Temperature, and Other External Factors. The influence of solvent properties on acylcarboxylic acid tautomeric equilibria has only rarely been studied. Polar solvents favor the cyclic form [31, 32, 68]. The mechanism of solvent effects on ring-chain equilibria **(1A)** ⇌ **(1B)** is highly complex. Most important is the stabilization of the ring or chain form by intermolecular hydrogen bonds with the solvent molecules. There are some approaches which may be helpful in special cases. A dilution of solutions [10, 30], a use of pyridine or triethylamine as solvent or their addition to solutions [30] shifts the equilibrium toward the open form. In acid media the equilibrium is shifted in favor of the less acid cyclic form [10, 16-18]. Raising the temperature favors the open form [65, 131].

2.1.2. Acyl Chlorides

Unlike the acylcarboxylic acids which form the ring tautomer as the result of an intramolecular nucleophilic addition to the C=O bond, the ring closure of acylcarboxyl chlorides takes place by intramolecular electrophilic addition to the keto or aldehydo group:

(2A)   (2B)

Structural factors, increasing electrophilicity of the COCl group and/or the nucleophilicity of the C=O group oxygen atom favor ring formation. The stability of cyclic chlorides of acylcarboxylic acids is mainly due to the high electrophilicity of the COCl group afforded by the $-I$-effect of the chlorine atom, when the steric structure of the chain connecting the interacting groups and the substituent R allow for a ring closure.

Bhatt and co-workers, investigating kinetically the solvolysis reactions of levulinic and 2-benzoylbenzoic acid chlorides [132] and some of their substituted derivatives [133], found no tautomeric equilibrium between the ring and chain forms of these acid chlorides.

IR spectra of the cyclic isomers show an intense lactone C=O band. Compared with a COCl band [134] it is slightly shifted toward high frequencies for five-membered chlorolactones, but toward low frequencies for six-membered analogues (see Table 6). Its frequency does not significantly depend on the state of the substance (solid or solution). However, the most important piece of evidence for the ring structure of acylcarboxylic acid chlorides is the absence of the aldehydo or keto group absorption in the range 1720-1660 cm^{-1} (Fig. 6).

A simple ^{13}C-NMR spectroscopic method has been suggested [135] for distinguishing between ring and chain isomers of ketocarboxylic acid chlorides and dicarboxylic acid dichlorides (see the discussion on page 47).

3-Acylcarboxylic acid chlorides, with the exception of a few special cases, have been obtained only as cyclic isomers **(2B)** [137], as has been shown for the chlorides of 3-formyl [30, 138], 3-acetyl [139-141], and other

Table 6. Chlorolactonic C=O Frequencies in the IR Spectra of
Acylcarboxylic Acid Chloride Ring Isomers

Formula	R	Solvent	$\nu_{C=O}$ cm^{-1}	Refs.
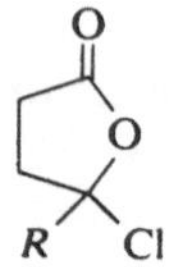	H	CCl$_4$	1812	30
	H Me Ph	CCl$_4$ CCl$_4$ Dioxan	1790 1798 1792	64 64 80
	H Me Ph	CCl$_4$ CHCl$_3$ CHCl$_3$	1818 1802, 1785 1780	35 20 109
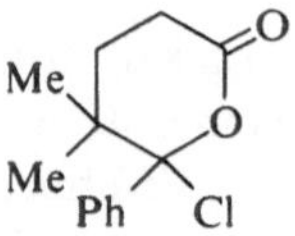		Nujol	1776–1754	136
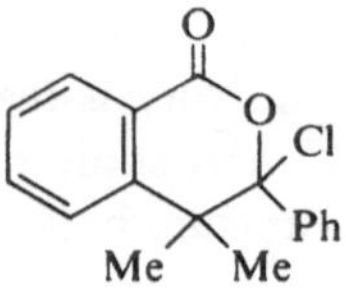		CCl$_4$	1750	120
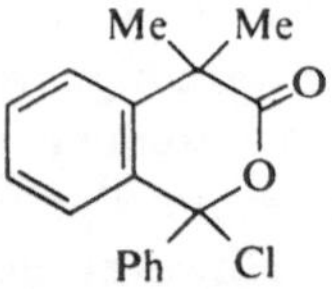		CCl$_4$	1770	119
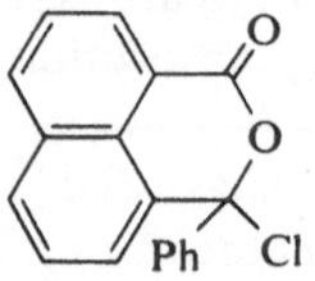		Dioxan	1731	128

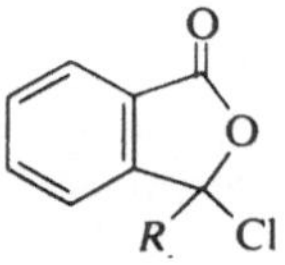

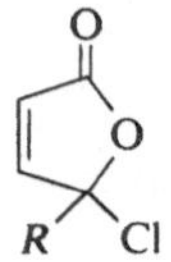

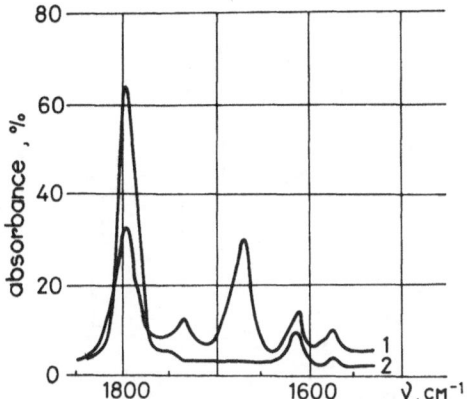

Figure 6. IR spectra of open and ring isomers of 2-aroylbenzoylchlorides in dioxan solution
($c = 5 \times 10^{-2}$ mol/l): 1—2-(2',4',6'-trimethylbenzoyl)benzoylchloride; 2—3-chloro-3-(2',4'-dimethylphenyl)phthalide.

3-acylpropionic [142, 143], 2-formyl [64, 144], 2-acyl (R = alkyl) [64, 145–149], and 2-aroylbenzoic [78–80, 146–148, 150–152], Z-3-formyl [35], Z-3-acyl (R = alkyl) [20, 111], and Z-3-aroylacrylic [9, 10, 109, 153–155] acids.

2-(2',4',6'-Trimethylbenzoyl)benzoyl chloride **(24)** [78, 81, 156], an exception among 2-acylbenzoyl chlorides, exists in an open structure due to steric hindrance of the keto group by two o,o'-methyl groups [157].

(24)
1) $\nu_{C=O}$ 1797, 1732†
2) $\nu_{C=O}$ 1673

Attempts to synthesize 2-(9-anthroyl)benzoyl chloride from the acid with thionyl chloride gave 9-chloro-9,10-dihydroanthrylidenephthalide **(26)** [158]. The phthalide $C_{(3)}$ atom in the intermediate carbocation **(25)** appears to be shielded by two peri-protons of the anthryl group hindering the chloride ion from adding to the cation at the $C_{(9)}$ atom. A similar shielding effect of an anthryl substituent on a keto group was observed [159] for

† IR C=O frequencies for dioxan solutions.

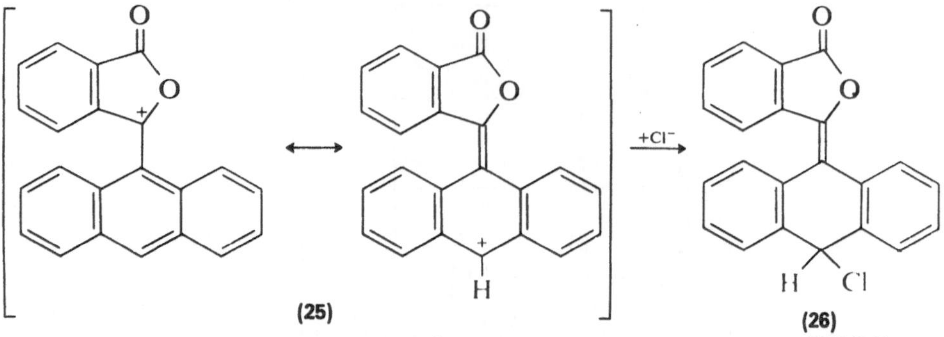

(25)

(26)
$\nu_{C=O}$ 1786 (6.8)

2-[o-(9-anthroyl)phenyl]-2-propanol, which exists exclusively as an open tautomer.

The pentacyclic-4-pyridyl of compound **(27)** in comparison with an anthryl substituent has a lesser influence on the keto group: the chloride **(27)** exists in the ring form [160].

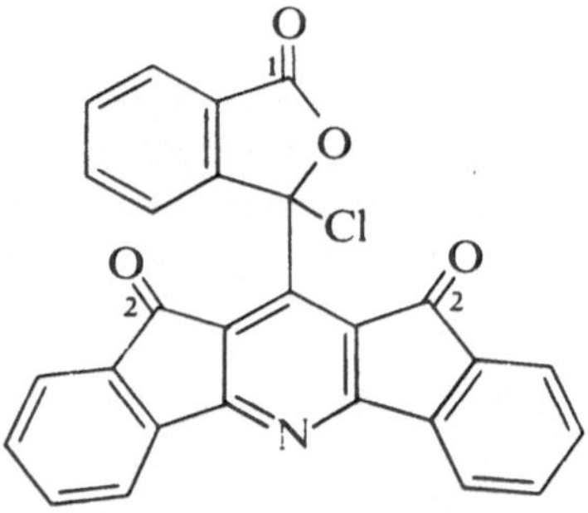

(27)
1) $\nu_{C=O}$ 1801
2) $\nu_{C=O}$ 1726

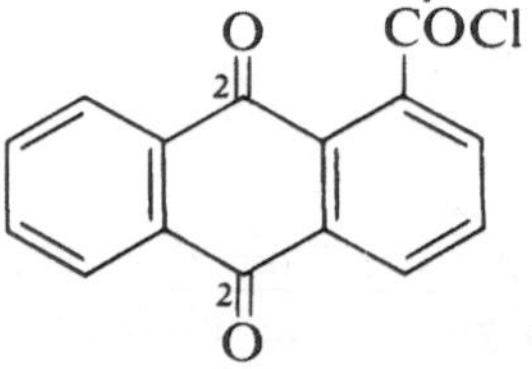

(28)
1) $\nu_{C=O}$ 1801
2) $\nu_{C=O}$ 1726

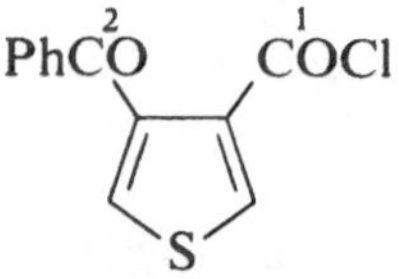

(29)
1) $\nu_{C=O}$ 1750
2) $\nu_{C=O}$ 1650

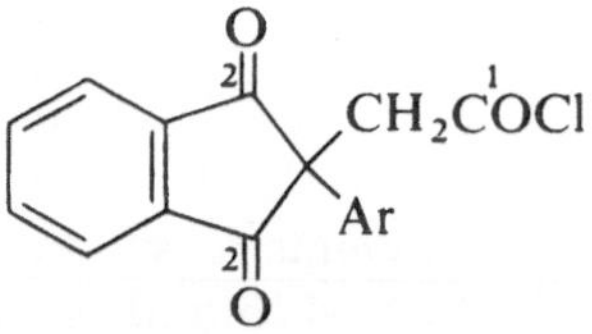

(30)
1) $\nu_{C=O}$ 1800–1770
2) $\nu_{C=O}$ 1740, 1700

Besides the shielding influence of the substituent at the keto group, the steric structure of the connecting link may prevent the chlorolactone ring from closing. Thus, no cyclic isomers of the chlorides of anthraquinone-1-carboxylic acid **(28)** [87], 4-benzoylthiophene-3-carboxylic **(29)** [161], and (2-aryl-1,3-indanedione-2-yl)acetic **(30)** [59] acids have been observed.

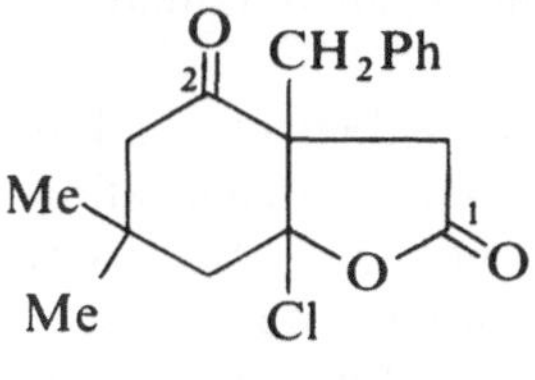

(31)
1) $\nu_{C=O}$ 1818 (4.0)
2) $\nu_{C=O}$ 1720 (1.7)

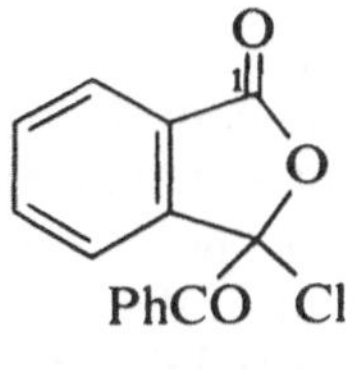

(32)
1) $\nu_{C=O}$ 1808 (4.6)
2) $\nu_{C=O}$ 1694 (1.2)

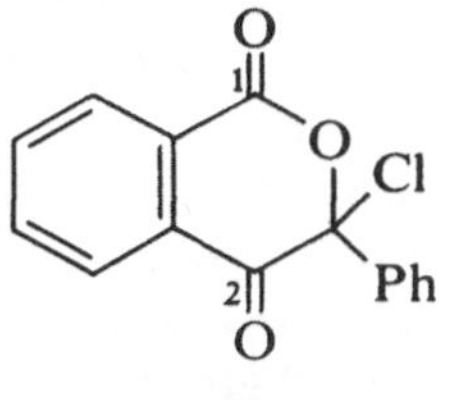

(33)
1) $\nu_{C=O}$ 1764 (3.6)
2) $\nu_{C=O}$ 1726 (1.7)

Attempts to prepare (2-benzyldimedone-2-yl)acetyl chloride have led to 9-benzyl-8-chloro-6,6-dimethyl-2,4-dioxohexahydrocoumarane **(31)** [162]. Here the five-membered chlorolactone ring is condensed with a six-membered cyclohexane ring (instead of a five-membered ring in the cyclic isomer of chloride **(30)**), which evidently favors the ring closure for steric reasons. Moreover, the keto groups of chloride **(30)** are conjugated with the aromatic ring. In a 1,3-cyclohexanedione derivative such conjugation is missing, which also favors the addition reaction to the C=O group of the 1,3-cyclohexanedione.

Two competitive intramolecular additions are possible for the chloride of benzil-o-carboxylic acid. Contrary to earlier studies [74] it was established by IR comparison [163] that both isomeric cyclic chlorides, i.e., 3-benzoyl-3-chlorophthalide **(32)** and 3-chloro-4-oxo-3-phenyl-3,4-dihydroisocoumarine **(33)** were observed. The six-membered chlorolactone easily undergoes isomerization involving ring contraction **(33)** → **(32)**.

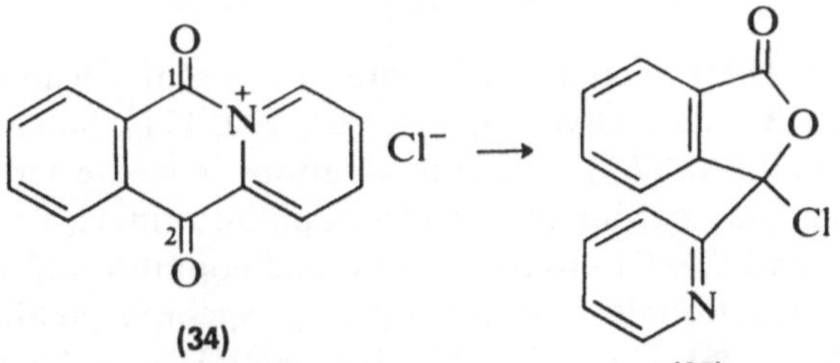

(34)
1) $\nu_{C=O}$ 1765
2) $\nu_{C=O}$ 1694

(35)
$\nu_{C=O}$ 1785 (Nujol)

It was recently shown [164] that the reaction of 2-(2-pyridylcarbonyl)benzoic acid with thionyl chloride leads to 4a-azoniaanthraquinone chloride **(34)**, which in polar solvents isomerizes into 3-chloro-3-(2-pyridylcarbonyl)phthalide **(35)**.

The chlorides of 4-acylcarboxylic acids have been comparatively rarely studied [137, 165]. By reaction of thionyl chloride and 4-benzoyl-4,4-dimethylbutyric acid Newman [136] obtained a mixture of open and cyclic isomeric chlorides. A similar mixture of isomers has been obtained for 4-benzoylbutyryl chloride [137].

It is impossible to obtain chlorides of 2-acylmethylbenzoic and o-acylphenylacetic acids since the reaction is accompanied by dehydration, leading to isocoumarines [120] and benzopyran-3-ones [119, 166, 167], respectively. But on replacement of the methylene hydrogen atoms by two methyl groups one succeeds in obtaining cyclic chlorides (for acids (20b, c) [120] and (21c, d) [119]). 8-Acyl-1-naphthoyl chlorides possess a ring structure for reasons discussed earlier [81, 124, 127, 168].

The chloride of 2'-benzoyldiphenyl-2-carboxylic acid was obtained only as an open isomer [169]. Cyclic isomers of 5-acylcarboxylic acid chlorides are not known.

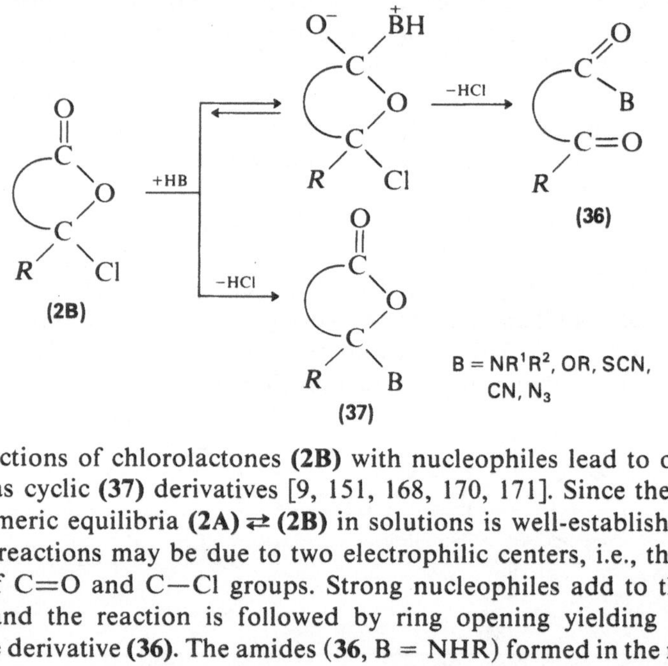

Reactions of chlorolactones (2B) with nucleophiles lead to open (36) as well as cyclic (37) derivatives [9, 151, 168, 170, 171]. Since the absence of tautomeric equilibria (2A) ⇄ (2B) in solutions is well-established [132], the two reactions may be due to two electrophilic centers, i.e., the carbon atoms of C=O and C—Cl groups. Strong nucleophiles add to the C=O group, and the reaction is followed by ring opening yielding an open structure derivative (36). The amides (36, B = NHR) formed in the reactions with ammonia and primary amines may isomerize further to give hydroxylactams (3B).

Weaker nucleophiles (aromatic amines, alcohols, urea, etc. [151]) replace the chlorine atom, leading to lactones (37). The direction of the reaction depends not only on the nucleophilicity of the reagent HB [151], but also on the structure of the chlorolactone (2B). Thus, substitution of the chlorine atom of 3-chloro-3-phenylperinaphthalide takes place even with aliphatic amines [168], which is quite unusual for a 3-chloro-3-phenylphthalide [146].

Dicarboxylic Acid Chlorides and Some Related Compounds. The electrophilic COCl group may intramolecularly add not only to the C=O bond of aldehydes and ketones, but also to other carbonyl groups. For example, some dicarboxylic acids may form two isomeric dichlorides of open (38A) and cyclic (38B) structures. These isomers are sometimes named as symmetrical and asymmetrical acid chlorides.

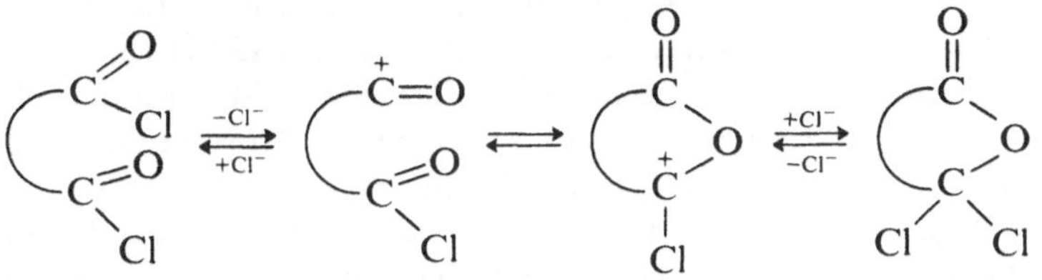

(38A)   (38B)

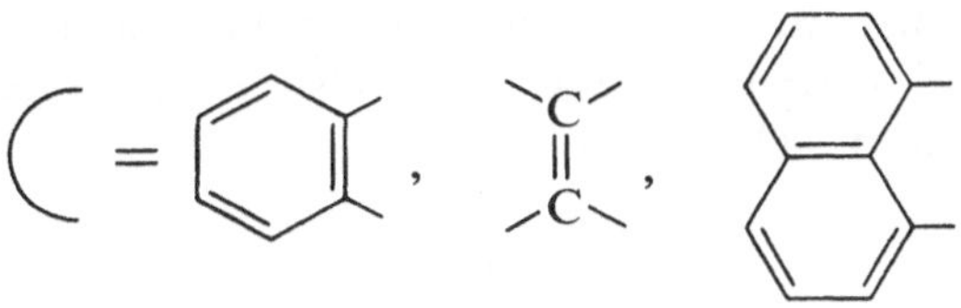

a   b   c

The most convenient way to identify sym- and asym-isomeric dichlorides is by means of ^{13}C-NMR [135]. Thus, for example, COCl carbon signals appear at $\delta 172.3$ ppm in succinyl dichloride and at $\delta 163.3$ ppm in sym-phthaloyl dichloride in CDCl$_3$. The 3,3-dichlorophthalide on the other hand shows signals at 169.1 ppm (C=O) and at 104.6 ppm (CCl$_2$). In addition three signals of benzene ring carbons appear in the spectrum of sym-phthaloyl dichloride due to the symmetry of the molecule, whereas the spectrum of 3,3-dichlorophthalide shows six benzene carbon peaks.

The isomerization of dicarboxylic acid dichlorides is an interesting case of intramolecular reaction between two identical functional groups (COCl), one of them acting as electrophile, the other as nucleophile. Evidently, a factor determining the stability of the cyclic isomers is the structure of the linking fragment. By means of spectroscopic methods the absence of a ring isomer in the solutions of succinic [135, 139, 142, 172] and glutaric [165] acid dichlorides was established. Cyclic isomers are formed from dichlorides of phthalic (38a) [135, 173], substituted maleic (38b) [173, 174], and 1,8-naphthalic (38c) [126] acids.

A condition for ring closure is the conformational rigidity of the linking fragment, which ensures the essential proximity of the COCl groups. Both isomers of phthalic acid dichloride have been obtained [175] and the isomerization **(38B)→ (38A)** proceeded on distillation. The reverse isomerization **(38A) → (38B̊)** takes place under the catalytic influence of Lewis acids (e.g., AlCl₃), which operate as chlorine anion transferrers.

The absence of a tautomeric equilibrium **(38A)** ⇌ **(38B)** for both isomers of phthaloyldichloride was established [176] by means of a kinetic study of their solvolysis in aqueous solutions of acetone, dioxan, and DMFA at 25–40°. Both isomeric phthaloyldibromides were also obtained [177].

The reactions of *O*-acetylsalicylic acid chloride **(39A)** with alcohols or aromatic hydrocarbons in the presence of Lewis acids produces 2-alkoxy or 2-aryl-2-methyl-1,3-benzodioxan-4-ones [178]. The authors [137] were unable to provide spectroscopic evidence for the presence of the cyclic isomer **(39B)**. This is in accordance with a kinetic study [137] of the solvolysis of **(39A)** in aqueous dioxan. The formation of cyclic products was attributed [179] to the capability of *O*-acetylsalicylic acid chloride to form carbonium ion **(40)**, confirmed by the isolation of the hexachloroantimonate **(40, X = SbCl₆)** [180]. A similar behavior was also observed for α-acyloxycarboxylic acid chlorides [180, 181].

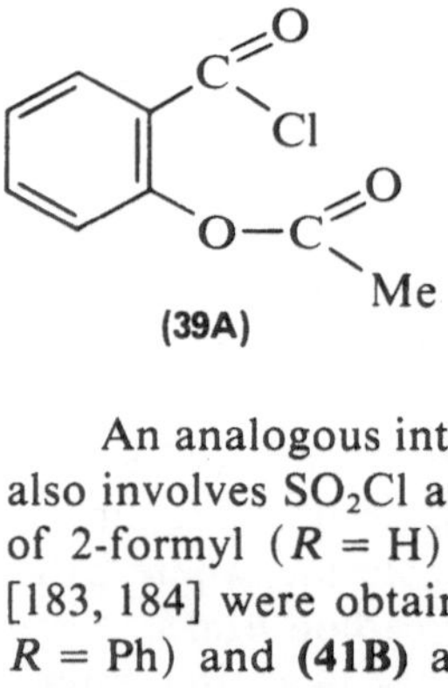

(39A) 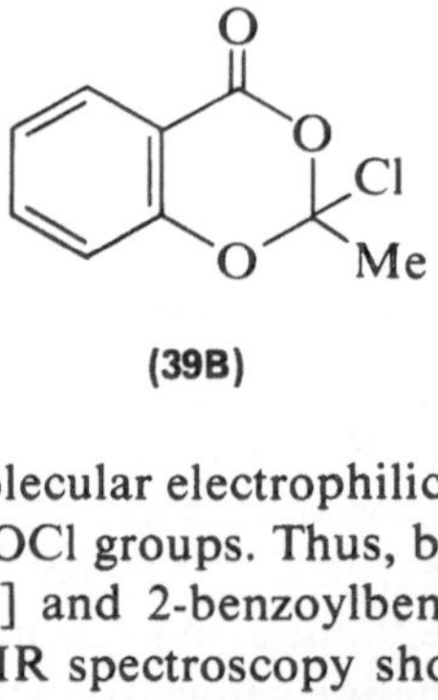   (39B) 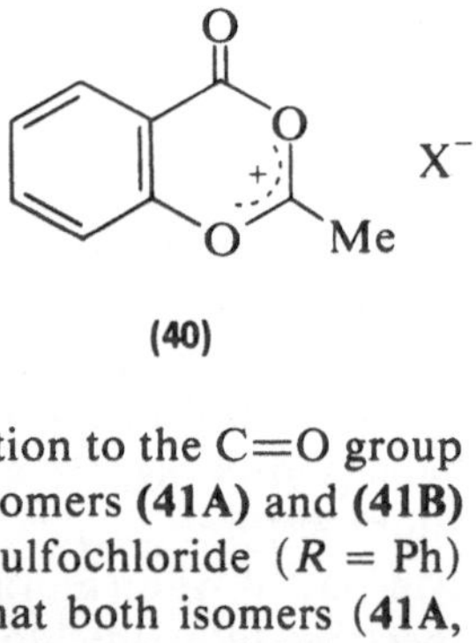   (40)

An analogous intramolecular electrophilic addition to the C=O group also involves SO₂Cl and SOCl groups. Thus, both isomers **(41A)** and **(41B)** of 2-formyl (*R* = H) [182] and 2-benzoylbenzenesulfochloride (*R* = Ph) [183, 184] were obtained. IR spectroscopy shows that both isomers **(41A, *R* = Ph)** and **(41B)** are stable in solutions and do not tautomerize. The mixture of isomers ([*A*] : [*B*] = 7 : 3) was obtained from **(41B)** by heating to 220°.

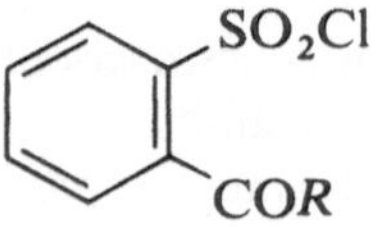

(41A) 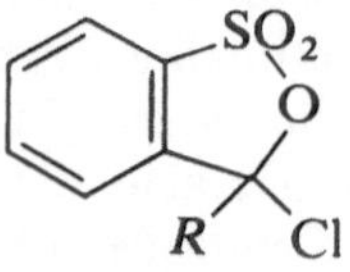   (41B)

Unlike the 2-formylbenzenesulfochloride isomers (**41A**, R = H) and (**41B**) [182], which are stable in solutions, 2-formylbenzenesulfinylchlorides (**42A**) are subject to a ring-chain equilibrium ($K_T \sim 0.25$) in CD_2Cl_2 solution as has been shown by ^{1}H-NMR investigations [185].

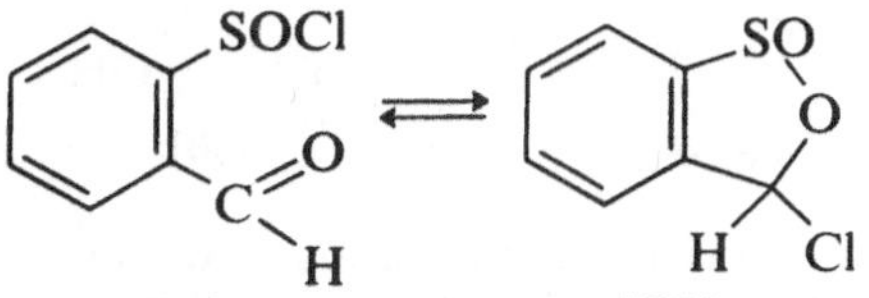

(42A)   (42B)

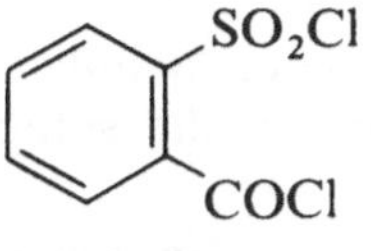

(43A) 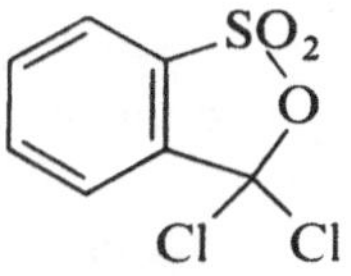   (43B)

Two isomeric 2-sulfobenzoic acid dichlorides (**43A**) and (**43B**) have been described [186]. Distillation leads to the isomerization (**43B**) → (**43A**) and the formation of the mixture of both isomers.

2-Sulfinobenzoic acid dichloride forms three isomers (**44A**), (**44B**) and (**44C**) [187, 188]. The isomerization (**44C**) → (**44A**) was carried out at 130°C in a vacuum.

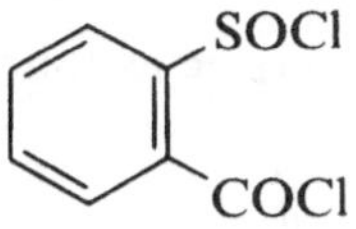

(44A) 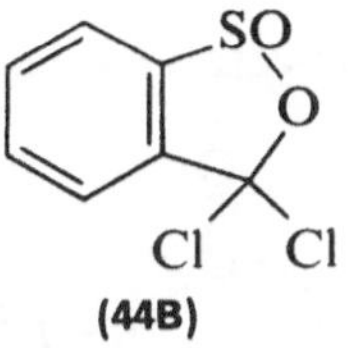   (44B) 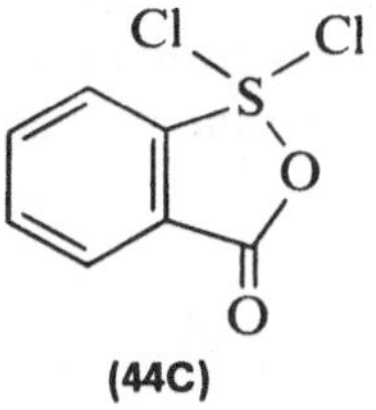   (44C)

2.1.3. Amides, Hydrazides, and Amidines

Contrary to the acyl carboxylic acids, which equilibrate in solutions (see Sec. 2.1.1.), *N*-unsubstituted and *N*-monosubstituted amides of these acids are more stable. In most cases the tautomeric equilibria (**3A**) ⇌(**3B**) do not occur in neutral aprotic solvents at room temperature. The acidities or basicities of amide groups, obviously, are insufficient for acidic or basic catalysis. Assistance of a base especially seems indispensable for an intramolecular addition to the keto group by a weak nucleophile such as the amide nitrogen atom. Acid on the other hand increases the electrophilicity of the keto group by protonation of the oxygen atom [189].

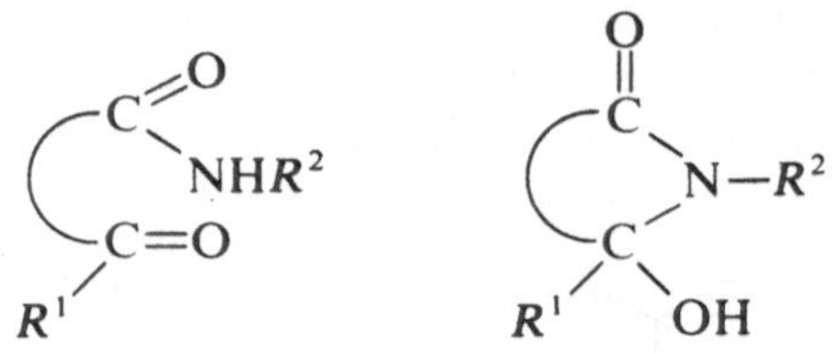

In protic solvents the interconversions of the keto amides usually take place at a measurable rate. Equilibria were obtained in methanol-d$_4$ [190] (for rate constants see Table 13) and ethanol [191]. The reaction takes place more rapidly in aqueous pyridine and a considerable acceleration may be achieved by adding sodium bicarbonate to the solution [192] (for rate constants see Table 10).

In aprotic solvents equilibration occurs after the addition of a proton acceptor to the solution. Thus, for example, N-(t-alkyl)-2-aroylbenzamides and their cyclic isomers, being stable in dioxan solution, equilibrate when triethylamine is added [193]. The equilibrium position (generally reached in 24–36 hours) is independent of the starting isomer (ring or chain).

In aprotic solvents ring-chain equilibria may be observed for ketoamides, which contain additional amino groups, the latter providing the basic catalyst. Ring-chain equilibria are described for N-unsubstituted, N-methyl, N-ethyl, and N-phenyl-2-(4-dimethylaminobenzoyl)benz-amides [191], and also for N-(t-alkyl)-2-(2-imidazolylcarbonyl)benzamides [194] (see Table 14).

The mechanism of the base-catalyzed equilibration may be as follows:

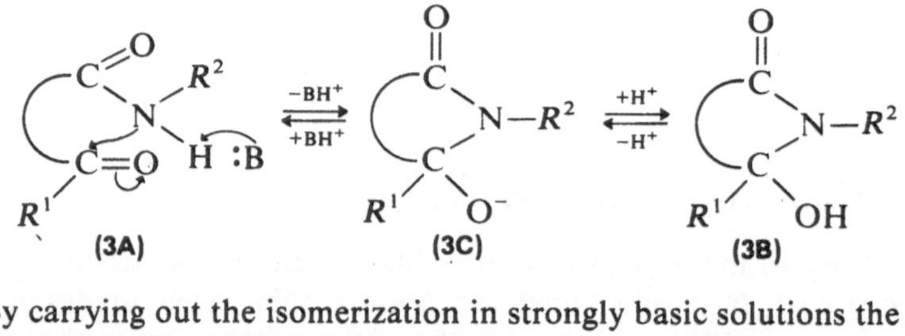

By carrying out the isomerization in strongly basic solutions the anion of the hydroxylactam (3C) is formed; hydroxylactams (3B) possess a stronger acidity ($pK_a \sim 12$) than amides (3A) ($pK_a = 16$–19). During acidification a fast protonization (3C) → (3B) takes place.

Since the energy of the hydrogen bond O—H$\cdots$B is greater than that of N—H$\cdots$B, the proton acceptors not only act as catalysts in the reversible isomerization, but, in greater concentration, also shift the equilibrium toward the cyclic isomer.

A confirmation follows from IR investigations [195, 196] showing that intramolecular hydrogen bonds O—H···N stabilize the cyclic isomers of N-(β-dialkylaminoethyl)amides of (2-phenyl-1,3-indanedione-2-yl)- and (2-benzyl-2-dimedonyl)acetic acids:

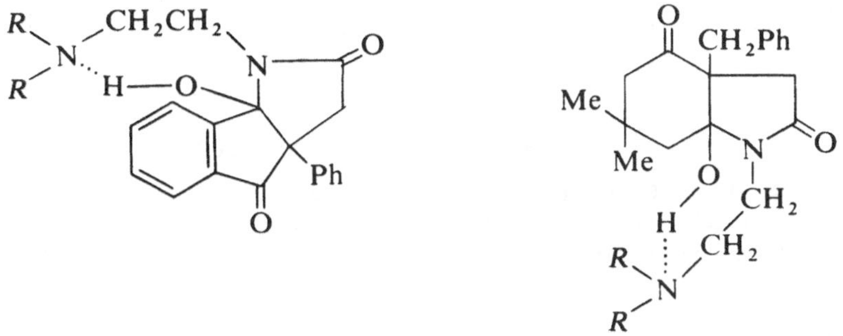

The IR method is most frequently used for a reliable determination of open **(3A)** and closed **(3B)** structures (see generalized data in [33]). Proof for the open structure of N-monosubstituted amides is the presence of an amide-II band ($\nu_{C-N} + \delta_{N-H}$) at 1570–1520 cm^{-1} [33, 146, 189, 197–200]. Since this band appears in the range which is usually transparent in the spectra of hydroxylactams **(3B)** it can be used as an analytical band for quantitative measurement of isomer proportions [146, 193]. However, in the spectra of nitro group-containing amides this band is useless since asymmetrical vibrational bands of nitro groups (ν_{NO_2} as) appear in the same region (1550–1515 cm^{-1}).

In the solid-state IR spectra of open amides **(3A)** one can easily detect the comparatively tight band of intermolecularly bonded N—H groups at 3400–3150 cm^{-1} and distinguish it from the broad bands of bonded O—H groups (O—H···O=C) appearing in the same range for cyclic isomers **(3B)** (see Fig. 7) [146, 200].

Open amides **(3A)** show C=O, amide-I, and amide-II bands in the double bond region. In solid-state spectra lowering of the amide-I band frequency and splitting of this band together with an increase of the amide-II band frequency takes place [68, 78, 146, 198] as contrasted with solution spectra. This is due to the intermolecular hydrogen bonds CONH···O=CNH, usually characteristic of amides in the solid state.

Intramolecular hydrogen bonds N—H···O=C(ketone) have been detected [202, 203] by means of the IR method in a number of N-aryl-3-acylpropionamides.

When interpreting the open isomer IR spectra difficulties sometimes appear; these are due to an overlap of ketone C=O and amide-I bands

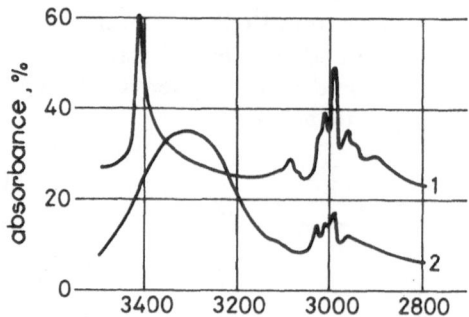

Figure 7. N—H and O—H bands in the IR spectra of ring and chain isomers of 2-acyl-benzamides in solid state: 1—*N*-(*t*-butyl)-2-(4-chlorobenzoyl)benzamide; 2—2-(*t*-butyl)-3-(4-chlorophenyl)-3-hydroxyisoindolinone [201].

[78, 79, 146, 197, 199]. In such cases measuring the integrated intensity of both band is recommended [204]. In this way the IR spectra of *N*-(*t*-butyl)-2-benzoyl-x-nitrobenzamides (x = 3, 4, 5, 6) were successfully interpreted [79]; two overlapping bands were observed: $\nu_{C=O}$ + amide-I at 1680–1670 cm^{-1} and $(\nu_{NO_2})_{as}$ + amide-II at 1540–1530 cm^{-1} (in dioxan) (see Fig. 8).

The lactam C=O band in the spectra of cyclic isomers of 3-acylcar-boxylic acid amides (**3Ba–c**) in solution is shifted to high frequencies by 30–40 cm^{-1} when compared with the amide-I band in the spectra of open isomers, this is due to the angle strain in the five-membered ring (see Table 7).

The use of C=O bands for the structure determination of acylcarboxylic acid amides is complicated because hydroxylactams (**3B**) can form inter-molecular hydrogen bonds O—H⋯O=C(lactam) in the solid state [20, 78, 146, 147, 162, 163, 170, 196, 198]. This causes a C=O band shift to low frequencies as compared with solution spectra or even a C=O band splitting due to the free and bonded C=O group absorption (see Table 7, Fig. 9). The hydrogen bonded C=O group bands of crystalline 2-benzoylbenzamides, strictly established as cyclic [33, 80, 146, 170, 205], have sometimes been erroneously attributed to the overlapping bands ($\nu_{C=O}$ diarylketone + amide-I) of open isomers (**3A**). This led to an incorrect conclusion [151, 191, 206] about the structure of these amides in solid state, the incorrectness of which is confirmed [207] by the absence of an amide-II band and the presence of a broad OH band in the spectra. For this reason the IR spectra of 2-phenylacetylbenzamides [208–210] and 2-α-(*N*-alkyl-amino)phenylacetylbenzamides [208] were interpreted erroneously as indicating an open structure. There is strong spectroscopic evidence

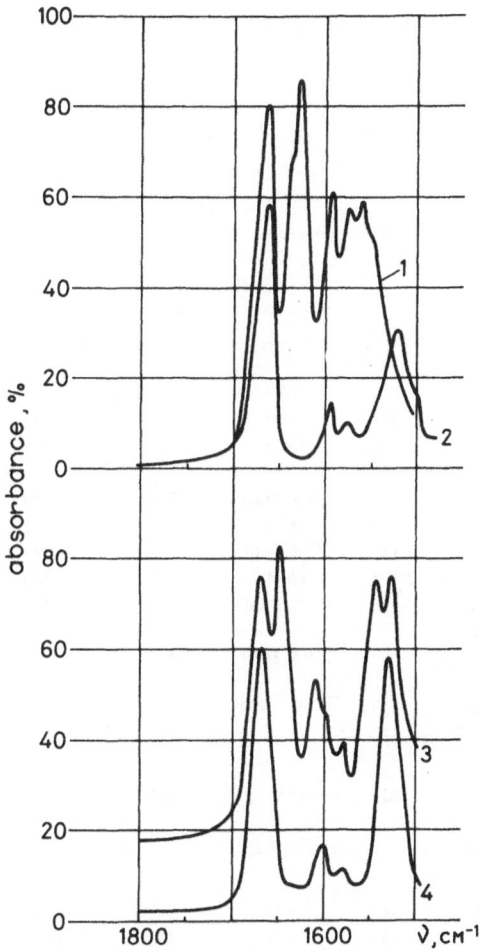

Figure 8. Partial IR spectra of 2-acylbenzamides: 1—N-isopropyl-2-benzoylbenzamide in the solid state; 2—its solution in dioxan; 3—N-(t-butyl)-2-benzoyl-4-nitrobenzamide in the solid state; 4—its solution in dioxan ($c = 5 \times 10^{-2}$ mol/l) [79, 146].

[147, 211–213] for the cyclic structure of the above-mentioned amides in the solid state.

It is recommended that frequencies and intensities of hydroxylactam **(3B)** C=O bands in proton acceptor solvents be measured. Dioxan is most suitable. As a result of the formation of intermolecular hydrogen bonds with the solvent molecule (O—H···dioxan) one can observe only the free C=O group absorption. In solvents of low polarity the hydroxylactam

Table 7. C=O and OH Band Frequencies in IR Spectra of Hydroxylactams (3B)

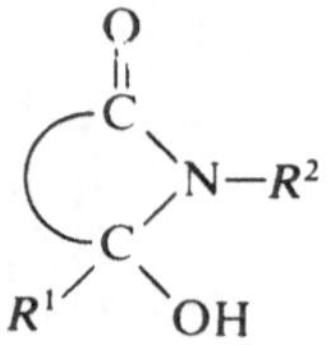

Ring	R^1	R^2	Solid state		Solution in dioxan		Refs.
			$\nu_{C=O}$	$\nu_{OH,NH}$	$\nu_{C=O}$	$A^\dagger_{C=O}$	
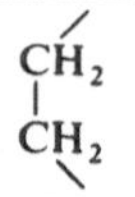 $\begin{array}{c}\text{CH}_2\\ \mid\\ \text{CH}_2\end{array}$	Me	H	1670	3280	1705		33
	Me	Me	1660	3250	1695		33
	Ph	Me	1680	3220	1698		33
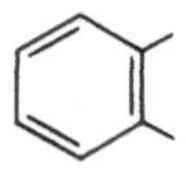	Ph	H	1703, 1661	3275	1720	5.5	146
	Ph	Me	1673	3259	1707	5.1	146
	Ph	i-Pr	1678, 1666	3169	1697	4.3	146
	4-ClC$_6$H$_4$	t-Bu	1674	3311	1700		201
	Ph	Ph	1675, 1660	3181	1707	4.7	170, 205
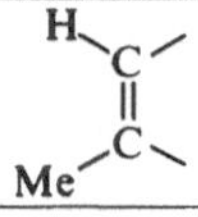 $\begin{array}{c}\text{H}\diagdown\text{C}\diagup\\ \parallel\\ \text{Me}\diagup\text{C}\diagdown\end{array}$	4-BrC$_6$H$_4$	H	1695, 1661	3390, 3344, 3279			9
	4-BrC$_6$H$_4$	Me	1698, 1658	3289			9
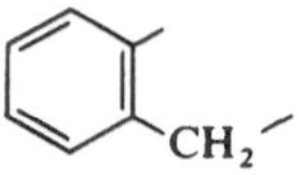 CH_2	Ph	Me	1638	3327	1661		214
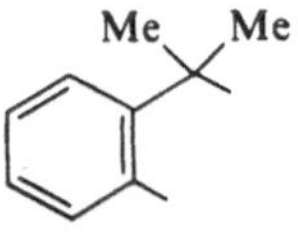 Me, Me	Ph	H	1640	3300	1667		171
	Ph	Me	1621	3317	1649		171
	Ph	Ph	1633	3295	1648		171
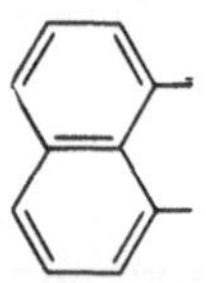	Ph	Pr	1637	3278	1651		168
	Ph	Ph	1661sh 1634	3281	1663	3.0	128

† Integrated intensities are given in units $10^4 \, l \cdot mol^{-1} \cdot cm^{-2}$ (ln).

C=O band shifts toward lower frequencies, increasing its intensity, and sometimes splitting simultaneously [198, 205]. These phenomena were erroneously said to indicate an equilibrium (3A) ⇌ (3B) [199].

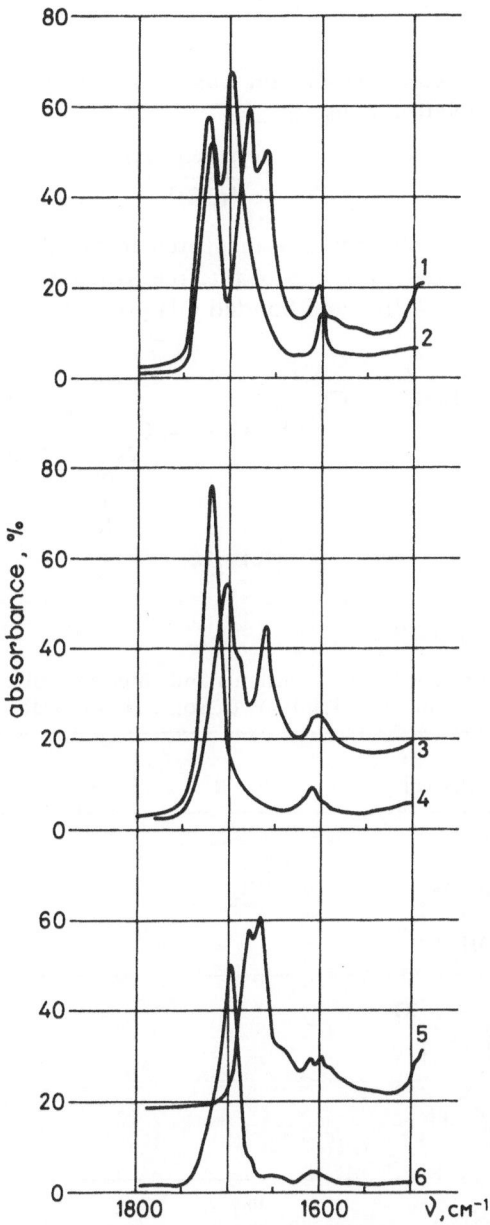

Figure 9. The splitting and the shift toward lower frequencies of the hydroxylactam C=O band in the IR spectra of ketoamide ring isomers in the solid state as compared with solution spectra: 1—2,3(CO)-benzoylen-1-ethyl-2-hydroxy-3-phenyl-5-pyrrolidone in the solid state; 2—its solution in dioxan; 3—3-hydroxy-3-phenylisoindolinone in the solid state; 4—its solution in dioxan; 5—3-hydroxy-2-isopropyl-3-phenylisoindolinone in the solid state; 6—its solution in dioxan [146, 198].

The hydroxylactam association was quantitatively evaluated [215] by means of an association constant

$$K_d = \frac{[(3B)\cdots(3B)]}{[(3B)]^2}$$

and hydrogen bond enthalpy (ΔH) measurements (Table 8). The concentrations of the associated form and of the free hydroxylactam were deduced from the intensities of free and bonded OH group bands in CCl_4 solution.

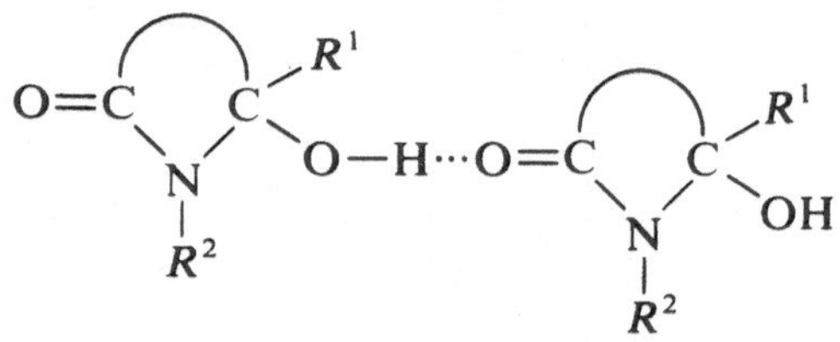

(3B)$\cdots$(3B)

Table 8. Dimeric Association Constants and Intermolecular Hydrogen Bond Enthalpies in CCl_4 Solutions of Some Hydroxylactams [215]

Hydroxylactam	R^1	R^2	K_d (l/mol)	$-\Delta H$ (kcal/mol)
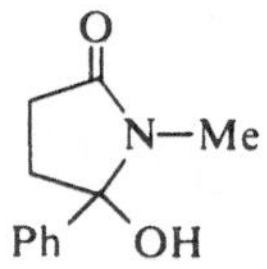			458	
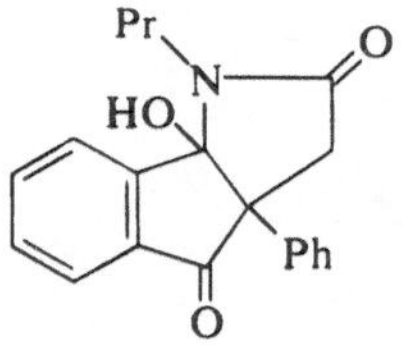			319	
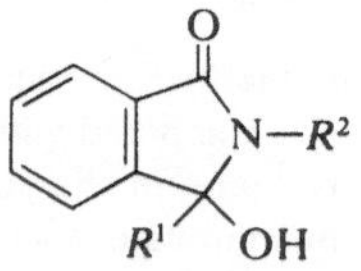	Ph	Pr	96	5.14
	Ph	i-Pr	27	4.63
	Ph	Ph	169	
	4-ClC$_6$H$_4$	PhCH$_2$	177	
	4-ClC$_6$H$_4$	Pr	126	5.33
	4-ClC$_6$H$_4$	i-Pr	63	4.74
	3-NO$_2$C$_6$H$_4$	Pr	145	5.36

In comparison with hydroxypyrrolidones, hydroxyisoindolinones (see Table 8) are less capable of association, which may account for the decrease of C=O group proton acceptor ability due to its conjugation with the benzene ring. Electron-withdrawing substituents R^1, which increase OH-acidity, favor association, while an increase in the steric bulk of substituent R^2 decreases the association constant.

IR spectroscopy has been applied to structure determination of ketoamide ring and chain isomers. Due to the association phenomena, many contradictory results have been published.

Apart from IR methods UV [9, 80, 194, 199, 216, 217], ¹H-NMR [33, 189, 190, 192, 199, 218–222], and mass spectrometric [189] methods were successfully applied to ketoamides. A detailed discussion of ¹H-NMR application is given by Flitsch [33].

We now consider the influence of the structure of the connecting link of a molecule and of substituents R^1 and R^2 on the stability of acylcarboxylic acid amide isomers (3Aa–h) and (3B) in the solid state and in solution, and also on their mutual interconversions.

3-Acylpropionamides. The isomeric structures (45A), (45B), and (46) of levulinic acid amide, obtained by Wolf from α-angelicalactone [223], were first suggested in Beilstein's handbook [224].

Lukes and Prelog [225–228] have systematically examined levulinic and 3-benzoylpropionic acid amides. However, due to the lack of a spectroscopic method they were unable to assign correctly the structure of these amides. A product obtained from the reaction of aniline and α-angelicalactone was reported to possess the open structure (45A, R^1 = Me, R^2 = Ph),

(45A)   (45B)   (46)

while a reaction of N-phenylsuccinimide with phenyl magnesium bromide was said to give the cyclic isomer (45B). Later Walton [229] ascertained that both reactions lead to the same product, to which he erroneously ascribed the ring structure (45B). Lukes [230], thirty years after his first investigations, using IR spectra, was able to show that the product formed in both reactions possesses the open structure (45A). The history of the investigations shows clearly that chemical methods, e.g., the synthesis of a compound, cannot be taken as an evidence of its structure, if the compound is capable of the isomeric interconversion.

Flitsch isolated [33, 231] both isomers (**45A**, R^1 = Me, R^2 = H) and (**45B**) of levulinic amide. The isomerization (**45B**) → (**45A**) was accomplished at 100°. A similar occurrence of equilibrium displacement toward the open form by heating a number of levulinic and 3,3-dimethyllevulinic amides was observed by other investigators [219, 232].

Keller and Prelog [189] isolated both levulinic anilide isomers and carried out the isomerization (**45A**) → (**45B**) by passing CHCl₃ solution through a column filled with acidic ion exchange resin. The reverse isomerization (**45B**) →(**45A**) proceeds when a methanolic solution is treated with silica gel. Both isomers are stable in the solid state and in neutral solvents. Over 10 days the formation of an equilibrium was observed in CDCl₃ solution in the presence of acidic (CF₃COOH) or basic (pyridine) catalysts.

Cromwell and Cook [197] comparing the IR spectra of a number of 3-benzoylpropionamides (**45**, R^1 = Ph, R^2 = H, Me, Ph, a.o.) with those of N,N-disubstituted amides found that in the solid state these amides with the exception of the N-methylamide possess an open structure. In the IR spectra of dioxan solutions of open forms the ketone C=O and the amide-I bands overlap leading authors [197] to propose incorrectly equilibrium (**45A**) ⇄ (**45B**) [199].

As a result of systematic IR comparisons of a large number of 3-acylpropionamides Chiron and co-workers [199, 200, 202, 203, 233] showed that all these amides exist as open isomers in the solid state as well as in solution. In some cases the open structure is stabilized by an intramolecular hydrogen bond N—H···O=C(ketone).

Only N-methylamides of 3-acylpropionic acids possess a cyclic structure, an exception being some 3-aroylpropionamides having an electron-donating substituent in the aroyl group.

The +I-effect of substituent R^2 in (**45**) increases the nucleophilicity of the nitrogen atom. Therefore the cyclic structure is stabilized. The +I and +C-effects of substituent R^1 at the keto group in (**45**) act in the reverse direction: the electrophilicity of the keto group decreases. This leads to the decrease in the ability of the keto group to add the nucleophilic nitrogen.

The increased stability of the cyclic isomer of N-methylamides may arise from hyperconjugation [199].

Another interpretation, however, has been suggested [214]. +I-Effect increases in a series of substituents: H < Me < Et < i-Pr; therefore in the same order substituents at a nitrogen atom should increase the stability of the cyclic form. However, since steric hindrance by a substituent at nitrogen destabilizes the cyclic structure [3, 70, 146, 198], the stability of the ring form must be lowered in the same order as that of the substituents. Both opposing effects may explain the increased stability of the ring isomer for N-methylamides only.

Table 9. The Isomeric Structure of Some 3-Acylpropionamides in the Solid State

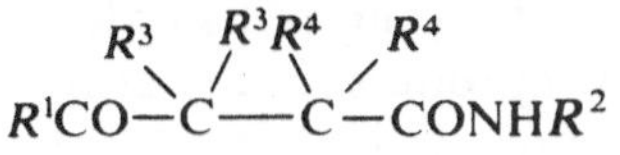

$$R^1CO-\overset{\overset{\displaystyle R^3}{|}}{C}-\overset{\overset{\displaystyle R^4}{|}}{C}-CONHR^2$$

(A)

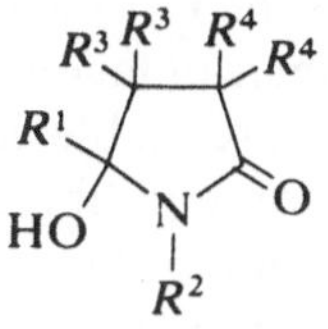

(B)

R^1	R^3	R^4	The structure amides, when $R^2 =$				Refs.
			H	Me	Ph	PhCH$_2$	
H	H	H	B	B			239, 240
Me	H	H	A, B	B	A, B	A, B	33, 189, 218, 220
Ph	H	H	A	B	A	A	33, 197, 199
Me	Me	H	B	B	B		231
Ph	H	Me		B		B	200
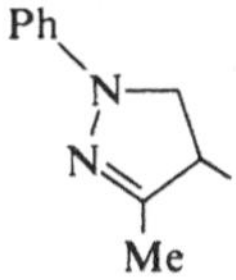 (pyrazole structure) Me	H	Me		B		A, B	238

A great number of levulinic [33, 189, 199, 218, 220, 230, 231, 234, 235], other 3-acylpropionic (R^1 = alkyl) [199, 220, 236, 237], 3-aroylpropionic [33, 197, 199, 202, 203], and methylsubstituted acylpropionic [141, 200, 221, 231, 237, 238] amides were synthesized and their structures strictly proven by means of spectroscopic methods. But one rarely succeeded in obtaining both individual isomers. This was the case for the N-unsubstituted amide [33], the N-benzylamide [218, 220], and the anilide [189] of levulinic acid, as well as for the N-benzylamide of 2,2-dimethyl-3-(3-methyl-1-phenyl-pyrazolyl-4-carbonyl)propionic acid [238] (see Table 9).

The N-monosubstituted amides of 3-formylpropionic [239-245] and 3-formyl-3-R-butyric [246, 247] acids were obtained in the ring form irrespective of the structure of the substituent at the nitrogen atom. Evidently this is due to the greater electrophilicity of the aldehyde group as compared with the ketone group.

3-Acylpropionic amides, bearing methyl groups in the chain, predominantly, also possess a cyclic structure [141, 200, 221, 231, 237, 238].

The data concerning tautomeric equilibria of solutions of 3-acylpropionic amides are contradictory. Again, conclusions stating the presence of equilibria **(45A)** ⇌ **(45B)** were sometimes based on incorrect interpretations of IR spectra [197, 199].

Both isomers of levulinic anilide [189] and *N*-benzylamide [220], which were isolated, are stable in neutral aprotic solvents at room temperature and do not form equilibrating mixtures. Using ¹H-NMR spectroscopy the presence of both isomers of levulinic anilide was demonstrated [33] in DMSO-d₆ solution.

Sheiman, Denisova, and Berezovskii [231, 248, 249] using the ¹H-NMR method have observed the tautomeric equilibria **(47A)** ⇌ **(47B)** which are displaced toward the cyclic tautomer in the case of *N*-unsubstituted and *N*-arylamides of 3,3,dimethyllevulinic acid. The situation is complicated by a dehydration of **(47B)** to give methylenepyrrolidone **(48)** which takes place under the condition of ring-chain equilibration in non-aqueous solutions.

X = NEt₂, OMe, H, Cl, Br, COOH, COOMe, CN, NO₂

Equilibrium constants were determined [249] using 50% aqueous pyridine. Table 10 shows that electron-withdrawing substituents X in the benzene ring destabilize the cyclic form and shift the equilibrium toward **(47A)** (see also [231]). The thermodynamic data ΔH and ΔS were calculated from the K_T temperature dependence.

Due to an increase of conformational freedom ring opening is accompanied by rising entropy. For most of the amides investigated ΔS values, being in the range 11–12 e.u., slightly depend on the substituent.

The introduction of strong electron-withdrawing substituents X into the benzene ring causes a simultaneous decrease of ΔH and ΔS values. The decrease in entropy is explained [249] as a decrease of conformational freedom of the open isomers **(47A)**. This should be due to the formation of stronger intramolecular hydrogen bonds N—H···O=C caused by electron-withdrawing substituents, increasing N—H acidity.

A good linear correlation $\log K_T/(K_T)_0 = -1.38\sigma^+$ (r = 0.96; s = 0.18) was found.

Solvents exert only a slight influence on the equilibrium position of **(47A)** ⇌ **(47B)**. However, the rate of isomeric interconversion strongly

Table 10. 3,3-Dimethyllevulinic N-(4-X-Phenyl)amide Ring-Chain Equilibrium (47A) $\rightleftarrows$ (47B) and Rate Constants, Thermodynamic and Activation Parameters, Determined by ^{1}H-NMR [192, 249]

X (R = 4-XC$_6$H$_4$)	K_T = [B]/[A] (34°C)	$-\Delta H$ (kcal/mol)	$-\Delta S$ (e.u.)	(A) → (B)			(B) → (A)		
				k_1 (s^{-1}, 50°)	$\Delta H^{\neq}$ (kcal/mol)	$\Delta S^{\neq}$ (e.u.)	k_2 (s^{-1}, 50°)	$\Delta H^{\neq}$ (kcal/mol)	$\Delta S^{\neq}$ (e.u.)
(R = H)									
NEt$_2$	14.7	5.3	11.9	0.31	25.7	18.7	0.031	30.1	27.6
OMe	16.4	5.2	11.2	2.34	19.3	2.7	0.19	24.6	14.2
H	17.2	5.1	11.1	54.6	19.1	8.7	14.2	23.9	20.9
Cl	5.85	4.7	11.8	75.9	18.7	7.8	18.0	24.7	23.5
Br	5.26	4.8	12.4	77.6	15.3	-2.4	23.4	19.6	8.1
I	5.0	3.9	9.6						
COOH	4.37	4.5	12.1	0.03	22.9	5.1	0.013	27.2	8.6
COOMe	2.13	3.2	9.3	195.0	14.6	-2.7	148.4	17.7	6.1
CN	1.66	3.1	9.5	512.9	17.9	11.3	478.6	21.1	18.9
NO$_2$	1.62	2.4	7.0						

depends on the properties of solvents. In the aprotic polar solvents (chloroform, pyridine, DMSO) the equilibrium is reached very slowly (usually in 11–26 days), in protic solvents the process proceeds more rapidly (~0.3–0.8 h). The addition of basic catalysts to the aqueous solution accelerates the equilibration even more. It may be possible that the reaction in aprotic solvents is catalyzed by traces of water, arisen from dehydration.

At elevated temperatures the equilibrium shifts toward the open form [192] (see also [219, 231]). The kinetics of equilibration was examined [192] by means of dynamic ^{1}H-NMR spectroscopy and the activation parameters of the tautomeric interconversions were determined (Table 10) using 50% aqueous pyridine containing 0.032 mol/liter sodium bicarbonate. The bicarbonate was added to bring the rates into an experimentally convenient range for all substituted derivatives. A linear correlation is observed between k_1 and k_2 and the concentration of bicarbonate.

Going from the electron-donating to electron-withdrawing substituents X rate constants k_1 and k_2 increase very strongly (Table 10). A good linear correlation is found between $\log k/k_0$ and σ or even better σ^+ coefficients of substituents X:

$$\log k_1/(k_1)_0 = 3.56\sigma^+, \qquad r = 0.997$$
$$\log k_2/(k_2)_0 = 4.68\sigma^+, \qquad r = 0.993$$

A large rate enhancement brought about by addition of bases led Sheiman and co-workers [250] to suppose that deprotonation is the rate-limiting step. This suggestion is in accordance with an acceleration of tautomeric interconversions by electron-withdrawing substituents X ($\rho > 0$).

On the basis of kinetic and thermodynamic data the authors [251] suppose that the conversion **(47A)** $\rightleftarrows$ **(47B)** proceeds via an open structure intermediate anion.

This conclusion contradicts the above-mentioned conjecture [251], that based on the larger acidity of hydroxylactams compared with open keto amides anions of ketoamides should possess a cyclic structure.

3-Cycloalkanonecarboxamides. 2-Decalone-9-carboxamides [252, 253] and analogous compounds of steroidic structure [254–256] possess cyclic structure **(49)** or **(50)**. Spectroscopic methods did not succeed in detecting the open isomers in solutions of these compounds.

(49)   (50)

Both isomers **(51A)** and **(51B)** of *N*-homoveratrylamide of 3-cycloheptanonecarboxylic acid were isolated [257]; however, their stability and the conditions or their interconversions were not studied.

The cyclic structure **(52)** was established [258] for *trans*-bicyclo[5.3.0]-6-decanone-1-carboxamide.

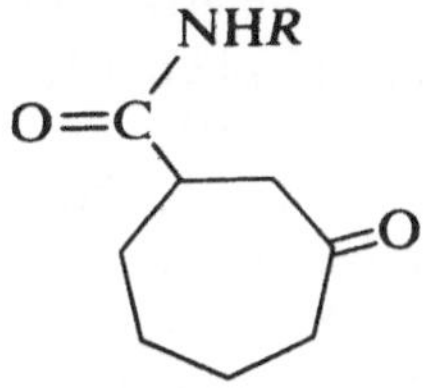

| (51A) | (51B) | (52) |

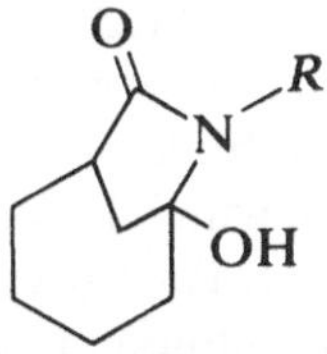

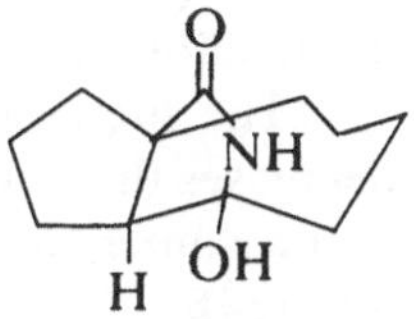

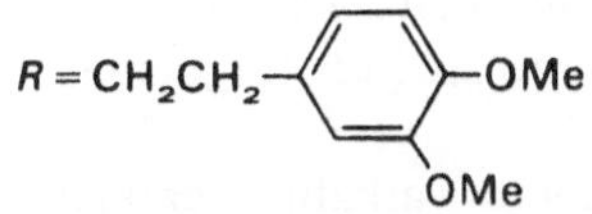

$R = CH_2CH_2$—⟨benzene ring⟩—OMe, OMe

(2-Phenyl-1,3-indanedione-2-yl)acetamides and Hydrazides. Either open amides **(54A)** or their cyclic isomers **(54B)** were formed, depending on the reaction conditions and on the structure of substituent R, by acylation of primary amines with the (2-phenyl-1,3-indanedione-2-yl)acetyl chloride **(53)** [198]. The probability of intramolecular cyclization decreases with increasing steric bulk of the substituents.

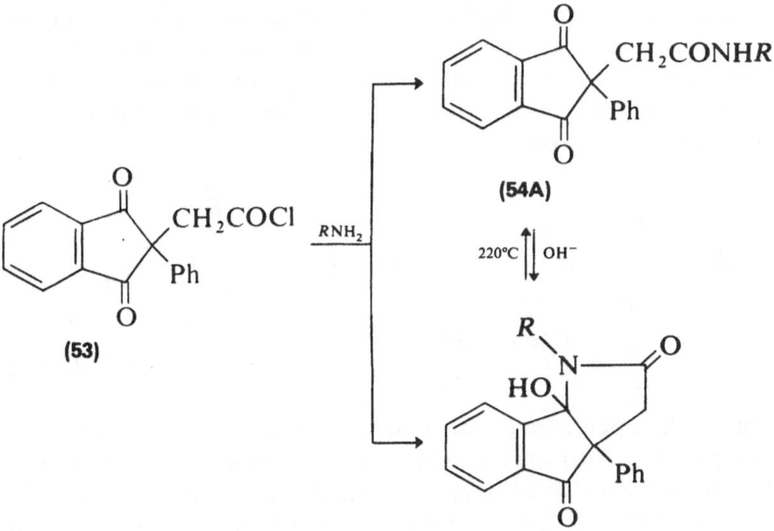

(54A)

(53)

(54B)

Cyclic isomers (**54B**, R = Et, Pr, Bu) show two melting points: the substance crystallizes after the melting and melts again at the higher temperature, which corresponds to the melting point of the open isomer (**54A**). This ring opening of hydroxylactams (**54B**) occurs at 200–220° and is the first reported example of a thermal ring opening of hydroxylactams.

Similar double melting points, indicating thermal isomerizations (**54B**) → (**54A**), were later observed [259] for some esters of (2-phenyl-1,3-indanedione-2-yl)acetylaminoacids, e.g., for (**54**, R = CH$_2$COOEt).

The increase of the steric bulk of substituents at the nitrogen atom decreases the stability of (**54B**), which is confirmed by a noticeable lowering of the isomerization temperature. The lower melting point of N-butylderivative (**54B**, R = Bu) is poorly defined. It is impossible to detect the true melting point of the N-benzylderivative (**54B**, R = PhCH$_2$) because the isomerization proceeds before the hydroxypyrrolidone melts. Therefore both isomers show the same experimental melting point and are indistinguishable by the method of mixed melting points.

Amides (**54A**, R = n-alkyl) under conditions of alkaline catalysis isomerize into hydroxypyrrolidones (**54B**), which is generally characteristic for addition reactions of weak nucleophiles to the C=O group.

Both series of isomers are stable in neutral aprotic solvents. Ring-chain equilibration is not observed even after 10 days at room temperature in dioxan or dichloroethane solutions.

Bulky substituents at the nitrogen atom prevent cyclization: amides (**54A**) with R = i-Pr, cyclo-Hexyl, t-Bu, and 1-Adamantyl do not undergo the isomerization. Sodium hydroxide in aqueous dioxan does not isomerize these amides. A hydrolytic opening of indanedione ring takes place, yielding N-(sec- or t-alkyl)amides of 3-(2-hydroxycarbonylbenzoyl)-3-phenylpropionic acid, which as a 2-acylbenzoic acid, form tautomeric equilibria (**55A**) ⇌ (**55B**) [71, 73].

Either the open isomers of hydrazides (**56A**) or the mixtures of both isomers (**56A**) + (**56B**) are formed in the reactions of chloride (**53**) with N,N-dialkylhydrazines [217]. The recrystallization of hydrazides (**56A**) from alcohols results in an isomerization (**56A**) → (**56B**), obviously due to the presence of an NRR group in the molecule, which in boiling alcohol acts

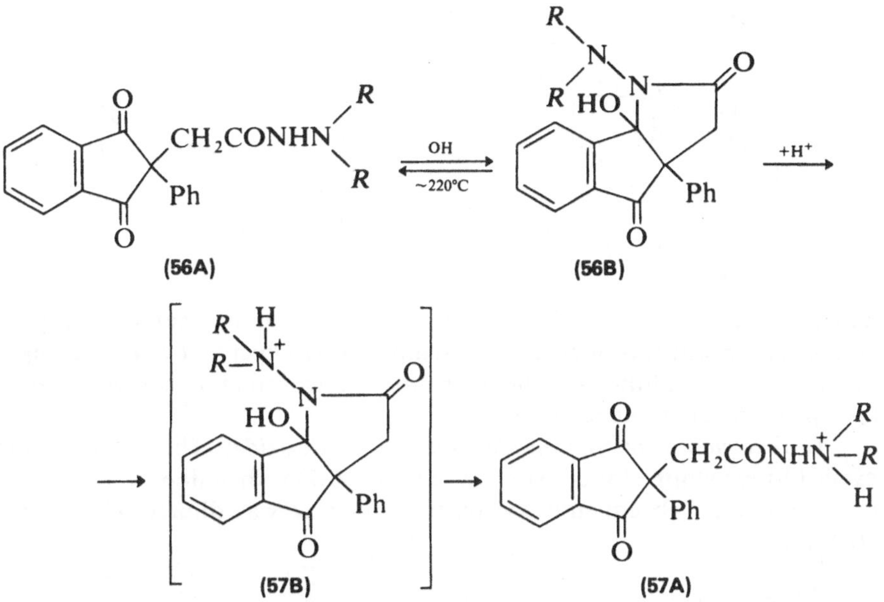

(56A)   (56B)

(57B)   (57A)

as a catalyst for the intramolecular nucleophilic addition of the NH group to the keto group.

True melting points of cyclic isomers (56B) cannot be obtained because of a thermal isomerization into open hydrazides.

On protonation of the NRR group of compounds (56B) ring opening occurs, which obviously is due to the action of a strong -I-effect of the substituent -$\overset{+}{N}HRR$, destabilizing the ring form. One may assume that protonation favors the ionization of the OH bond in (57B) shifting the electron density in the C—N bond toward the nitrogen atom simultaneously and finally promotes the heterolytic breaking of this bond and subsequent isomerization (57B) → (57A). The deprotonation of hydrochlorides (57A) results in a complete or partial isomerization to the initial hydroxypyrrolidones (56B).

Hydrazides and their cyclic isomers in solution slowly form equilibrating mixtures (56A) ⇌ (56B). Equilibrium is attained from both sides within 5–10 days. Isomeric interconversions are apparently facilitated by catalytic influences of the NRR group being able to transfer protons. The equilibrium of the hydrazide (56, R = Me) in dioxan (K_T = 9.1) and acetonitrile is shifted toward the ring form; in CHCl$_3$ (K_T = 0.54) and nitromethane, however, the open form is more favored. Proton acceptor solvents (dioxan, acetonitrile) form intermolecular hydrogen bonds with the pyrrolidone hydroxy group (58) thus stabilizing the ring form. In proton

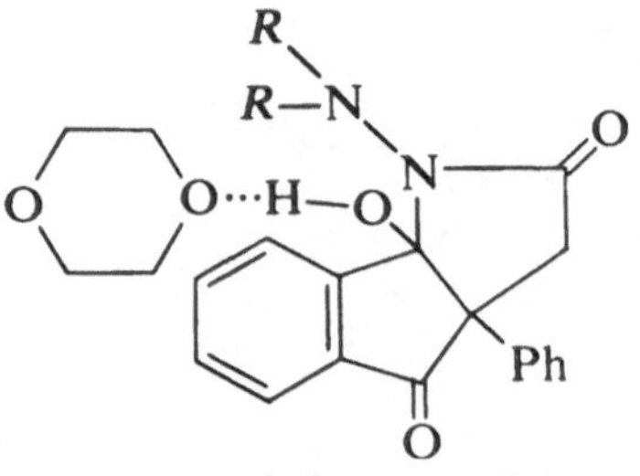

(58)

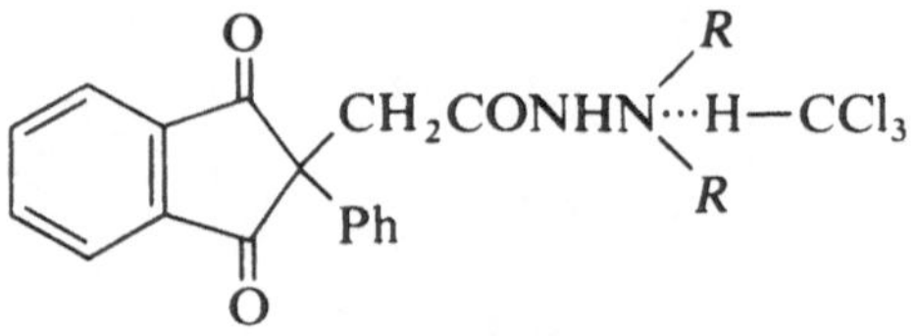

(59)

donating solvents (chloroform, nitromethane) intermolecular hydrogen bonds of type **(59)** may appear strengthening the -I-effect of the NRR group; hence, the nucleophilicity of the hydrazide α-nitrogen atom is lowered and the open isomer stabilized [251].

(2-Benzyldimedone-2-yl)acetamides and Hydrazides. The reactions of cyclic (2-benzyldimedone-2-yl)acetyl chloride **(31)** with amines were carried out in dioxan. Subsequently the reaction mixture was diluted with water [162, 196].

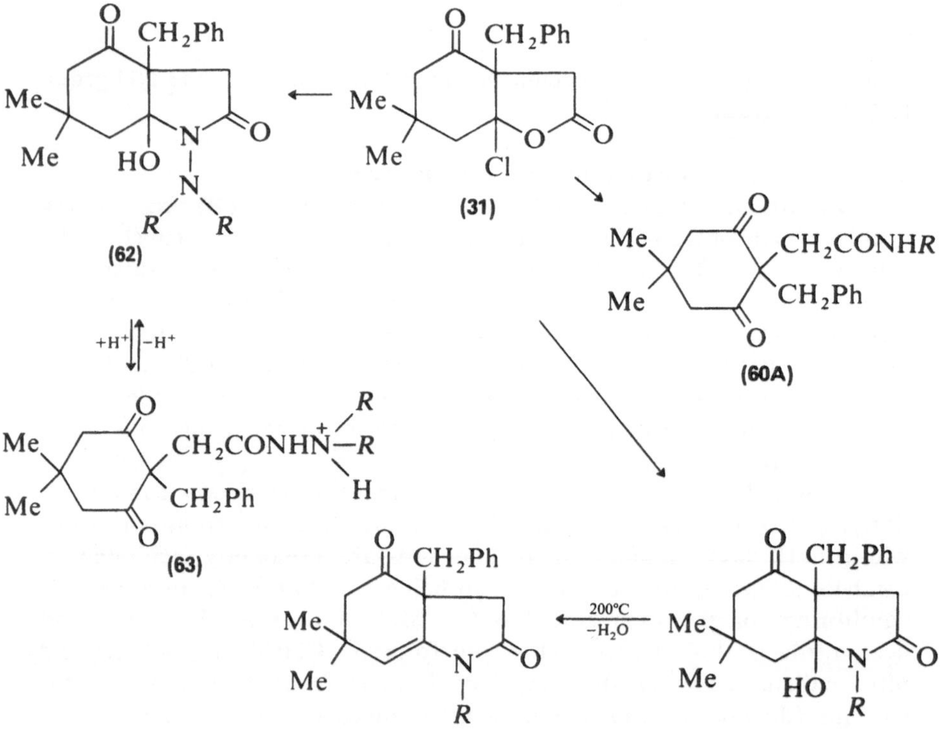

(62)

(31)

(60A)

+H$^+$ ‖ −H$^+$

(63)

(61)

200°C
−H$_2$O

(60B)

The *n*-alkylamines gave the cyclic isomers **(60B)**; isopropyl and *t*-butylamines, however, gave open amides **(60A)**, which do not isomerize [**(60A)** → **(60B)**] under the influence of alkaline catalysts. This happens in a reaction of chlorolactone **(31)** with ammonia and propylamine in benzene yielding open isomers of amides (**60A**, R = H, Pr), which isomerize to hydroxylactams **(60B)** under basic conditions.

On heating, cyclic isomers **(60B)** dehydrate forming 1-alkyl-9-benzyl-6,6-dimethyl-2,4-dioxo-2,3,4,5,6,9-hexahydroindoles **(61)**, and this conversion excludes the thermal isomerization **(60B)** → **(60A)**.

N,N-Dialkylhydrazides **(62)** are obtained only as ring isomers. Protonation of the dialkylamino group opens the ring of hydroxypyrrolidone **(62)** → **(63)**; deprotonation gives rise to the reverse conversion.

Contrary to indanediones, the keto groups of cyclohexanedione are not conjugated with the benzene ring and are therefore more electrophilic. Amides **(60)** as compared to the (2-phenyl-1,3-indanedione-2-yl)acetamides **(54)** display a greater capability of forming cyclic isomers: *N*-unsubstituted amide forms ring isomer, *N*-methyl and *N*-ethylamides are obtained as ring isomers even if their synthesis is carried out in benzene solution; *N,N*-dialkylhydrazides exist only in the cyclic form.

2-Acylbenzamides. In the first investigations published in the last century [260–263] the open structure **(64A)** was erroneously ascribed to the 2-acylbenzamides.

However, in 1896 Graebe and Ullmann [264] proposed that 2-benzoylbenzamide (**64**, R^1 = Ph, R^2 = H) represents a mixture of isomers **(64A)** and **(64B)**. Later Sachs and Ludwig [265] obtained a number of 2-acylbenzamides by the action of alkyl and phenyl magnesium bromides on the *N*-ethylphthalimide. They concluded that these compounds possess cyclic structure **(64B)**.

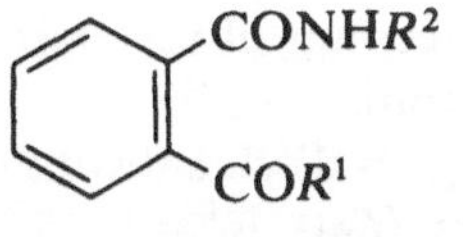

(64A) 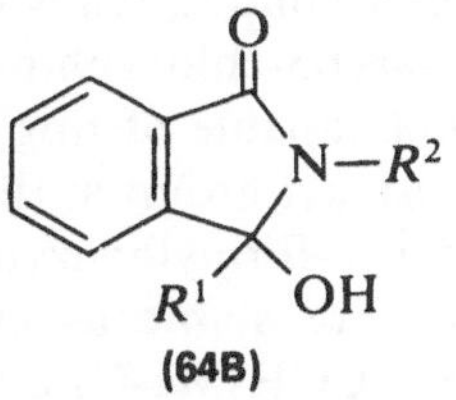   (64B)

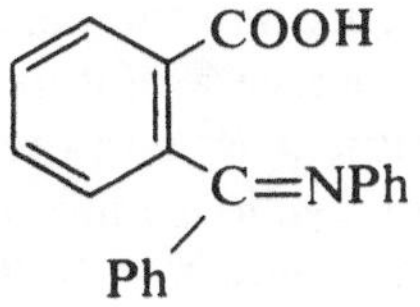

(65A) 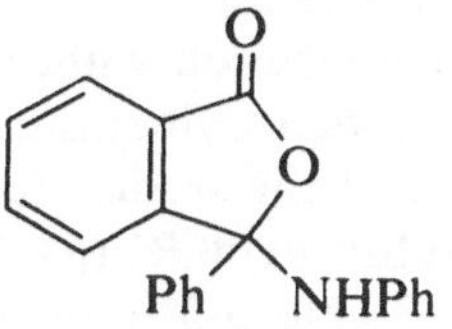   (65B)

The history of the determination of the structure of the products obtained from 2-benzoylbenzoic acid and its chloride with aniline provides an instructive example of the insufficiency of chemical methods for an elucidation of the structure of ring-chain isomers. Meyer [266] obtained two products with m.ps. of 195 and 221°, to which on the basis of chemical properties he assigned the structures (64A) and (64B).

Later it was concluded from polarographic data [267] that the compound having m.p. 195° possesses structure (64A, $R^1 = R^2 = $ Ph) but that the other one with m.p. 221° should be a mixture of both 2-benzoylbenzoic acid anil (65A) and 3-phenylamino-3-phenylphthalide (65B).

These results were refuted by IR spectroscopy [33, 170, 205] demonstrating that the compound with m.p. 195° possesses a cyclic structure (64B) in the solid state as well as in solutions. The amide-II band does not appear, but the isoindolinone C=O band does at 1710–1699 cm^{-1} in solutions. The frequency lowering and the splitting of the latter band in the solid state spectra, as well as the band splitting in dichloroethane solution spectra are caused by intermolecular hydrogen bonds O—H···O=C. The absorption of the bonded OH group in the solid state spectra appears as a broad band at 3180 cm^{-1}.

It was confirmed [170] that the compound with m.p. 221° really represents a mixture of (65A) and (65B) with a considerable predominance of the latter.

For the first time the IR method was used by Graf and coauthors [80] for establishing the structure of 2-acylbenzamides. All amides examined by these authors exist in the ring form (64B, $R^1 = $ aryl, $R^2 = $ alkyl).

It has been shown subsequently by many investigators that almost all the 2-acylbenzamides exist as 3-hydroxyisoindolinones in the solid state as well in solutions (see Table 11).

2-Aroylbenzamides, really existing as open amides, were first obtained [146] from 3-aryl-3-chlorophthalide and t-alkylamines (Table 12). These amides are not capable of ring closure on account of the steric hindrance of the bulky t-alkyl group at the nitrogen atom.

Note that 2-formylbenzamides having t-alkyl substituents at the nitrogen atom are stable as cyclic isomers (64B, $R^1 = $ H, $R^2 = t$-Bu, 1-adamantyl, $Me_3CCH_2Me_2C$) and do not undergo a thermal isomerization into open amides [222].

Introducing electron-withdrawing substituents into the benzoyl group of N-(t-alkyl)-2-benzoylbenzamides one can observe an interesting combined influence of the steric effect of the substituent R^2 and the electronic effect of the substituent R^1 (64) [193, 201]. Only cyclic isomers (64B) were obtained from 3-chloro-3-(X-phenyl)phthalides (where X = 3-NO_2, 4-NO_2, 4-Cl) with t-butylamine, when the reaction was carried out in dioxan with

Table 11. References for the Proof of Cyclic Structure (64B) for
2-Acylbenzamides

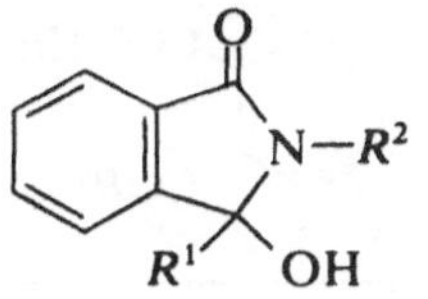

R^2 = H, Me, Ph, etc.

R^1	References
H	33, 148, 220, 222, 248, 268–270
Me	33, 271–274
cyclo-Pr	275
PhCH$_2$	147, 211, 212, 276, 277
t-Bu	148
Other alkyls and substituents of more complicated structure	33, 146, 213, 236, 276, 278–284
Ph and other aryls	33, 78–80, 146, 148, 170, 201, 220, 277, 285–291
PhCO	163
R^1R^2NCO	292
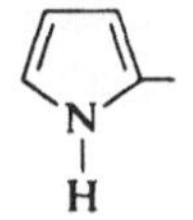	33
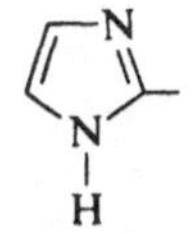	293

a subsequent dilution of the reaction mixture with water. A reaction of strictly equimolar quantities of reagents (chlorophthalide, *t*-butylamine, triethylamine) in benzene afforded an open isomer when R^1 = 4-ClC$_6$H$_4$, but the ring isomer when R^1 = 4-NO$_2$C$_6$H$_4$. Hence, an increasing electron-

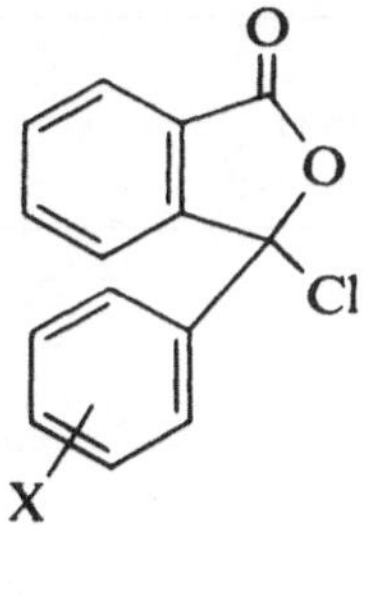

Table 12. N-Monosubstituted 2-Acylbenzamides Possessing Open
Structure **(64A)**

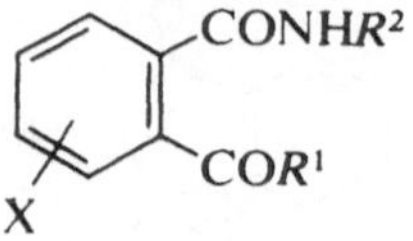

R^1 $(X)^\dagger$	R^2	Isomeric structure in solid state	Refs.
Ph	i-Pr t-Bu $Me_3CCH_2Me_2C$ 1-adamantyl	A, B A A A	146
$2,4\text{-}Me_2C_6H_3$	i-Pr t-Bu	A, B A	78
$PhCH_2$	t-Bu	A	147
$PhCH_2(Ph)CH$	t-Bu $Me_3CCH_2Me_2C$ 1-adamantyl	A A A	146
$4\text{-}Me_2NC_6H_4$	1-adamantyl	A	290
$3\text{-}NO_2C_6H_4$	t-Bu $Me_3CCH_2Me_2C$ 1-adamantyl	A, B A A	201
$4\text{-}NO_2C_6H_4$	t-Bu	A, B	201
$4\text{-}ClC_6H_4$	t-Bu $Me_3CCH_2Me_2C$ 1-adamantyl	A, B A A	201
$4\text{-}BrC_6H_4$	t-Bu $Me_3CCH_2Me_2C$ 1-adamantyl	A, B A A	193
$3\text{-}NO_2\text{-}4\text{-}ClC_6H_3$	t-Bu 1-adamantyl $Me_3CCH_2Me_2C$	A, B A, B A	193
Ph $(X = 3\text{-}NO_2, 4\text{-}NO_2,$ $5\text{-}NO_2, 6\text{-}NO_2)$	t-Bu	A	79

continued

Table 12. (*cont.*)

R^1 (X)[†]	R^2	Isomeric structure in solid state	Refs.
Ph (X = 3-Cl)	t-Bu	A, B	
	1-adamantyl	A, B	193
	Me₃CCH₂Me₂C	A	
Ph (X = 4-Cl)	t-Bu	A	193
	H, Me, Pr, i-Pr, PhCH₂, Ph	A	78
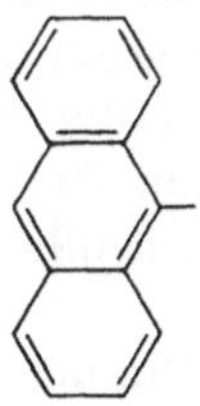	H, Me, Pr, i-Pr, t-Bu, PhCH₂, Ph	A	158
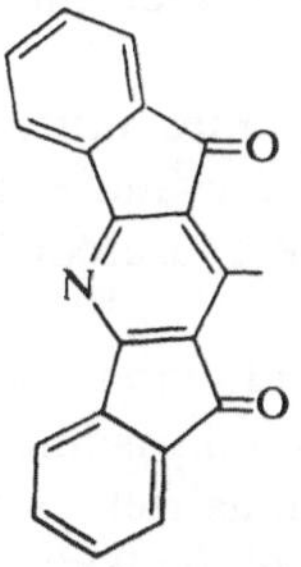	H, Et	A	160
	t-Bu	A	
	1-adamantyl	A	293
	Me₃CCH₂Me₂C	A	

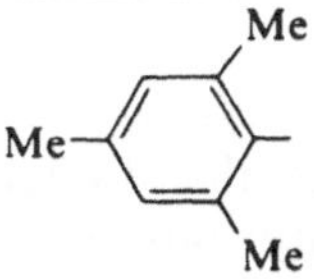

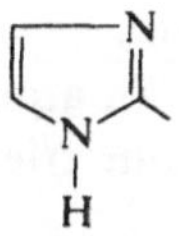

[†] When X is not noted then X = H.

withdrawing effect of substituents R^1 may favor the cyclic isomer (**64B**) even in the presence of such a sterically bulky substituent at the nitrogen atom as *t*-butyl. However, additional enhancement of the steric bulk of substituent R^2 (**64**) prevents the cyclization: *N*-(1,1,3,3-tetramethylbutyl)- and *N*-(1-adamantyl)amides of 2-(4-chloro or 3-nitrobenzoyl)benzoic acids exist exclusively in the open form. Nevertheless, an accumulation of two electron-withdrawing substituents in the benzoyl group (R^1 = 3-NO_2-4-ClC_6H_3 in (64)) permits isolation of the cyclic isomer of N-(1-adamantyl)-amide.

$$R = t\text{-Bu, 1-adamantyl}$$

Cyclic isomers have also been obtained for N-(t-butyl) and N-(1-adamantyl)-2-benzoyl-3-chlorobenzamides. In addition to the -I-effect of the chlorine atom increasing the C=O group electrophilicity, most probably a considerable role is played by a steric effect [28] of the chloro atom on the C=O group in the open structure which brings the C=O group near to the amide group. This is corroborated by the fact that the N-(t-butyl)-2-benzoyl-4-chlorobenzamide exists exclusively as the open isomer. The shift of the chlorine atom from position 3 to position 4 cancels the steric influence leaving the electronic effect nearly unchanged.

The melting points of 2-(t-alkyl)-3-aryl-3-hydroxyisoindolinones are poorly defined since these compounds thermally isomerize into open amides (64B) → (64A).

Ring and chain isomers of N-(t-alkyl)-2-aroylbenzamides differ in reactivity: thionyl chloride transforms the open isomers into 2-cyanobenzophenones [294]; the cyclic isomers, however, into 2-(t-alkyl)-3-chloro-3-arylisoindolinones under the same conditions [295].

Treating 2-(t-butyl)-3-(4-chlorophenyl)-3-hydroxyisoindolinone with CF_3COOH at room temperature makes it possible to carry out the isomerization (64B) → (64A). In concentrated sulfuric acid the open as well as the cyclic isomers of N-(t-alkyl)-2-aroylbenzamides undergo N-dealkylation forming the corresponding 3-aryl-3-hydroxyisoindolinones [296].

The reverse isomerization (64A) → (64B) is carried out in an aqueous dioxan solution of potassium hydroxide or by boiling in ethanolic triethylamine. The last method is more convenient preparatively.

Both series of isomers are stable in dioxan at room temperature. Equilibration occurs on addition of triethylamine to a dioxan solution [193] and also in methanol [190].

The signals of the t-butyl protons in ^{1}H-NMR spectra of CD_3OD solutions differ for ring and chain isomers. Consequently, the time dependence of the intensity ratio and rate constants for the isomeric interconversions have been determined (Table 13).

Table 13. Ring-Chain Equilibrium and Rate Constants of
N-(t-Butyl)-2-acylbenzamides in CD_3OD Solution as Determined by
^{1}H-NMR Method [190]

	t-Bu protons δ, ppm					Temperature
R	(A)	(B)	K_T	$k_1 \times 10^4\,s^{-1}$	$k_2 \times 10^4\,s^{-1}$	(°C)
$PhCH_2$	1.40	1.76	0.72	2.0	2.8	24
Ph	1.19	1.50	0.56	2.9	5.2	27
$4\text{-}ClC_6H_4$	1.12	1.43	0.92	4.5	4.9	28
$4\text{-}BrC_6H_4$	1.12	1.43	0.98	5.9	6.3	23
$4\text{-}Cl\text{-}3\text{-}NO_2C_6H_3$	1.22	1.52	3.55	260	73	24

Table 13 shows that the introduction of electron-withdrawing substituents into the aryl group not only shifts the equilibrium towards the cyclic form but also accelerates tautomerization.

In aprotic solvents ring-chain equilibria were observed for 2-acylbenzamides containing a proton acceptor group in the acyl substituent. Thus, for example, by means of the UV method ring-chain equilibrium constants were measured [194] for N-(t-alkyl)-2-(2-imidazolylcarbonyl)benzamides in various solvents (Table 14). For a discussion of a dependence of these equilibria on the substitution pattern see p. 76.

Table 14 shows that in proton donating solvents the equilibrium is shifted to the cyclic form. One may suppose that the ring form is stabilized by intermolecular hydrogen bonds $ROH\cdots N$ (imidazole) thus increasing the electron withdrawing effect of the imidazole group. An analogous solvent shift of the equilibrium in ethanol is observed for N-unsubstituted, N-methyl, N-ethyl, and N-phenyl-2-(4-dimethylaminobenzoyl)benzamides [191].

Heating transforms 2-isopropyl-3-hydroxy-3-phenylisoindolinone into N-isopropyl-2-benzoylbenzamide [196]. On cooling slowly a mixture is obtained in which the cyclic form predominates. The bulky N-isopropyl substituent stabilizes the open structure; hence, it is possible to freeze the equilibrium, which is shifted in favor of the open form by an increase in the temperature. On rapid cooling a mixture is obtained which contains a considerable proportion of the open isomer. Attempts to carry out an

Table 14. Ring-Chain Equilibrium Constants for
N-(t-Alkyl)-2-(2-imidazolylcarbonyl)benzamides
Determined by a UV Method [194]

		K_T with $R =$	
Solvent	t-Bu	1-Adamantyl	Me$_3$CCH$_2$Me$_2$C
Ethanol	1.13	0.65	0.30
Methanol	0.79	0.75	0.23
Chloroform	0.20	0.16	0.15
Dioxan	0.18	0.16	0.15
Acetonitrile	0.08	0.06	0.03

analogous isomerization of 2-(n-alkyl or phenyl)-3-hydroxy-3-phenyliso-
indolinones failed.

Recently Sheinker, Anisimova, and Valters [297] have reported the
results of a mass spectrometric investigation of 2-acylbenzamides and their
cyclic isomers. The open isomer is characterized by a fragment ion [M-
NHR^2]$^+$ (see formula (64)) the fragmentation of the ring isomer yields
mainly the ion [M-OH]$^+$ and, to a lesser degree [M-R^1]$^+$ (see also [189]).
The proportions of the intensities of peaks

$$\frac{I_{[M\text{-}OH]}}{I_{[M\text{-}NHR^2]}} \quad \text{and} \quad \frac{I_{[M\text{-}R^1]}}{I_{[M\text{-}NHR^2]}}$$

quickly fall in a series of compounds (64, R^1 = Ph, R^2 = H > Me > Et >
Pr > Ph > t-Bu). The proportions of the cyclic form accordingly decrease
in gaseous phase (at 180°). An examination of the intensities of the fragmentation peaks at various temperatures (40–140°) showed that the amount
of the cyclic form decreases with increasing temperature.

The intensities of the fragmentation peaks do not give completely
correct quantitative data concerning the proportion of the tautomeric forms
in the gas phase because the intensity of any peak depends not only on the
concentration of a specific isomer, but also on the ease of bond breaking
leading to the formation of fragments and also on the further break up of
these fragments [297].

Irrespective of the substituent at the nitrogen atom, 2-mesityloyl [78] and 2-(9-anthroyl)benzamides [158] exist only in the open form (see Table 12]. The two ortho-methyl groups or two peri-protons in these amides hinder the intramolecular attack of the nucleophilic nitrogen at the C=O group perpendicular to its plane. Evidently for the same reason the open structure of 2-[2,3(CO),6,5(CO)-dibenzoylenisonicotinoyl]-benzamides is more stable [160] (Table 12).

Summarizing, stable open isomers of 2-acylbenzamides can be obtained only by an introduction of bulky substituents either at the nitrogen atom or at the keto group.

Valters and Karlivans [290, 293, 298] presented two examples studying the influence of a protonation of an amino group which is involved in the ring-chain interconversion process of 2-acylbenzamides.

Using IR, UV, and ^{1}H-NMR methods they were able to show [290, 298] that contrary to earlier results of Indian chemists [191] N-unsubstituted and N-monosubstituted 2-(4-dimethylaminobenzoyl)benzamides in the solid state possess a cyclic structure (**64B**, $R^1 = 4$-Me$_2$NC$_6$H$_4$; $R^2 = $ H, Me, Et, PhCH$_2$, Ph), N-(1-adamantyl)amide **(66)** being an exception.

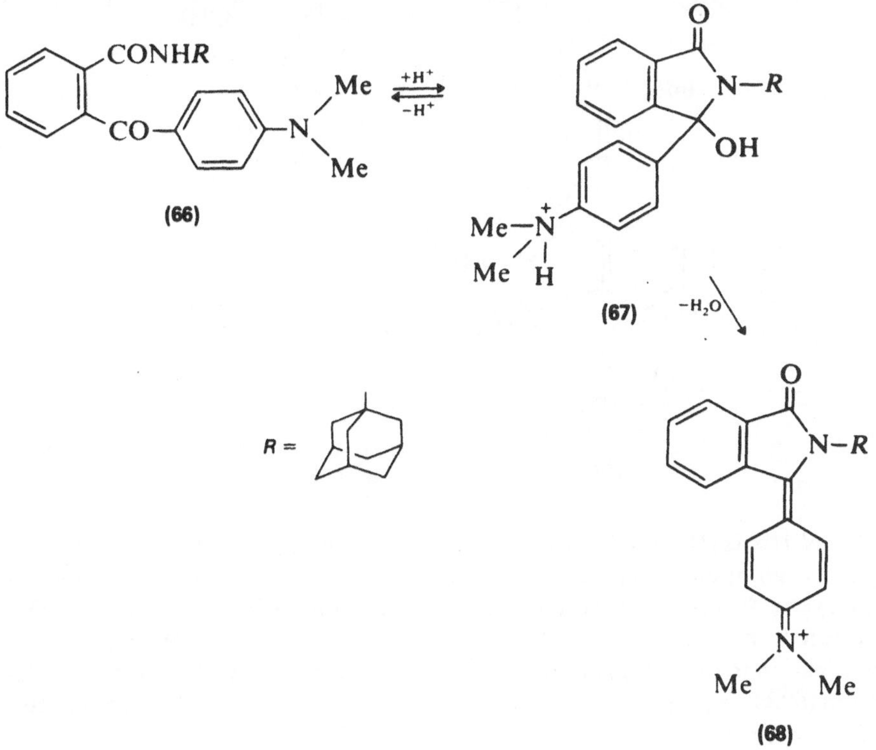

The protonation of the dimethylamino group in the amide (66) leads to a cyclization giving (67). Due to the electron-withdrawing influence of the $\overset{+}{N}HMe_2$ group the electrophilicity of a C=O carbon atom is increased so much that an intramolecular addition of an amide group bearing a bulky 1-adamantyl substituent becomes possible. According to ^{1}H-NMR spectroscopy the hydroxyisoindolinone (67) dehydrates in CF$_3$COOH forming the quinoide compound (68) [290].

The reaction of ammonia and primary amines with imidazolo[1,2-b]isoquinoline-5,10-dione gives 2-(2-imidazolylcarbonyl)benzamides (69A) [299]. A more detailed spectroscopic examination showed [293] that these amides, except N-(t-alkyl)substituted, exist in the cyclic form (69B, R = H, Me, Et, Pr, i-Pr, PhCH$_2$, Ph, 4-MeC$_6$H$_4$) in the solid state and in DMSO solution.

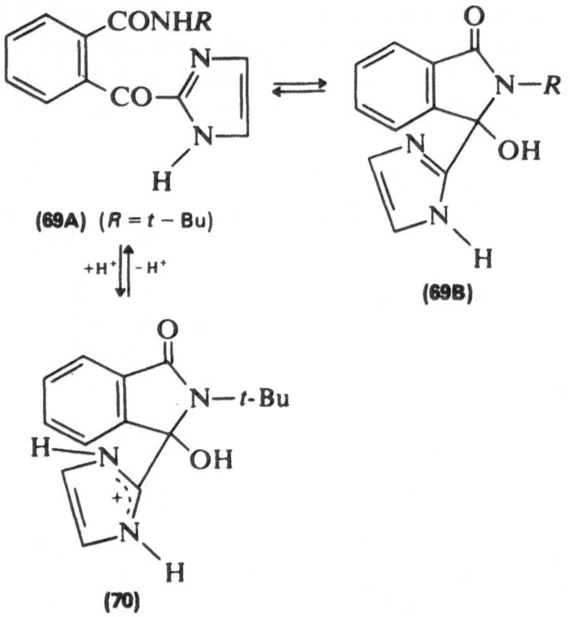

(69A) (R = t − Bu)

(69B)

(70)

N-(t-Alkyl)amides (69, R = t-Bu, 1-adamantyl, Me$_3$CCH$_2$Me$_2$C) possess an open structure in the solid state, but in solution the equilibrium (69A) ⇌ (69B) is observed (see Table 14 and the discussion on p. 73). The protonation of the imidazole ring of N-(t-alkyl)amides (69A) promotes cyclization to (70); deprotonation reverses the process. Evidently, the cyclization (69A) → (70) occurs for reasons similar to those for the interconversion (66) → (67).

2-Acylbenzhydrazides. An analogous but contrary influence of protonation on cyclization takes place in 2-dialkylamino-3-hydroxy-3-phenylisoindolinones (**71**, R^1 = Ph, R^2 = R^3 = Alkyl) [162]. Protonation of the dialkylamino group gives rise to ring opening (**71**) → (**72**). Deprotonation reverses the transformation.

The *N*-phenylhydrazide of 2-phenylacetylbenzoic acid also possesses a cyclic structure (**71**, R^1 = PhCH$_2$, R^2 = Ph, R^3 = H) [300].

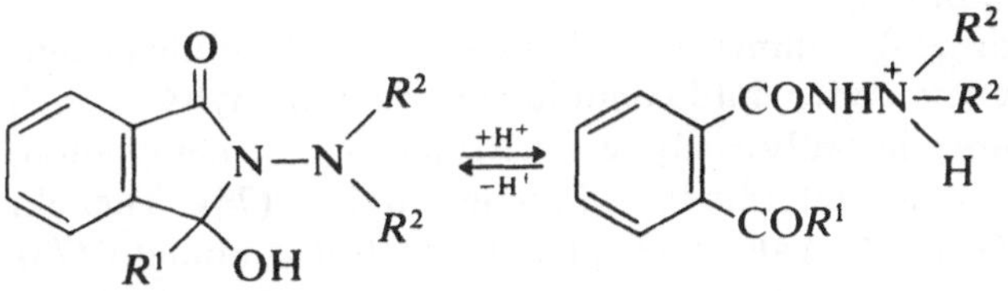

In the case of *N,N'*-dimethyl-2-acylbenzhydrazides, ring closure is possible owing to the intramolecular addition of the hydrazide β-nitrogen to the keto group.

N,N'-Dimethylhydrazides of 2-benzoyl and 2-(2,4-dimethylbenzoyl)benzoic acids exist as 4-aryl-4-hydroxy-2,3-dimethyl-3,4-dihydrophthalazones (**73**) [301], but the 2-mesityloylbenzoic acid derivative has open structure (**74**), presumably due to steric hindrance of the mesityl group.

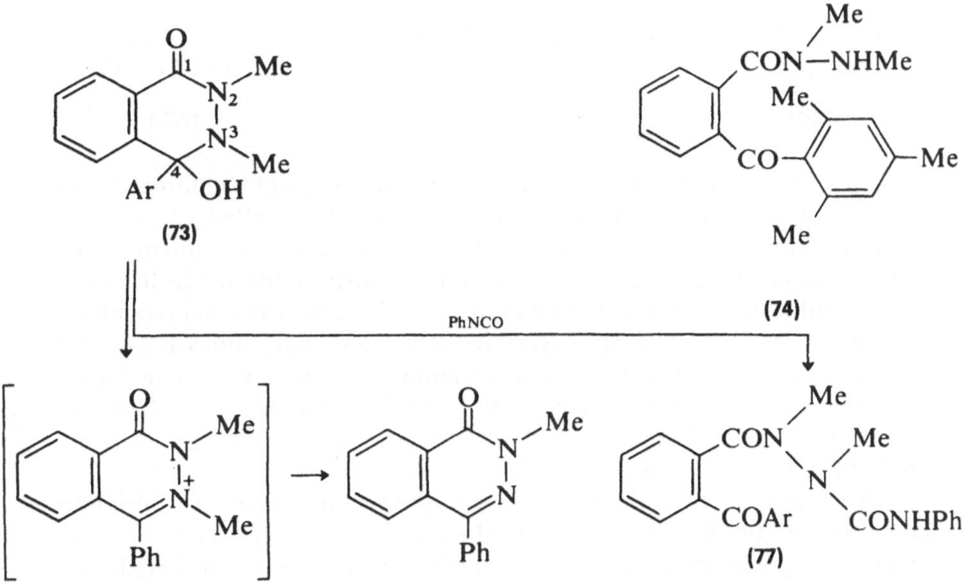

Compounds of open (74) or cyclic (73) structure in a reaction with phenylisocyanate form identical products (77). The formation of the open-structured (77) (but not the corresponding N-phenylurethane) from (73) may be explained by the presence of rapidly equilibrating ring and open tautomers with an open form reacting faster with phenylisocyanate than the cyclic form. Alternatively, a hydroxyphthalazone ring opening may occur in the course of the reaction with an initial attack of phenylisocyanate on the $N_{(3)}$ of (73).

4-Hydroxy-2,3-dimethyl-4-phenyl-3,4-dihydrophthalazone and electrophile agents under mild conditions gives 2-methyl-4-phenylphthalazone (76). This unexpectedly easily accomplished $N_{(3)}$-demethylation, apparently, proceeds via an intermediate immonium ion (75). The elimination of methanol (73) → (76) also takes place by heating compound (73). An attempt to thermally isomerize compounds (73) into the open hydrazides failed.

Benzil-o-carboxamides. Benzil-o-carboxylic acid represents an interesting example to study the dependance of ring isomer stability on ring size. N-Monosubstituted amides of this acid are expected to form three isomeric structures: open amides (78A), five-membered hydroxylactams (78B) and six-membered hydroxylactams (78C). Both chlorides of benzil-o-carboxylic

(78C)   (78A)   (78B)

acid, i.e., the five-membered (32) and the six-membered (33) chlorolactones in reactions with primary amines form exclusively 3-benzoyl-3-hydroxy-2-R-isoindolinones (78B, R = alkyl, Ph) irrespective of the substituent at the nitrogen atom [153]. Hydroxyisoindolinones (78B) are formed in the reactions of chlorolactones (32) or (33) even with t-butyl or 1-adamantylamines. This indicates that the ring closure of benzil-o-carboxamides is favored compared with that of other 2-acylbenzamides which may account for an increased electrophilicity of the C=O group due to the -I- and -C-effects of the neighboring benzoyl group. Attempts to obtain the open isomers (78A) of benzil-o-carboxamides failed.

A reaction of chlorolactone (32) with propylamine gave in addition to compound (78B, R = Pr) its six-membered isomer (78C), which at 200°C or in boiling ethanol isomerized under ring contraction to the 3-hydroxy-isoindolinone (78B).

This rearrangement corroborates the greater stability of a five-membered hydroxylactam ring compared with that of a six-membered isomer (see also [302]).

An analogous ring contraction of a six-membered hydroxylactam has been reported [303] for the more complicated 8,13-dioxo-14-hydroxy-canadine **(79C)** → **(79B)**.

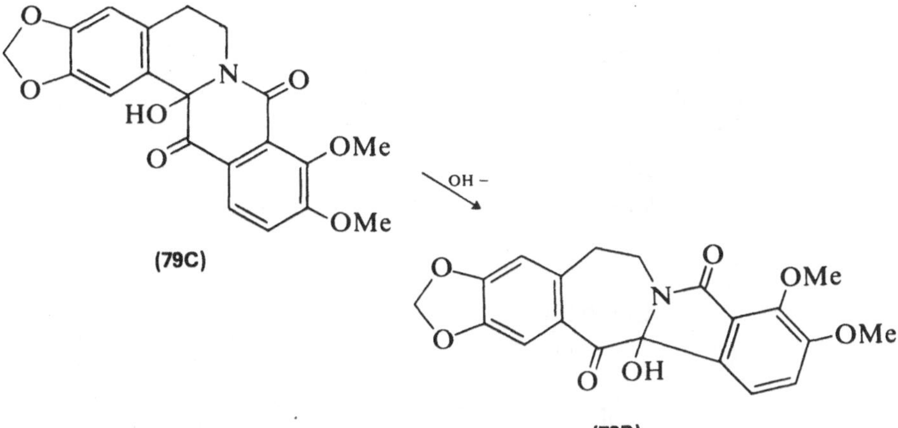

Similar regularities have been observed investigating monoimines of benzil-*o*-carboxylic acid. Usov and Freimanis [269] prepared 3-aryl-3-hydroxy - 1 - oxo - 2 - phenyl - 4 - phenylimino - 1,2,3,4 - tetrahydroisoquino-lines **(80C)** and showed that in polar solvents these compounds isomerize rapidly and irreversibly into isoindolinones **(80B)**. Here ring contraction also takes place although the CONH*R* group adds to a C=N rather than to a carbonyl group.

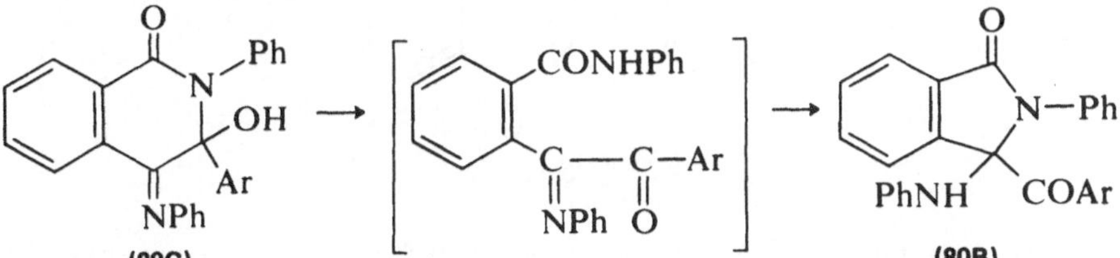

Anthraquinone-1-carboxamides, Amidines, and Hydrazides. For stereoelectronic control of an intramolecular nucleophilic addition of an NH group to a keto group it is necessary for the latter to rotate around the CO—Ar bond and to take a conformation which allows the nitrogen atom to approach the C=O group nearly perpendicular to its plane. In derivatives of anthraquinone-1-carboxylic acid **(81A)** this possibility is lacking and therefore they do not form cyclic isomers **(81B)** [87, 304].

For the same reason *N*-butylamide **(82)** exists only in the open form [305]. Here, due to the bond angles in the five-membered ring CONHBu and C=O groups are still more removed from each as in the amides **(81A)**.

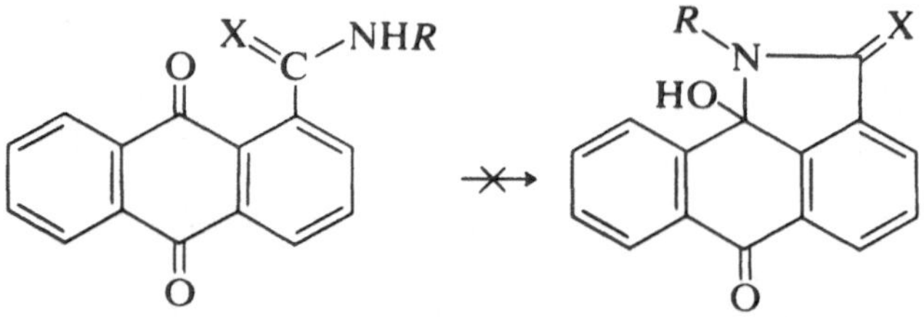

(81A)   X = O, N*R*   **(81B)**

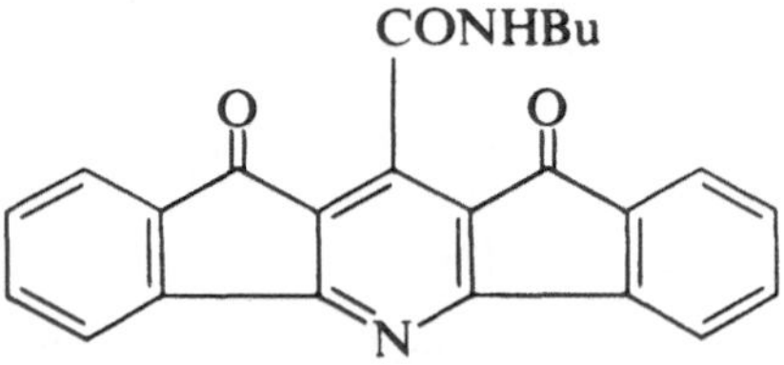

(82)

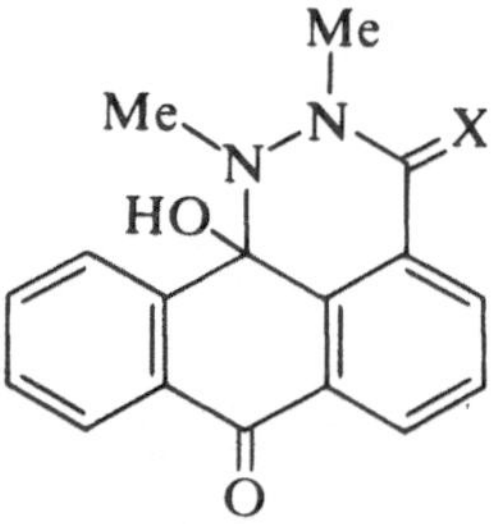

(83)
X = O, NPh

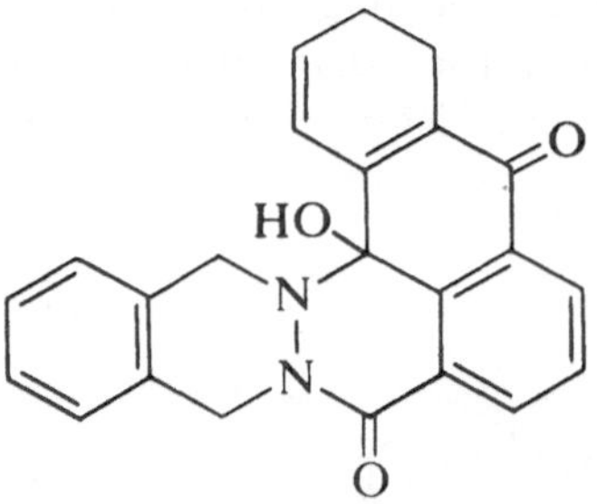

(84)

An exception among anthraquinone-1-carboxylic acid derivatives is *N,N'*-dimethylhydrazide and *N,N'*-dimethyl-*N''*-phenylamidrazone existing only in the cyclic form **(83**, X = O, NPh) [87, 304]. *N*-(Anthraquinone-1-carbonyl)tetrahydrophthalazine also possesses ring structure **(84)** [306]. The formation of stable cyclic isomers **(83)** is favored by the greater nucleophilicity of the β-nitrogen atom of hydrazides as compared with the α-nitrogen or with the amide nitrogen atoms of compounds **(81)**. This is confirmed by the fact that N'-acylhydrazides of anthraquinone-1-carboxylic acid do not cyclize [307].

Another very important factor favoring cyclization may be the enhanced conformational mobility of the chain $-CX-N-NH$ as compared with $-CX-NH$ giving rise to conformations which are more favorable stereoelectronically for an intramolecular attack of the nitrogen atom on the C=O group.

Z-3-Acylacrylamides. Following spectroscopic investigations amides of Z-3-formyl [240], Z-3-acetyl [20, 308–310], Z-3-acyl (R^1 = alkyl) [311–314], and Z-3-aroylacrylic [9, 10, 109, 216, 308, 315–317] acids are cyclic **(85B)** irrespective of the substituents at the nitrogen atom and at the ethylene bond.

Using the example of Z-3-(4-bromobenzoyl)-3-methylacrylic acid Lutz and coauthors [9] have shown that cyclic amides (**85B**, R^2 = H, Me, Ph) are formed: a) by the action of ammonia or primary amines on the corresponding chlorolactone; b) by the action of ammonia or amines on the cyclic as well on the open methyl ester of these acids; c) by the sunlight irradiation (E-Z-isomerization) of E-3-(4-bromobenzoyl)-3-methylacrylamides. These data confirm the extraordinary great tendency of Z-3-acylacrylamides to form cyclic isomers†. Communications [312, 318, 319] which claim the existence of open Z-3-acylacrylamides should be revised (see [207]).

Recently Japanese authors [320] using ^{1}H-NMR spectroscopy were able to show that the introduction of a benzyloxycarbonyl group at the nitrogen atom of Z-3-aroylacrylamides (**85**, R^2 = PhCH$_2$OCO) results in ring-chain equilibria which depend on the substituent in the aroyl group (**85**, R^1 = XC$_6$H$_4$): electron-donating substituents stabilize the open form. Evidently, here the electron-withdrawing PhCH$_2$OCO group at the nitrogen atom decreases its nucleophilicity thus favoring the stability of the open form. Moreover, it was shown [321] by IR and ^{1}H-NMR methods that E-3-bromo-3-(4-pentoxybenzoyl)acrylamides (**85**, R^1 = 4-C$_5$H$_{11}$OC$_6$H$_4$,

$R^2 = $, , R^3 = H, R^4 = Br)

with bulky 1-ethoxycarbonyl-1-cyclopentyl and -1-cyclohexyl substituents at the nitrogen atom possess an open structure.

Z-3-Benzoyl-2,3-diphenylacrylhydrazide having cyclic structure **(86B)** dehydrates upon heating and transforms into the pyridazone **(87)** [109]. This transformation suggests the presence of equilibrium **(86B)** $\rightleftarrows$ **(86A)** in the course of the reaction.

---

† See also A. A. Jakubowski, F. S. Guziec, Jr., M. Sugiura, C. Ch. Tam, and M. Tishler, *J. Org. Chem.* **1982**, *47*, 1221.

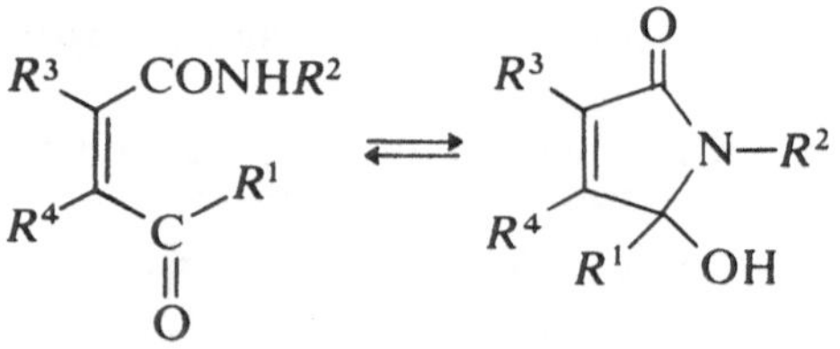

(85A)   (85B)

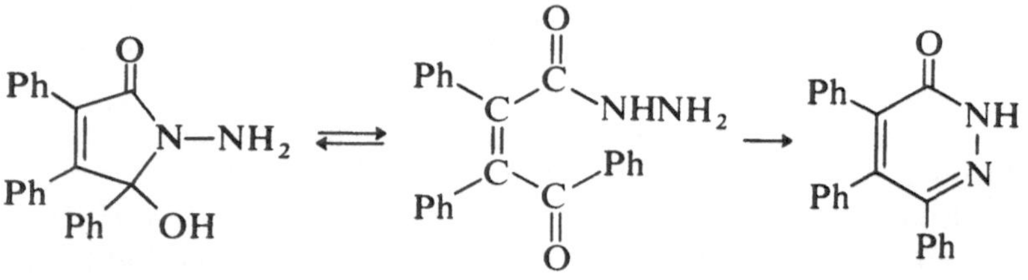

(86B)   (86A)   (87)

4-Acylbutyramides. Lukes and coauthors [322, 323] have obtained *N*-methylamides of 4-acylbutyric acids by the action of Grignard reagent on *N*-methylglutarimide **(88)**. They erroneously suggested that the compounds possess cyclic structure **(89B)**. It was shown later by IR spectroscopy [324] that these compounds exist in open form **(89A)**.

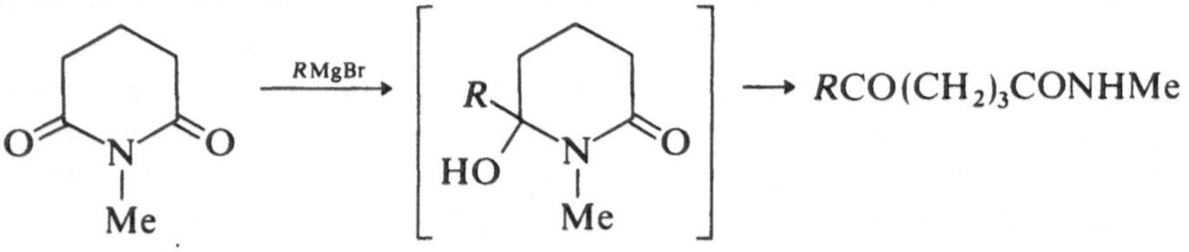

(88)   (89B)   (89A)

This seems to be the case more generally [200, 325, 326]. Evidently the chain between CONHR and C=O groups containing three sp^3-carbon atoms does not create favorable steric conditions for the cyclization.

3-(2-Phenyl-1,3-indanedione-2-yl)propionamides also exist only in an open form [68, 195, 198].

Some 4-formylbutyramides [244] and their chain substituted derivatives [327, 328] possess a cyclic structure, probably due to an increased electrophilicity of the formyl group as compared with the keto group.

4-Aroyldimethylbutyramides **(90)** [200] and **(91)** [329] are cyclic. In this case the cyclization is favored by the Thorpe–Ingold effect [47–49] (see p. 27 for a discussion).

2-Acylmethylbenzamides. *N*-Monosubstituted 2-phenacyl and 2-(4-methyoxyphenacyl)benzamides are obtained by the interaction of 3-

arylisocoumarines with primary amines [214, 330, 331]. This reaction was first carried out by Gabriel [332], who erroneously suggested that all the

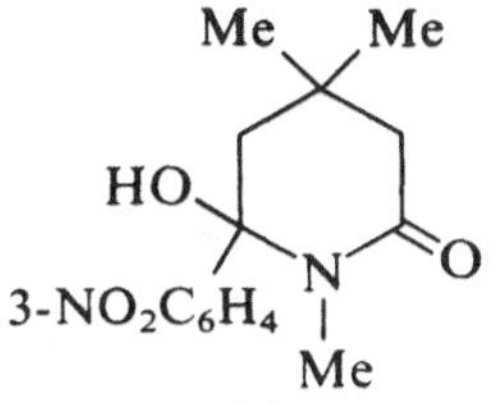

(90)

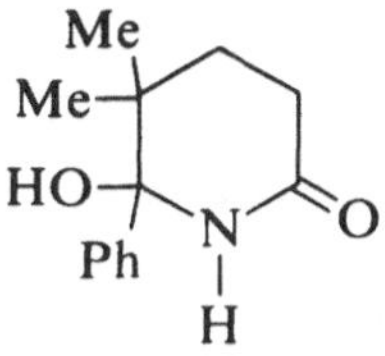

(91)

product amides are open structured. However, IR shows [214, 330, 331] that 2-phenacylbenzamides having N-(n-alkyl) substituents exist in the cyclic form (92B), but N-isopropylamides possess an open structure (92A) in the solid state as well as in solution. N-Monosubstituted 2-(4-methoxyphenacyl)benzamides, with the exception of N-methylamide (92B, R^1 = 4-MeOC$_6$H$_4$, R^2 = Me), are open structured (92A).

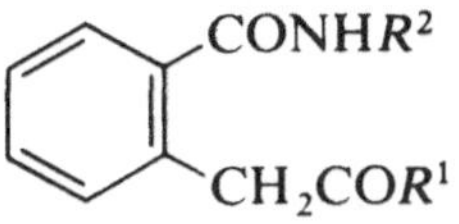

(92A)

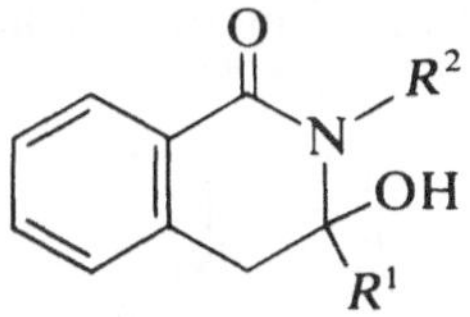

(92B)

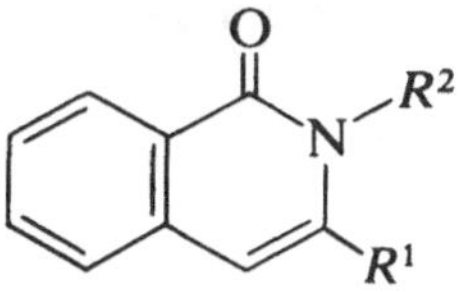

(93)

The reduced tendency of 2-(4-methoxyphenacyl)benzamides to form the cyclic isomer is mainly due to a decreased electrophilicity of the keto group brought about by the electron-donating methoxy group.

The cyclic isomers (92B, R^1 = aryl) very easily dehydrate forming 1-isoquinolones (93) [333]. It is therefore impossible to carry out thermal isomerization (92B) → (92A).

Unlike aryl substituted ring isomers (92B, R^1 = aryl) the corresponding methylderivatives (R^1 = Me) cannot be obtained from a reaction of 3-methylisocoumarine with n-alkylamines since they spontaneously dehydrate yielding only 1-isoquinolones (93). The dehydration is favored by +I-effect of the methyl group.

N-Isopropyl-2-acetonylbenzamide, however, has an open structure (92a, R^1 = Me, R^2 = i-Pr) and does not undergo spontaneous dehydration [334].

Generally N-alkyl-2-acylmethylbenzamides as compared to 4-acyl-butyramides show an increased tendency to form the cyclic isomers.

o-Acylphenylacetamides. By action of ammonia on methyl 2-acylphenyl-acetates Jones [335] obtained 3-isoquinolones (95) instead of the expected amides **(94A)**, which, obviously, are formed as a result of a dehydration of the cyclic amide isomers **(94B)**. For the majority of the known N-unsubstituted and N-monosubstituted o-acylphenylacetamides the open structure

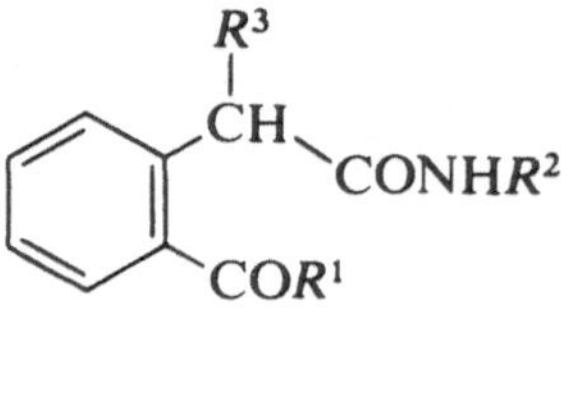

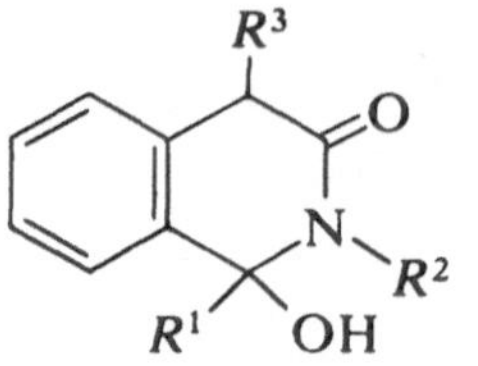

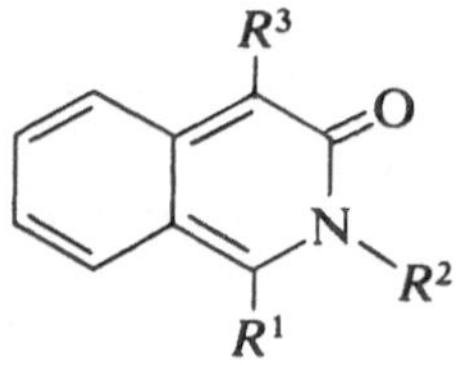

(94A)   **(94B)**   **(95)**

(94A) was established [166, 336]. A rare exception is the cyclic 1-hydroxy-6,7-dimethoxy-2-methyl-1-veratryl-1,4-dihydro-3-isoquinolone **(96)** [336]. However, all these amides in an acidic medium or on heating very easily dehydrate and transform into 3-isoquinolones **(95)** [166, 336–338], which hampered the investigation of isomeric interconversions **(94A)** $\rightleftarrows$ **(94B)**. For this reason 2-benzoylphenyl-α,α-dimethylacetamides **(97)** were investigated

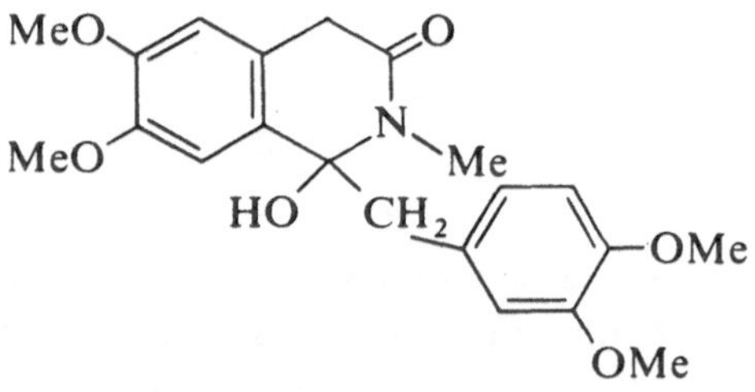

(96)

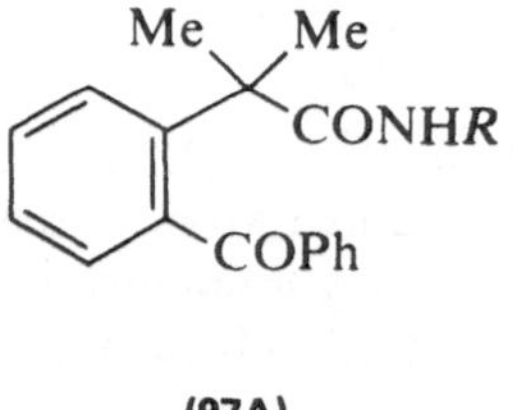

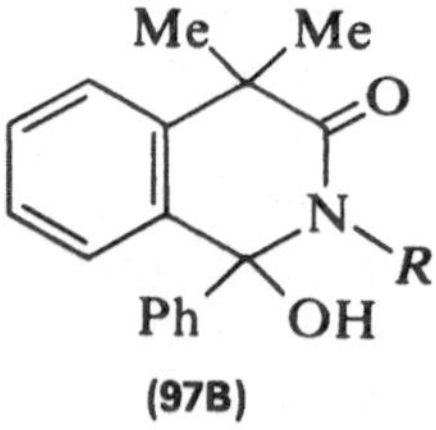

(97A)   **(97B)**

[171], since a 1,4-dehydration of the cyclic isomers is impossible. Under the influence of the Thorpe–Ingold effect these amides exist exclusively as cyclic isomers **(97B**, R = H, Me, Pr, i-Pr, Ph) which do not undergo thermal isomerization into open forms. The reverse is true only for N-(t-butyl)amide which, possessing an open structure, does not isomerize **(97A**, R = t-Bu) $\rightarrow$ **(97B)** even under conditions of alkaline catalysis.

On the whole *o*-acylphenylacetamides display a low tendency to form cyclic isomers compared with 2-acylmethylbenzamides.

8-Acyl-1-naphthamides. 8-Aroyl-1-naphthamides, obtained by the interaction of 3-chloro-3-phenylperinaphthalide with amines or by the isomerization of 3-alkyl(or aryl)amino-3-arylperinaphthalides, possess the cyclic structure **(98B)** [128, 168, 339].

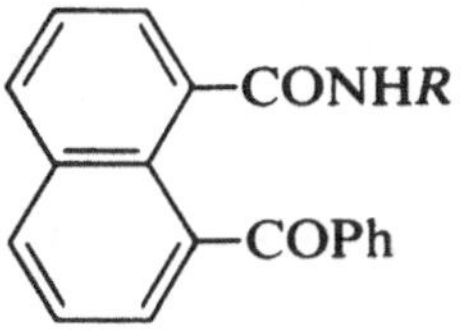

(98A) 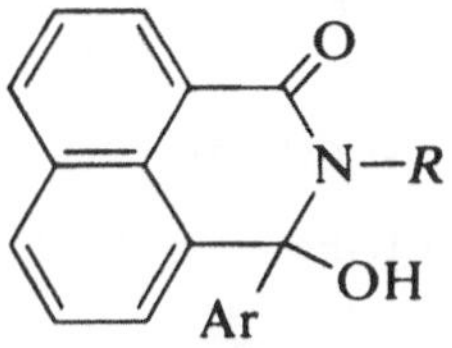   (98B)

All attempts to carry out the thermal isomerization of **(98B,** $R = i$-Pr, Ar = Ph) to **(98A)** failed. Hence it follows the cyclic form of *N*-alkyl-8-benzoyl-1-naphthamides is more stable than that of *N*-alkyl-2-benzoylbenzamides.

The only amide of 8-benzoyl-1-naphthoic acid obtained as an open isomer is the *N*-(*t*-butyl)amide **(98A,** $R = t$-Bu), which is formed in the reaction of the methyl ester of this acid with an excess of *t*-butylamine in an autoclave at 230°. The alkaline isomerization **(98A,** $R = t$-Bu) → **(98B)** does not take place.

2-Benzoyldiphenyl-2'-carboxamides possess the open structure **(99)** and do not ring close even under alkaline catalysis [81, 169].

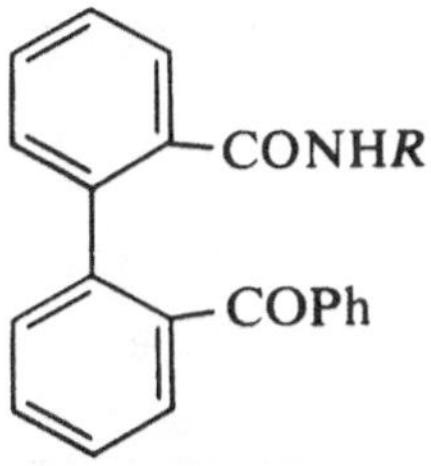

(99)

Summary of Structural Influences on the Relative Stability of Open and Cyclic Isomers of Acylcarboxamides. The structural factors on which the relative stability of ring or chain isomers depends may be divided into three groups.

1. The influence of the structure of the chain connecting the groups CONH*R* and C=O. Increasing the rigidity of the chain favors the proximity of mutually interacting groups and, hence, the stability of cyclic form rises.

Table 15. Influence of Keto Amide Structure and of Steric Bulk of a Substituent at Nitrogen on the Existence of Open (**A**) and Closed (**B**) Isomers

Keto amide structure	Favored isomer when $R =$				
	H	Me	n-Alkyl	sec-Alkyl	t-Bu
$PhCH_2CH_2CONHR$	A	B	A	A	
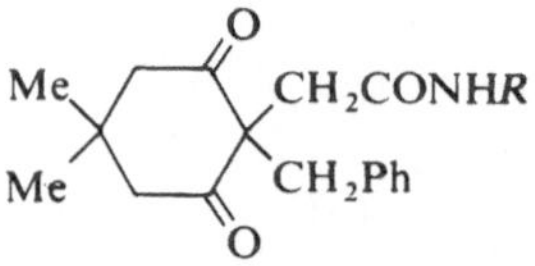 Me, Me cyclohexanedione with CH_2CONHR and CH_2Ph	A, B	B	A, B	A	A
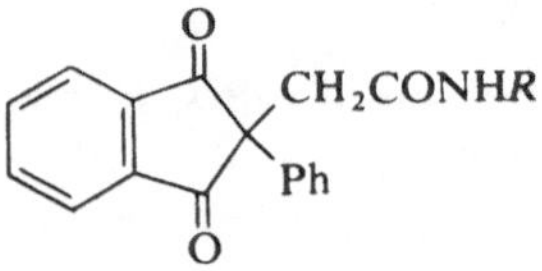 indandione with CH_2CONHR and Ph	A	A, B	A, B	A	A
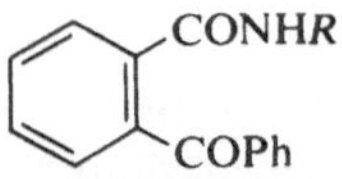 benzene with $CONHR$ and $COPh$	B	B	B	A, B	A
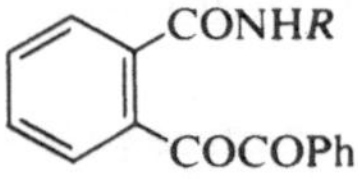 benzene with $CONHR$ and $COCOPh$	B	B	B	B	B
$PhCOCH_2CH_2CH_2CONHR$	A	A	A	A	
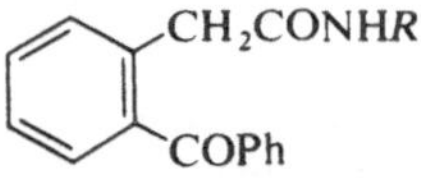 benzene with CH_2CONHR and $COPh$	A		A		
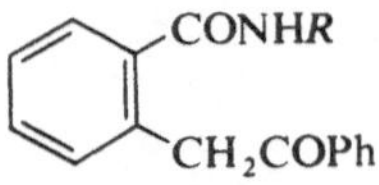 benzene with $CONHR$ and CH_2COPh		B	B	A	
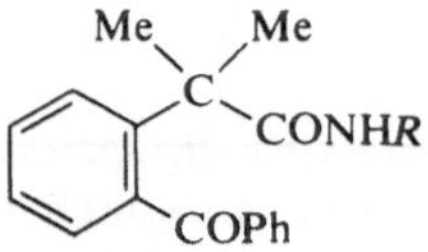 benzene with $C(Me)_2CONHR$ and $COPh$	B	B	B	B	A
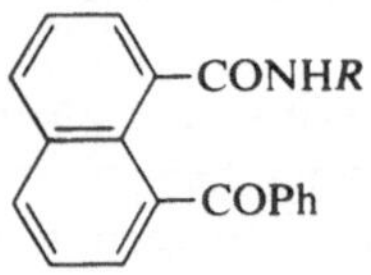 naphthalene with $CONHR$ and $COPh$	B	B	B	B	A

The formation of a five-membered ring is favored over that of a six-membered one. As an example anthraquinone-1-carboxamides having a rigid arrangement of the keto group in a conformation, which does not allow for a stereoelectronically favored nearly perpendicular approach of the nitrogen atom to the C=O group, does not undergo cyclization.

2. *The steric influence* of substituents at a nitrogen atom and/or keto group. The ring form is destabilized as the bulk (branching) of the substituent at the nitrogen increases in a series: Me < n-alkyl < sec-alkyl < t-alkyl. Substituents (mesityl, 9-anthryl) at the keto group which hinder the approach of the amide group act similarly.

3. *The electronic influence* of substituents. The electron-withdrawing effect of substituents at a keto group increases its electrophilicity thus increasing the stability of the cyclic form; electron-donating substituents operate in the opposite direction. Electron-withdrawing substituents at the nitrogen atom decrease its nucleophilicity and stabilize the open form.

It is shown in Table 15 that as the rigidity of the chain connecting the groups CONHR and C=O increases keto amides form cyclic isomers with very bulky substituents at the nitrogen atom.

Numerous examples of electronic influences on ring-chain equilibria have been discussed in the preceding chapters. In addition to 2-acylbenzamides, discussed on p. 68, N-(t-alkyl)-2-acylbenzamides [193, 201] may be cited as an example where the increasing electrophilicity of a keto group by an electron-withdrawing substituent outweighs the steric hindrance of an N-(t-alkyl) substituent. Thus, N-(t-butyl)-2-acylbenzamides having alkyl or phenyl substituents at the keto group exist only in the open form and are not capable of hydroxyisoindolinone ring closure even under alkaline catalysis. However, the introduction of electron-withdrawing substituents X into a benzoyl group allows cyclization of N-(t-butyl) and even N-(1-adamantyl)-2-(X-benzoyl)benzamides into hydroxyisoindolinones.

2.1.4. Esters

The esters of 3-acylpropionic [21, 30, 139, 140, 340], 2-acylbenzoic [28, 29, 32, 36, 74, 80–83, 151, 341–345], Z-3-acylacrylic [9, 10, 15, 22, 27, 101,

108, 109, 346-348], and 4-acylcarboxylic [40, 119, 120, 345, 349-355] acids were obtained in an open (4A) as well as in a cyclic (4B) form. Both isomers are stable in neutral solvents at room temperature. The equilibrium (4A) ⇌ (4B) may be observed only under the conditions of acid or alkaline catalysis [28, 36, 108, 140, 356] or by heating [346].

The determination of the structures of the isomers was carried out by UV [10, 29, 30, 82, 216, 346], IR [10, 27, 36, 40, 62, 82, 109, 151, 346, 357], and ^{1}H-NMR methods [27, 29, 30, 36, 38, 40, 346]. Quantitative measurements were made by gas–liquid chromatography (GLC) [36], UV [30, 356], IR [82] and ^{1}H-NMR spectroscopy [28, 29, 36, 38].

In the IR spectra of five-membered lactones (4Ba–c) the bands of the lactonic C=O group appear in the range of 1790–1760 cm^{-1} (see Table 16), which is transparent for the open isomers (4A). This band can be used as an analytical tool for quantitative measurements. It is not applicable for

Table 16. Frequencies of C=O Bands in the IR Spectra of Open (4A) and Cyclic (4B) Acylcarboxylic Acid Esters in Solution (cm^{-1})

Structure of open isomers		R	Solvent	Open isomer		Cyclic isomer C=O	Refs.
				COR	COOMe		
CH₂–COOMe / CH₂–COR		H	CCl₄	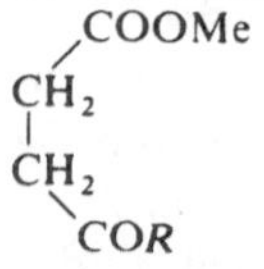 1717	1750	1790	30
benzene-COOMe / COR		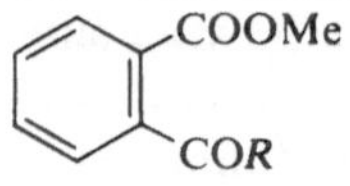 H	Dioxan	1700	1727	1782	26
		Me	Dioxan	1706	1727	1777	26
		Ph[a]	Dioxan	1680	1727	1783	26
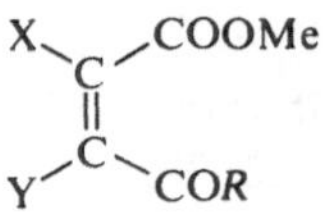 X–C(=C–Y)–COOMe / COR	X H, Y H	H	Pure liquid	1695	1733	1795, 1661	35
	Me H	Me	CHCl₃	1695	1724	1761	101
	H Me	Ph	Dioxan	1676	1727	1777	27
	Cl Cl	H	CHCl₃	1706	1732	1795	346
	Br Br	H	CHCl₃	1714	1731	1796	346
naphthalene–COOMe / COR		H CCl₄		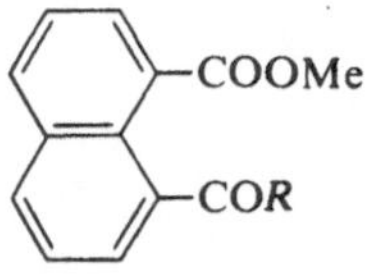 1704	1734	1742	38
		Me CCl₄		1693	1722	1730	38
		Ph CCl₄		1669	1739, 1726	1737	38

[a] For the IR spectra of 2-aroylbenzoic acid esters substituted in phthaloyl group see [37].

esters of 4-acylcarboxylic acids because the C=O bands of six-membered alkoxylactone and of ester groups differ only slightly.

In the ^{1}H-NMR spectra of open structured acylcarboxylic esters the signals of alkyl substituents at the keto group (R^1 in formula (4A)) and the alkoxygroup (R^2) are shifted toward lower fields as compared with those of cyclic isomers (4B) (see Table 17). This characteristic has been used for quantitative experiments. For a structure determination of aroylcarboxylic acid esters the ^{1}H-NMR method is of little use.

Esterification methods of acylcarboxylic acids generally lead to one specific isomer [1]. The open derivatives are formed by the action of alkyl halides on the silver salt of acylcarboxylic acids [10, 35, 101], and by esterification of acylcarboxylic acids with ethyl orthoformate or with diazomethane [10, 30, 40, 82, 101, 109, 119, 346], but there are some exceptions to the last reaction [83, 120].

Table 17. Chemical Shifts of Protons in ^{1}H-NMR Spectra of Open (4A) and Cyclic (4B) Isomers of Acylcarboxylic Acid Esters

Structure of open isomer (solvent)	R^a	Open isomer		Cyclic isomer		Refs.
		$\delta_{(R)}$	$\delta_{(OMe)}$	$\delta_{(R)}$	$\delta_{(OMe)}$	
RCOCH$_2$CH$_2$COOMe	**H**	9.63	3.60	5.90	3.7	30
	CH$_3$	2.11	3.60	1.54	3.29	139
COOMe / COR (in CD$_3$OD)	**CH$_3$**	2.55	3.87	1.82	3.06	36
	CH$_2$CH$_3$	1.18	3.88	0.83	3.04	
	CH(CH$_3$)$_2$	1.17	3.87	0.81	2.99	
	C(CH$_3$)$_3$	1.24	3.90	1.03	2.99	
	CH$_2$Ph	4.07	3.97	3.35	2.99	
	Ph	—	3.44	—	3.16	
X C COOMe Y C COR (in CDCl$_3$) X Y	**CH$_3$** (H H)	2.27	3.72	—	—	102
	Ph (H H)	—	3.58	—	3.27	102
	CH$_3$ (Me H)	2.22	3.79	1.60	3.20	101
	H (Cl Cl)	10.02	3.93	5.76	3.54	346
COOMe / COR (in CCl$_4$)	**H**	10.21	3.87	6.32	3.65	38
	CH$_3$	2.62	3.83	1.87	3.20	
	CH$_2$CH$_3$	1.23	3.85	0.82	3.20	
	CH(CH$_3$)$_2$	1.25	3.80	0.88	3.16	
	Ph	—	3.32	—	3.28	

^{a} The chemical shifts of protons printed in bold are given in the columns $\delta_{(R)}$.

Beska, Rapos and Winternitz [346] have observed that esterification of 3-formyl-2,3-dihalogenoacrylic acids with diazomethane at 0° yields open isomers but with an increasing proportion of ring isomers at higher temperatures. The open isomers of these acid esters undergo thermal isomerization into the cyclic ones.

The 2-formyl and 2-acetylbenzoates of open structure were obtained [343, 358] by the esterification of the acids with methyl iodide in the presence of potassium carbonate.

Alkoxylactones have been formed mainly by the alcoholysis of chlorolactones (2B) usually in the presence of pyridine [9, 10, 81, 109, 151, 350]. Urea has also been used instead of pyridine [359]. The cyclic isomers of levulinic ester were obtained [340] by the alcoholysis of α-angelicalactone.

The esterification of acylcarboxylic acids with alcohols in the presence of hydrogen chloride (Fischer-Speier method) or other acidic catalysts usually leads to a mixture of isomers (4A) and (4B), the proportion of which depends not only on the structures of the acid and the alcohol, but also on the reaction time (kinetic or thermodynamic control). The ratio of products is determined by the following equilibria [28, 29, 36, 360] (intermediates are not shown):

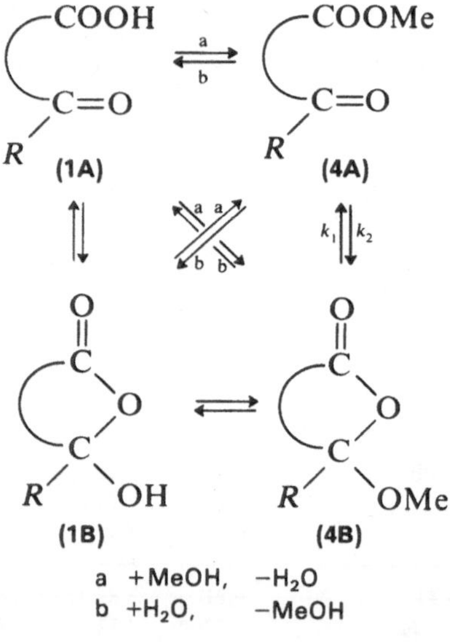

a +MeOH, $-H_2O$
b +H_2O, $-$MeOH

The scheme reflects the possibility of a nucleophilic attack of MeOH at both electrophile centers of tautomers (1A) and (1B).

A number of kinetic investigations have been published [342, 343, 351–354, 361–367] proving that the neighboring C=O group accelerates the hydrolysis of aldehydo or keto carboxylic acid esters by intramolecular catalysis with neighboring group participation [49,368]. In this case the reaction is initiated by an addition of the nucleophile to the keto group. The mechanism of the alkaline hydrolysis may be given as an example:

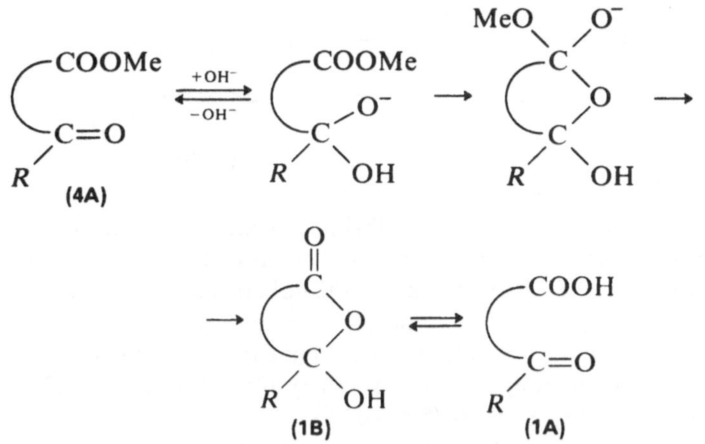

The result of thermodynamically controlled Fischer–Speier esterifications is determined by the equilibrium (4A) ⇌ (4B).

Using GLC and ^{1}H-NMR methods Bowden and Taylor [36] have measured equilibrium and rate (k_1 and k_2) constants (see Table 18) for ring–chain interconversions (100A) ⇌ (100B) of methyl 2-acylbenzoates in

Table 18. Equilibrium and Rate Constants for Ring-Chain Interconversions of Methyl 2-Acylbenzoates (100A) ⇌ (100B) in the Presence of Acid or Alkaline Catalysts [36] in Methanol

| | Equilibrium constants $K_T = [B]/[A]$ | | | | Rate constants $1 \cdot mol^{-1} \cdot s^{-1}$ | |
| | ^{1}H-NMR method, 36° | | GLC method, 40° | | GLC method, 40°C acid-catalyzed | |
R	Acid catalyzed	Alkaline	Acid	Alkaline	$k_1 \times 10^3$	$k_2 \times 10^3$
Me	1.9	2.2	1.7	1.8		1.03
Et	2.2	2.1	2.6	2.3	1.43	0.560
i-Pr	3.0	3.5	4.5	4.9	1.10	0.242
t-Bu	8.1	6.1	8.1	8.1[a]	0.273	0.0337
PhCH$_2$	2.8	2.6				
Ph	0.02	0.02				

[a] The rate constants under alkaline catalysis at 20°C are $k_1 \times 10^3 = 3.08$; $k_2 \times 10^3 = 0.380$.

methanol in the presence of acidic (hydrochloric acid) or alkaline (sodium methoxide) catalysts.

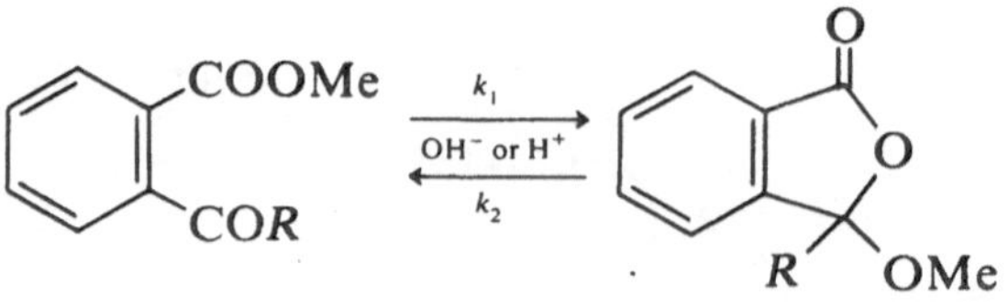

Table 18 shows that the equilibrium constants for acid catalyzed and base catalyzed equilibrations are equal within the limits of experimental error. K_T increases with an increase in the bulk of substituents R: Me < Et < i-Pr < t-Bu. The open form is destabilized by bulky substituents turning the C=O group out of the plane of the benzene ring (see [61]). Equilibrium constants correlate roughly with Taft's steric substituent coefficients E_s: log $K_T/(K_T)_0 = \delta E_s$, $\delta = -0.35 \pm 0.05$ [36].

The rate constants k_1 and k_2 for acid catalyzed isomerizations were obtained from GLC experiments. They decrease with increasing steric bulk of substituents R, k_2 being affected more ($\delta = 0.9$ in log $k_2/(k_2)_0 = \delta E_s$) than k_1 ($\delta = 0.5$).

Bowden and El-Kaissi [356] have suggested the following mechanisms of isomeric interconversions:

1) acid catalyzed:

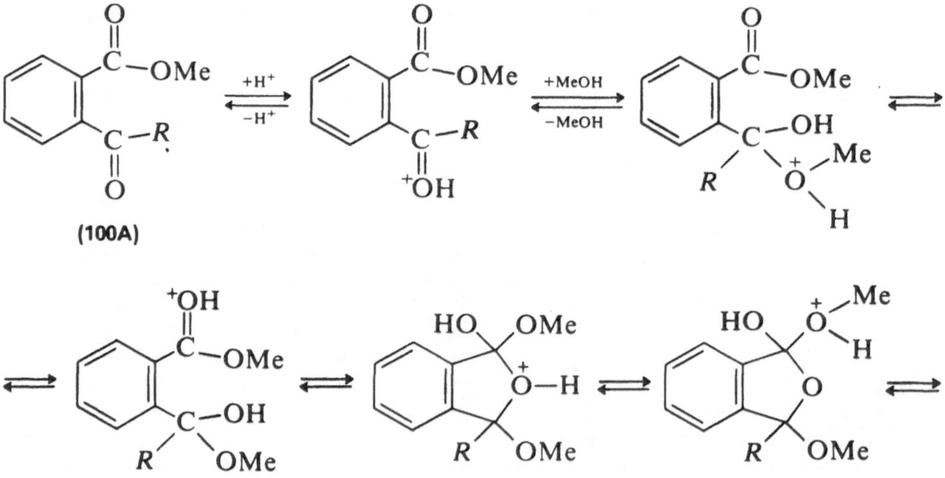

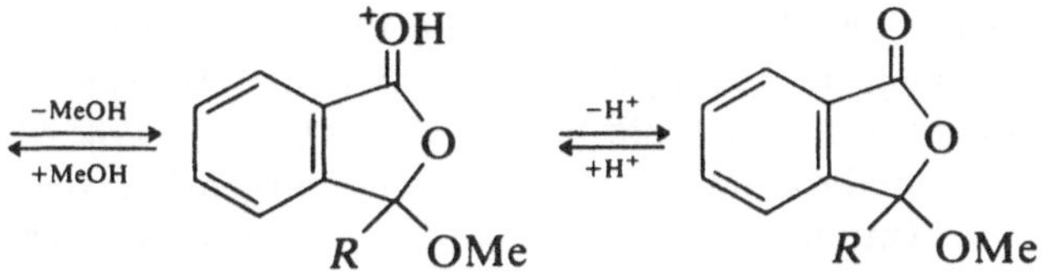

(100B)

2) methoxide catalyzed:

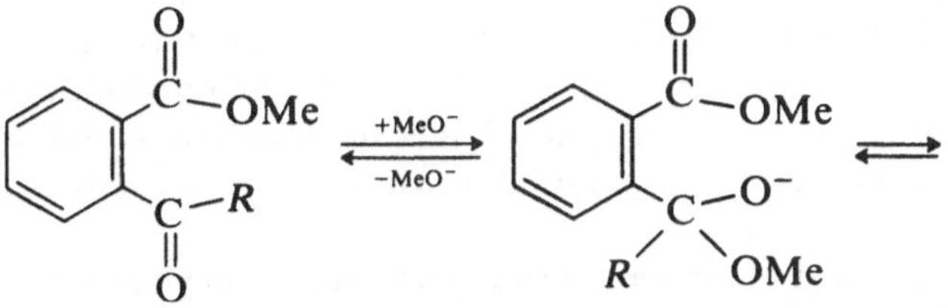

(100A)

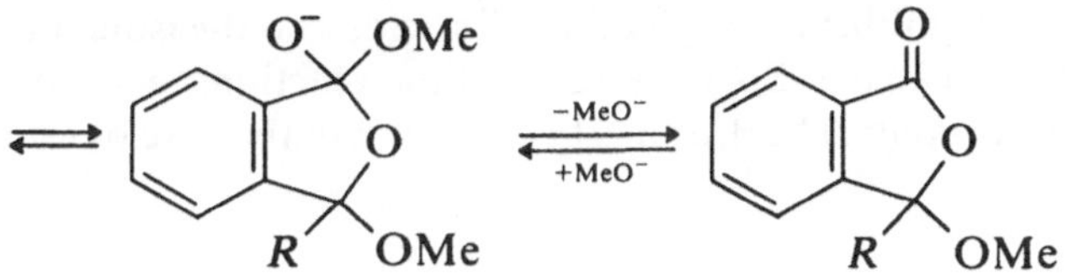

(100B)

The authors [356] carried out a kinetic investigation of base and acid catalyzed isomerization of the cyclic isomers of methyl 2-aroylbenzoates (**100B**, $R = 3$ or $4\text{-}XC_6H_4$) to **(100A)** and 8-aroylnaphthoates substituted in aroyl group **(101B)** to **(101A)**.

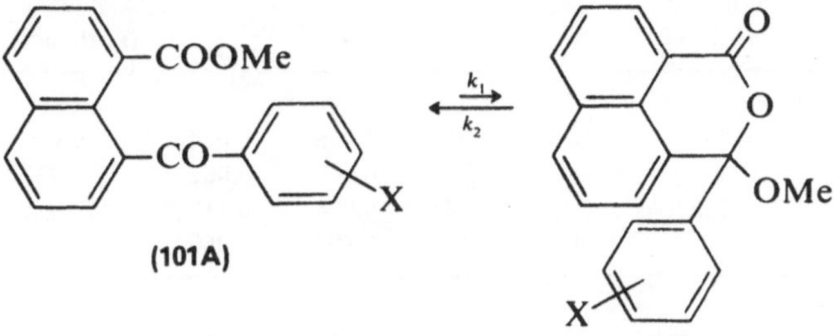

(101A)

(101B)

The isomerizations proceed according to the depicted mechanisms. In the rate-limiting step of the methoxide catalyzed isomerization a tetrahedral intermediate forms (for **(100B)**) or decomposes (for **(101B)**).

The equilibria **(100A)** ⇄ **(100B)** and **(101A)** ⇄ **(101B)** are almost fully displaced in favor of the open isomer ($K_T < 0.02$). The isomerization **(100B)** → **(100A)** proceeds more rapidly than **(101B)** → **(101A)**. Especially in the case of methoxide catalysis this difference arises from the lower enthalpy of activation (see Table 19) which is due to the greater strain in the five-membered ring. For these isomerizations (methoxide catalysis) a good linear correlation with σ or even better with σ^n values of substituents X has been obtained.

In acid catalyzed isomerizations substituents X exhibit an insignificant influence on the rate constants. Based on substituent, solvent, and solvent isotopic effects it has been concluded [356] that the rate-limiting step of the reaction sequence given on page 92 is a nucleophilic attack of methanol at the protonated C=O group of alkoxylactones or the subsequent proton transfer.

Generally ring-chain equilibrium constants of acylcarboxylic esters are governed by the same structural factors as equilibria of the corresponding free acylcarboxylic acids. Thus, there is a good correspondence between the tautomeric equilibria of acylcarboxylic acids and the isomer distribution of products of Fischer–Speier esterification reactions carried out under thermodynamic control [29], as has been shown in the case of esterifications

Table 19. Rate Constants, Correlation[a], and Activation Parameters of Acid- and Base-Catalyzed Isomerizations (**100B**, $R = XC_6H_4$) to **(100A)** and **(101B)** to **(101A)** in Methanol [356]

Initial compound	k_2 $1 \cdot mol^{-1} \cdot s^{-1}$ (X = H, at 60°C)	σ or σ^n	ρ	r	s	X = H, at 30°C	
						$\Delta H^{\neq}$ (kcal/mol)	$\Delta S^{\neq}$ (e.u.)
Methoxide catalyzed							
(100B),	4.05	σ	0.952	0.996	0.074	9.4	−28
$R = XC_6H_4$		σ^n	1.087	0.996	0.062		
(101B)	0.125	σ	2.041	0.996	0.112	12.6	−25
		σ^n	2.108	0.999	0.060		
Acid catalyzed							
(100B),	2.74×10^{-3}	σ	−0.172	0.771	0.054	11.3	−36
$R = XC_6H_4$		σ^n	−0.182	0.731	0.064		
(101B)	3.25×10^{-4}	σ	−0.123	0.878	0.030	12.2	−38
		σ^n	−0.125	0.878	0.032		

[a] For **(100)** $n = 9$, X = H, 4-Me, Me$_2$CH, Me$_3$C, MeO, F, Cl, I, 3-NO$_2$; for **(101)** $n = 7$, X = H, 4-Me, Cl, Br, 3-Me, Cl, CF$_3$.

of levulinic [140, 340], 2-aroylbenzoic [28, 29, 81, 82, 342], Z-3-aroylacrylic [27, 348], and 4-benzoylbutyric [350] acids mainly yielding open isomers.

2-Acylbenzoic (**1b**, *R* = H, alkyl) [26, 29, 36], Z-3-acylacrylic (**1c**, *R* = H, alkyl) [96, 101, 346], and 8-acyl-1-naphthoic [40] acids for which the ring-chain equilibrium is shifted toward a cyclic form (see Table 3) in the esterification mainly form cyclic esters. The isomerizations (rearrangements) (**4B**) to (**4A**) (levulinic [140, 340], 2-aroylbenzoic [80, 356], Z-3-aroylacrylic [10, 108], and 8-aroyl-1-naphthoic [356] esters) proceeding in the presence of acid or alkaline catalysts are reflecting the thermodynamically more stable isomer of the equilibrium (**4A**) ⇌ (**4B**).

2.1.5. Mixed Anhydrides

The identification of acylcarboxylic acid mixed anhydrides (**5A**) or their more widespread cyclic isomers, the acyloxylactones (**5B**), may be

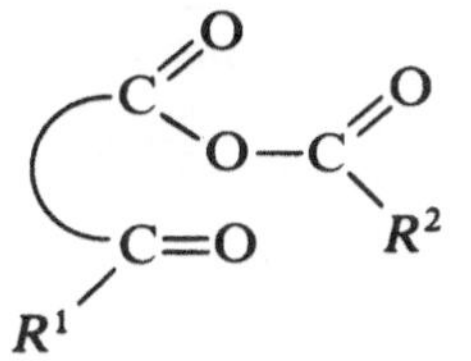

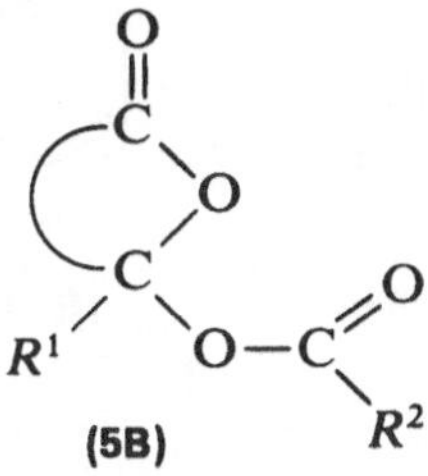

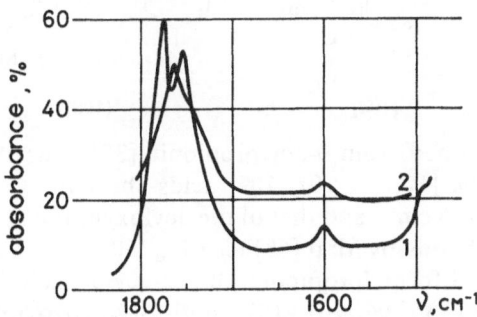

easily carried out using IR comparisons. In the IR spectra of open isomers (**5A**) anhydride C=O bands appear at 1820 and 1750 cm^{-1}, being clearly separated from the vibrational modes of the aldehydic or ketonic C=O groups at 1725–1660 cm^{-1}. Hence, absence of any absorption in the range of 1725–1660 cm^{-1} is good evidence for the cyclic structure (**5B**). Acyloxy-

Figure 10. C=O Bands in IR spectra of 3-acetoxyphthalides in the solid state: 1—3-acetoxy-3-(1,2-diphenylethyl)phthalide; 2—3-acetoxy-3-(1-phenylpropyl)phthalide [68].

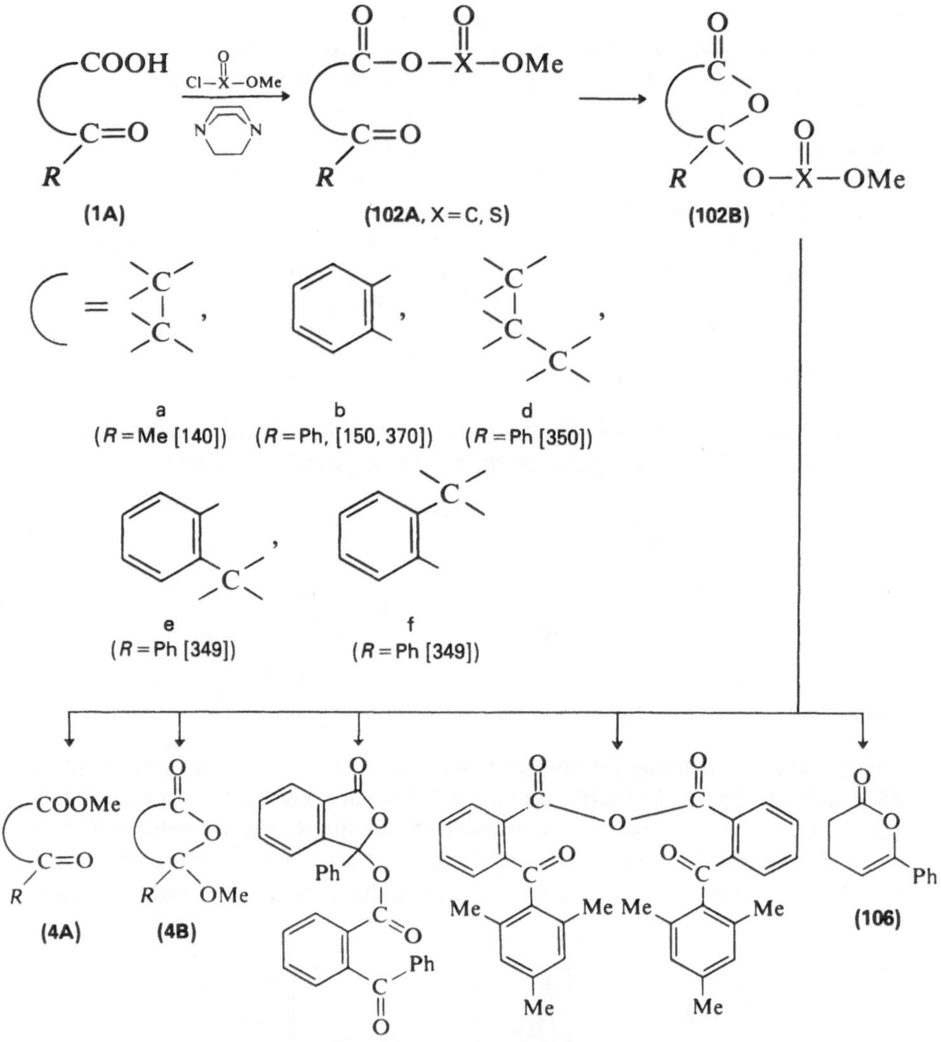

(1A)   (102A, X = C, S)   (102B)

a
(R = Me [140])

b
(R = Ph, [150, 370])

d
(R = Ph [350])

e
(R = Ph [349])

f
(R = Ph [349])

(4A)   (4B)   (103)   (105)   (106)

lactones (5B) obtained from 3-acylpropionic [30], 2-acylbenzoic [66, 68], and Z-3-acylacrylic [10, 35, 101, 109] acids show the C=O band of the lactone at 1810–1775 cm^{-1} and that of the acyloxygroup at 1770–1745 cm^{-1}. Sometimes these bands overlap [68] (see Fig. 10).

Acylation of 3-formylpropionic [30], 2-acylbenzoic [66–68, 81, 369], Z-3-acylacrylic [10, 35, 96, 101, 109], and 2-(2-benzoylphenyl)-2-methyl-propionic [121] acids with acetic anhydride [30, 67, 96, 101, 369], a mixture of acetic anhydride and pyridine [66, 68, 81, 109, 121] or sodium acetate

[66], acetic acid in the presence of sulfuric acid [10] or acetyl and benzoyl chlorides [35], gave acyloxylactones (5B). A more specific synthesis is the reaction of chlorolactones (2B) with silver acetate [10, 369]. Mixed anhydrides of open structure (5A) were obtained by action of ketene on 3-formylpropionic acid [30], by acetylation of 3-benzoyl-2,3-diphenylpropionic acid with acetic anhydride [109], and by the action of acetylchloride on the silver salt of benzil-o-carboxylic acid [369].

Newmnan and others [140, 150, 156, 350, 370] have obtained mixed anhydrides of 3- and 4-acylcarboxylic acids from methylchlorocarbonate or methylchlorosulfite in the presence of base, e.g., 1,4-diazabicyclo[2.2.2]-octane or by using the sodium salts of these acids.

Heating transforms mixed anhydrides (102A) into cyclic isomers (102B), which were not isolated in every case. These cyclic isomers on pyrolysis at higher temperature fragmented yielding mixtures of open (4A) and cyclic (4B) esters. Pyrolysis of mixed anhydride (102Ab, X = C, R = Ph), obtained from 2-benzoylbenzoic acid, gave in addition to esters (4A) and (4B) 2-benzoylbenzoic acid anhydride (103), a compound consisting of one open-structured acid moiety and a second cyclic part.

The anhydride (103) has been obtained also by the action of ethoxyacetylene on 2-benzoylbenzoic acid, of 3-chloro-3-phenylphthalide on 2-benzoylbenzoic acid in the presence of pyridine [150] and of 3-chloro-3-phenylphthalide on sodium nitrate [151].

(104)

On the basis of UV spectroscopic [369] and polarographic [371] data the structure (104) was first erroneously assigned to the anhydride (103).

The pyrolysis of 2-mesityloylbenzoic acid mixed anhydride (102Ab, X = C, R = 2,4,6-Me$_3$C$_6$H$_2$) gives the anhydride (105) as the only product, but from the mixed anhydrides of 4-benzoylbutyric acid (102Ad, R = Ph) and some gem-dimethyl derivatives unsaturated lactones (106) are obtained in addition to the esters (4B) [350].

Schmid and co-workers [369] have shown that 3-acetoxy-3-phenylphthalide (107) thermally isomerizes into mixed anhydride (108), i.e., a thermal isomerization takes place which is the reverse of the above-mentioned ((102A) → (102B)).

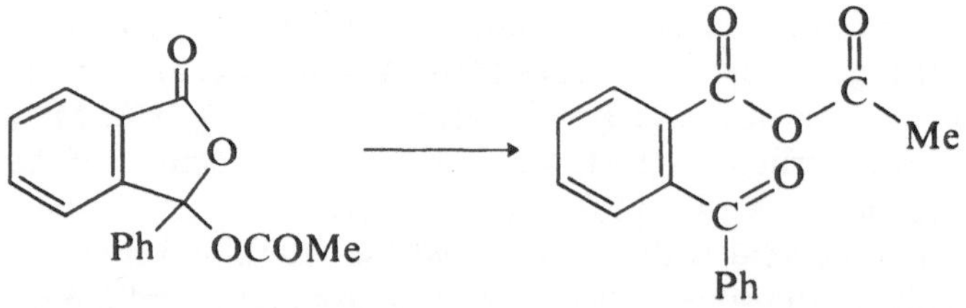

Ph   OCOMe

(107)   **(108)**

The available investigations do not allow conclusions to be reached concerning the influence of the structure of the acyl group (R^2 in formula **(5)**) on the stability of open and cyclic isomers of mixed anhydrides and on the conditions of their interconversions.

2.2. HYDROXY ALDEHYDES AND KETONES AND RELATED COMPOUNDS

2.2.1. Derivatives Containing a Hydroxy Group at Carbon (C—OH)

Ring-chain tautomerism in which a ring forms by intramolecular addition of a hydroxy group to a C=O bond occurs widely in organic chemistry. The structures are sometimes called ketolo-lactolic (as well as keto-lactolic) or oxo-cyclotautomeric [1]; ring isomers are sometimes named hemiacetals or hemiketals [372]. Commonly it is possible to isolate one of the isomers, but in solutions equilibration occurs. Both isomers have been isolated only rarely.

Mutarotation of aldoses and ketoses is a very important example of ring-chain interconversion of hydroxyaldehydes and hydroxyketones. Since a series of reviews [373–377] is dedicated to this problem we shall not discuss it here. Significant progress has been made recently in the study of the kinetics of tautomeric interconversions of sugars [378].

Here we shall discuss ring-chain isomeric interconversions of hydroxyaldehydes and hydroxyketones where the hydroxy group is located at an aliphatic, aromatic, or unsaturated (enols) carbon atom. Mainly those tautomeric systems will be considered which allow one to draw more general conclusions about the influence of structure on the relative stability of isomers and on the positions of equilibria in solutions. Unfortunately, precise quantitative data concerning tautomeric systems are rarely available. In most cases the influence of structural modifications on the relative stability of isomers was discussed only qualitatively.

Aliphatic hydroxyaldehydes and hydroxyketones provide the simplest examples for a study of ring-chain interconversions **(109A)** $\rightleftarrows$ **(109B)**. 4-Hydroxybutanal and 5-hydroxypentanal (**109**, R = H, n = 3, 4) forming five- and six-membered rings on cyclization possess the ring structure of hemiacetals **(109B)**. In the neat state [379–381], and in their solutions the equilibrium is strongly displaced toward the cyclic form (see Table 20). In larger rings the equilibrium shifts to the open form [382, 383].

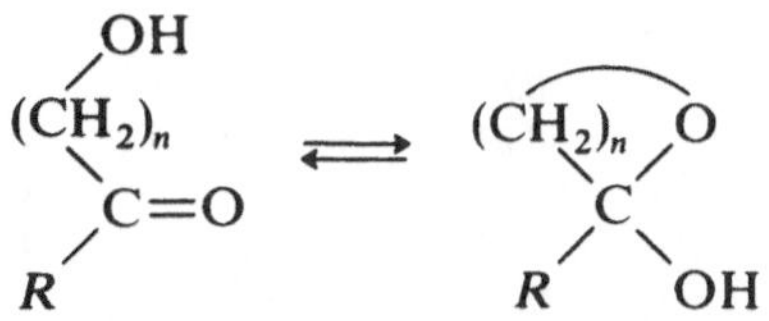

$$(CH_2)_n \overset{OH}{\underset{C=O}{\big|}} \quad \rightleftarrows \quad (CH_2)_n \overset{O}{\underset{C}{\diagdown}}$$

R	$R \qquad OH$
(109A)	**(109B)**

There is a striking difference between hydroxyaldehydes and hydroxyketones: in solutions of 5-hydroxy-2-pentanone (**109**, R = Me, n = 3) and 6-hydroxy-2-hexanone (**109**, R = Me, n = 4) the equilibrium is shifted toward the open form though five- and six-membered rings are formed [379, 384–386]. Equilibrium constants were determined [387] using the intensities of the signals of the methyl group in both tautomeric forms (slow equilibration on the ^{1}H-NMR time-scale), as well as by measurement of band intensities at 265–280 nm.

Table 20. Ring-Chain Equilibrium Constants of
Hydroxyaldehydes (**109A**, R = H) $\rightleftarrows$ **(109B)** in
Solutions [382, 383]

	$K_T = [B]/[A]$	
n	75% Aqueous dioxan[a] [382]	Toluene-d_8[b] [383]
3	7.8	
4	15.4	9
5	0.18	0.17
6		0.19
7	0.25	0.43
8		0.19
9	0.10	0.69
10		0.47
12		0.33
14		0.54

[a] at 25°C by a UV method
[b] at 70°C by an ^{1}H-NMR method

Table 21. Ring-Chain Equilibrium Constants in Solutions of
5-Hydroxy-2-pentanone and 6-Hydroxy-2-hexanone
(**109A**, R = Me, n = 3, 4) ⇌ (**109B**) [387]

	$K_T = [B]/[A]$			
	¹H-NMR method at 37°C		UV method at 25°C	
Solvent	$n = 3$	$n = 4$	$n = 3$	$n = 4$
Pure liquid	0.83	0.85		
Cyclohexane[a]	0.81	0.95	0.78	0.87
Dioxan			0.83	0.74
Carbon tetrachloride	0.73	0.95	0.75	0.96
Chloroform[a]	0.82	0.85	0.75	0.81
Acetone[a]	0.82	0.82		
Ethanol[a]	0.83	0.62	0.80	0.54
Methanol[a]			0.81	0.51
Acetonitrile[a]	0.82	0.79	0.79	0.61
DMSO[a]	0.64	0.81	0.56	0.77
Water[a]	0	0	0	0

[a] ¹H-NMR measurements are made in deuterated solvents.

Increasing temperature [386, 387] and solvent polarity displaces the equilibria in favor of the open form. Equilibrium constants in various solvents, except water, differ only slightly (see Table 21). This can be explained by the presence of similar proton-donating groups (O—H) in both forms.

In aqueous solution the cyclic form is absent. Addition of water to solutions of hydroxyketones in dioxan, ethanol or DMSO abruptly shifts the equilibrium toward the open form. This was explained [387] by a hydration effect on the carbonyl group. Thermodynamic characteristics of the equilibrium (**109A**, R = Me, n = 3, 4) ⇌ (**109B**) have been determined in different solvents.

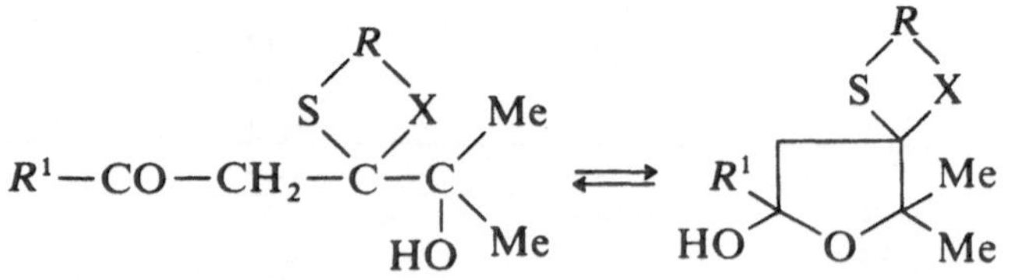

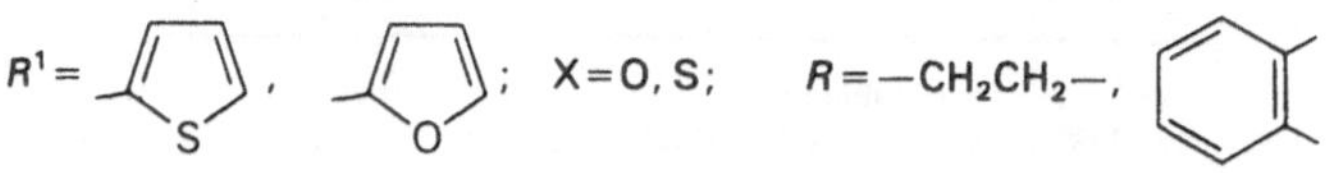

Curiously, the ring-chain equilibrium of more complicated compounds **(110A)** ⇌ **(110B)** shows a reverse solvent effect. Increasing the solvent polarity results in a shift of the equilibrium toward **(110B)** and equilibrium constants vary within larger limits [388]. Addition of sodium bicarbonate to an aqueous pyridine solution accelerates equilibration. The investigation of equilibria in aprotic solvents is complicated by the dehydration of the ring **(110B)**.

The greatest stability of a six-membered ring isomer was observed [389, 390] in the series of 2-(ω-hydroxyalkyl)cyclohexanones: 2-(3-hydroxypropyl)cyclohexanone possesses hemiketal structure (**111B**, $n = 3$), but its homologues (**111A**, $n = 4, 5$) are not capable of cyclization.

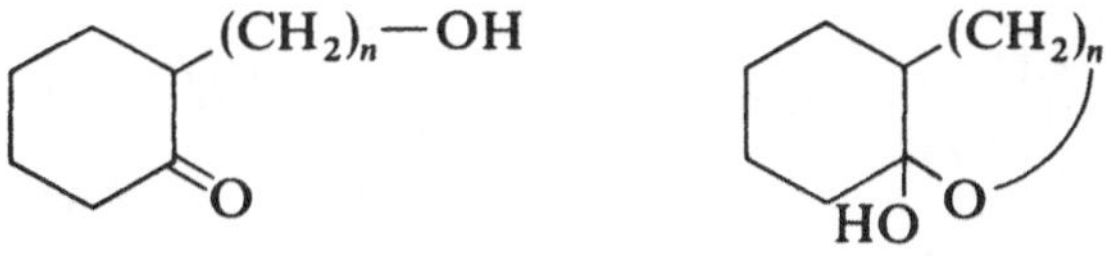

(111A)   (111B)

Compared with cyclohexanones the corresponding 2-(2-hydroxyethyl) and 2-(2 or 3-hydroxypropyl)cyclopentanones are less prone to form cyclic hemiketals [390].

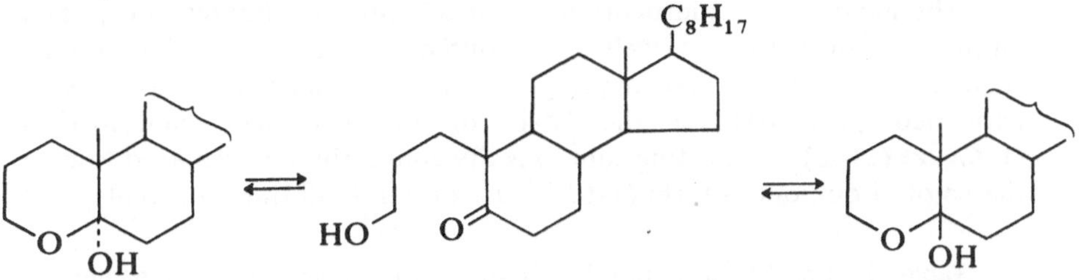

(112B)   (112A)   (112B′)

5-Oxo-3,5-seco-A-norcholestan-3-ol forms an equilibrium mixture **(112B)** ⇌ **(112A)** ⇌ **(112B′)** in carbon tetrachloride and chloroform. The equilibrium is displaced in favor of the epimeric cyclic forms ($K_T =$ [(112B)] + [(112B′)]/[(112A)] > 9) [391].

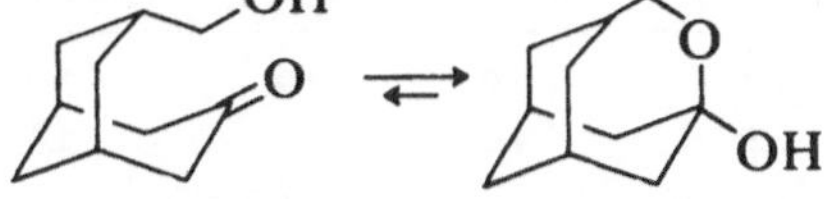

(113A)   (113B)

The equilibrium **(113A)** ⇌ **(113B)** of 7-hydroxymethylbicyclo[3.3.1]nonan-3-one in CDCl$_3$ is in favor of the lactol [392].

Sterically interacting groups in the open form shift the equilibrium to cyclic isomers. Thus, for a series of N-(2-hydroxyalkyl)-N-(2-oxoalkyl)amines **(114A)** $\rightleftarrows$ **(114B)** [393–398] the introduction of a third substituent at nitrogen (R^3 = alkyl) and in particular the protonation of the nitrogen atom [399] stabilizes the cyclic structure. Electron-donating substituents at the keto group (R^1 = 4-MeOC$_6$H$_4$) displace the equilibrium **(114A)** $\rightleftarrows$ **(114B)** toward the open form; electron-withdrawing substituents (R^1 = 4-ClC$_6$H$_4$) act in the reverse direction. Esterification in acidic media led to 2-alkoxymorpholines **(115)**.

| (114A) | (114B) | (115) |

The ring-chain equilibrium is also observed in the series of S-(2-hydroxyethyl)-S-(2-oxoalkyl)mercaptans [400] and two enantiomeric cyclic isomers have been detected.

Thermochromy, piezochromy, solvatochromy, fluorescence, and amphoteric properties as well as a double-step curve of polarographic reduction of 2-[N-(2-hydroxyethyl)-N-alkyl]aminobenzoquinones were explained [401–404] on the basis of the ring-chain tautomerism **(116A)** $\rightleftarrows$ **(116B)**. On heating solutions the equilibrium is displaced toward the colored quinonoid form **(116A)** [402, 403]. Equilibrium constants were

Table 22. Ring-Chain Equilibrium, Rate Constants, and Thermodynamic Parameters for 2-[N-(2-Hydroxyethyl)-N-alkyl]aminobenzoquinone **(116A)** $\rightleftarrows$ **(116B)** at 25°C^a [402]

| | | | Methanol | | | Phosphate buffer (pH = 7)-methanol (5:1) | | |
R	R^1	R^2	$K_T = [B]/[A]$	$-\Delta H$ (kcal/mol)	$-\Delta S$ (e.u.)	$K_T = [B]/[A]$	$k_1\,(s^{-1})$	$k_2\,(s^{-1})$
Me	Me	Me	1.1	2.9	9	2.0		
Me	H	Me	2.9	4.5	13	5.3		
Me	H	H	22	5.7	12.8	39	95	2.5
Et	H	H	53	6.1	12	83	76	0.9
Pr	H	H	67	6.3	12.5	100	73	0.7

^{a} The measurements are carried out at concentrations 2.5×10^{-4}–1.5×10^{-3} mol/l.

deduced from UV spectra of methanolic solutions and the rate constants of isomeric interconversions were determined polarographically (see Table 22). Thermodynamic parameters were calculated.

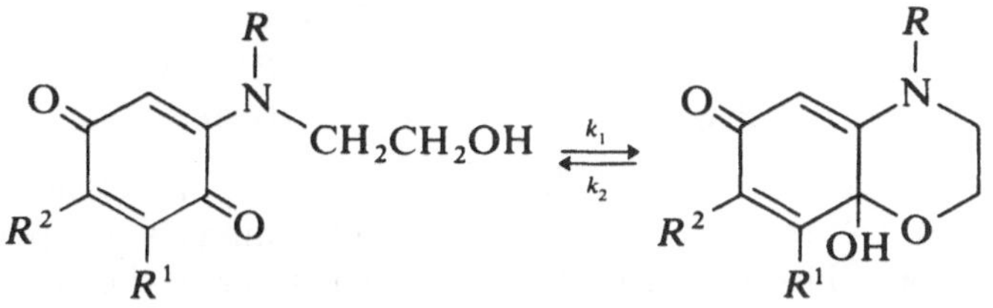

(116A)   (116B)

N,N-Disubstituted derivatives (116, R = H) are not at all capable of cyclization, and in N,N,N-trisubstituted derivatives (116, R = alkyl) the equilibrium shifts toward the cyclic form as the volume of the alkyl substituent at the nitrogen atom increases (Me < Et < Pr, see Table 22). The reason for this is the approach of the interacting groups due to a decreasing C—N—C bond angle and a hindering of the free rotation as a result of an increased steric bulk of substituent (R in (116)) at the nitrogen atom. In the benzoquinone ring, methyl groups (R^1 and R^2 in (116)) lower the electrophilicity of the C=O group thus shifting the equilibrium toward the open form. Isomeric interconversions proceed comparatively slowly.

Summarizing, one can state that a decreasing conformational mobility of the chain connecting the OH and C=O groups stabilizes the cyclic structure.

The ring-chain equilibrium (117A) ⇌ (117B) constants for 2-hydroxymethyl and 2-(2-hydroxyethyl)benzaldehydes in aqueous solution have been established by the UV method ($n = 1$, $K_T = 6.7$; $n = 2$, $K_T = 20$). The pertinent rate constants were measured by a pH jump method between pH 1 and 8 [405].

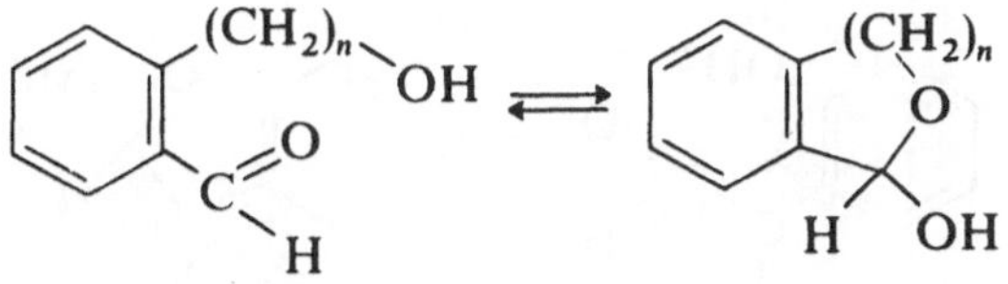

(117A)   (117B)

o-Acylphenylcarbinols with methyl or phenyl substituents in the methylene group (R^2 in formula (118)) exist in a stable cyclic form as 3-hydroxyphthalanes (118B) [159, 406–409].

The equilibrium (118A) ⇌ (118B) of a solution of 5-chloro-1-hydroxy-1-methyl-3, 3-diphenylphthalane in CDCl$_3$ ($K_T = 2.4$) was investigated [274].

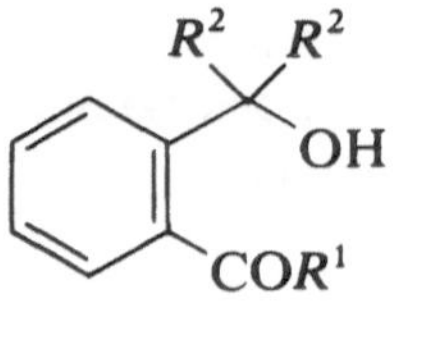

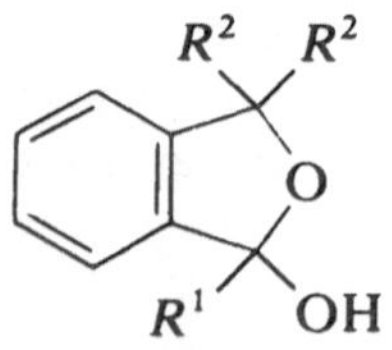

(118A)   (118B)

By introducing a mesityl [408, 410] or 9-anthryl [159] substituent to the ketogroup (R^1 in (118)) one succeeds in obtaining stable open isomers (118A, R^2 = Me, Ph). In this case the steric shielding of the keto group by the substituents (mesityl o,o'-methyl groups, anthryl peri-protons) prevents addition of the OH group. An analogous effect of these two substituents was noted also for 2-acylbenzoic acid derivatives [28, 78, 81, 82, 158].

α,β-Unsaturated γ-hydroxyketones possess mostly the ring structure of 2-hydroxy-2,5-dihydrofurans [406, 411]. 8-Acyl-1-naphthylcarbinols [126] also exist in a stable cyclic form (119).

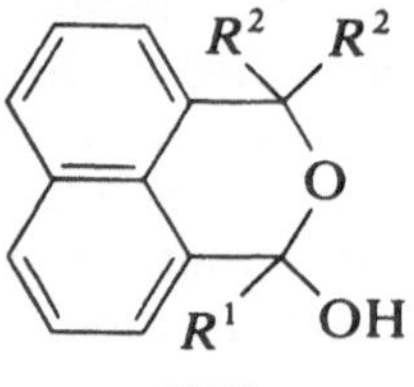

(119)

There are many examples of intramolecular additions of a phenolic hydroxyl group to aldehyde or ketone groups including the closure of a five-membered as well as a six-membered ring.

A preference of five-membered cyclic isomers was observed in the case of 2-hydroxyphenylglyoxal (120B, X = O) [412], 2-(2-hydroxy-phenyl)isobutyric aldehyde (120B, X = Me$_2$) [413], and ketones of similar structure [414, 415].

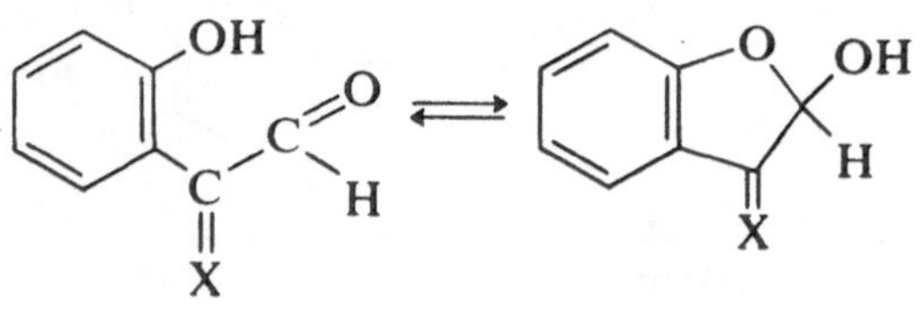

(120A)   (120B)

Six-membered cyclic isomers of 2-(2-hydroxybenzyl)cyclohexanone (121B, X = Y = H$_2$) [416, 417], 4-(2-hydroxybenzoyl)-1,3-cyclohexanedione (121B, X = Y = O) [418]), a series of the derivatives of 1-(2-hydroxyphenyl)-1,3-propanedione (122) [419–422], and similar compounds

[423] have been obtained. For (122, R^1 = Ph, R^2 = Br, OH, PhCOO) in chloroform equilibria (122A) ⇌ (122B) were observed (R^2 = Br: K_T = 3.7; R^2 = OH: K_T = 0.5) [419, 420]. Apparently, substitution of the hydroxyl group by a bromine atom increases the electrophilicity of the keto group, and the equilibrium is displaced toward the ring form. The compounds (122, R^1 = H, R^2 = H, Me) [421] and (122, R^1 = CF$_3$, R^2 = H) [422] exist exclusively in the cyclic form. Solutions of (122, R^1 = H, R^2 = Me) in acetone in the presence of a trace of t-BuONa show equilibria (122A) ⇌ (122B), the ring isomer (122B) existing as a mixture of cis- and trans-isomers, the interconversions of which proceed via the open form (122A).

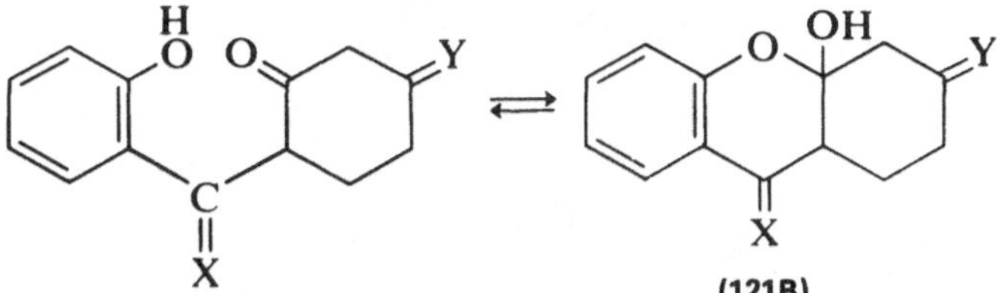

(121A) (121B)

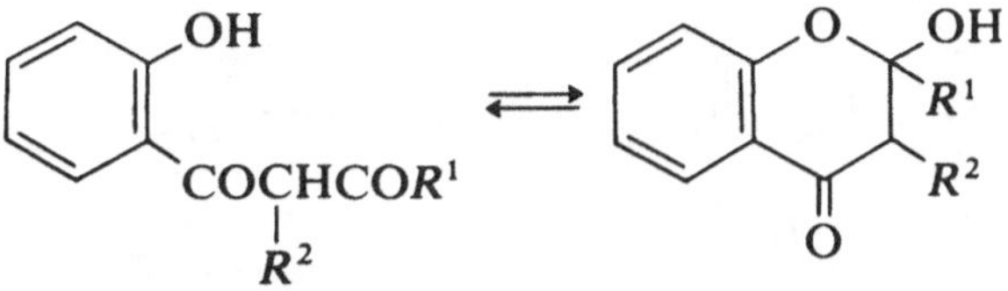

(122A) (122B)

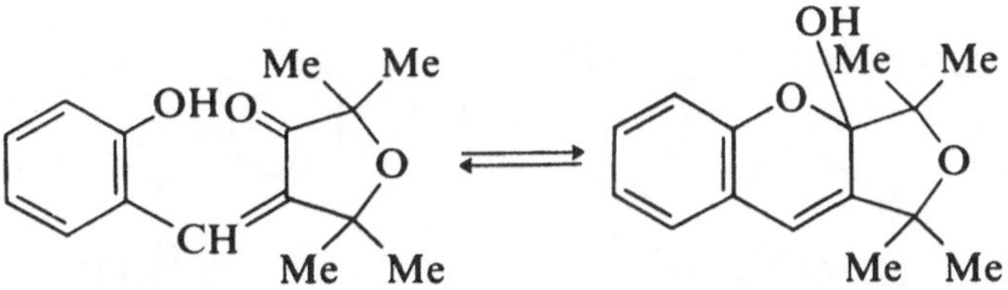

(123A) (123B)

Korobitsyna and co-workers [424] isolated both isomeric 4-(2-hydroxy-benzylidene)-2,2,5,5-tetramethyl-3-furanidones (123A) and (123B). The isomerization (123A) → (123B) proceeded on saturation of the ethereal solution of (123A) with dry hydrogen chloride. By dissolution of the hemiketal (123B) in sodium hydroxide solution and subsequent neutralization the isomerization was accomplished in the reverse direction (123B) → (123A). Evidently, under basic conditions the phenolate anion of (123A) is formed. The hemiketal (123B) can also be obtained by the addition of water to the corresponding furanidinobenzopyrylium salt which is formed by the action of strong acids on the hydroxyketone **(123A)**.

The condensation of ninhydrin with 2-amino-phenol and -thiophenol gives the ring structure **(124, X = O, S)** as shown by [425, 426] means of IR, ^{1}H-NMR, and ^{13}C-NMR spectroscopy, thus correcting incorrect structure assignments [427]. The introduction of substituents R^1 and R^4 into molecules of ninhydrin condensation products with polyphenols shifts the equilibrium **(125A) ⇄ (125B)** in favor of the cyclic isomer [428].

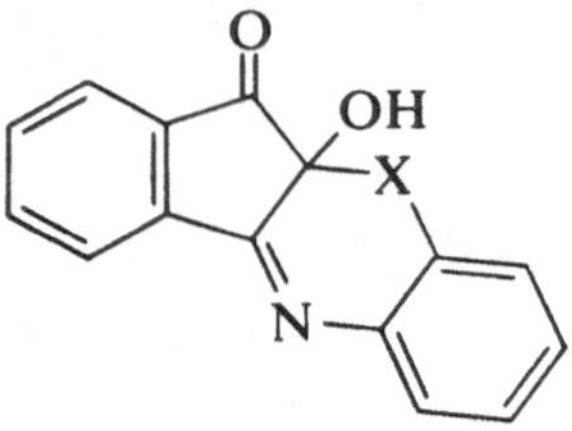

(124)

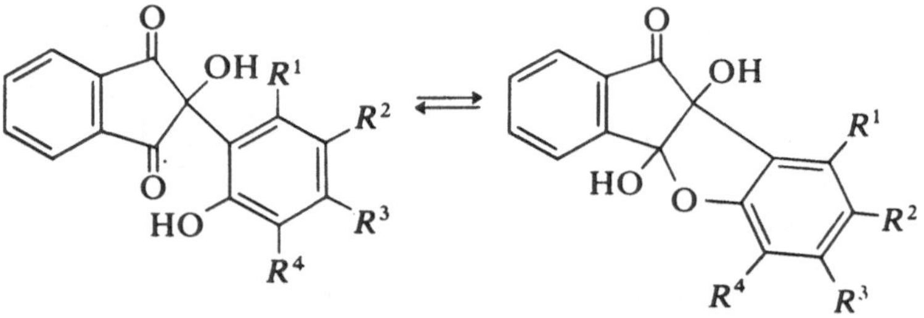

(125A)   **(125B)**

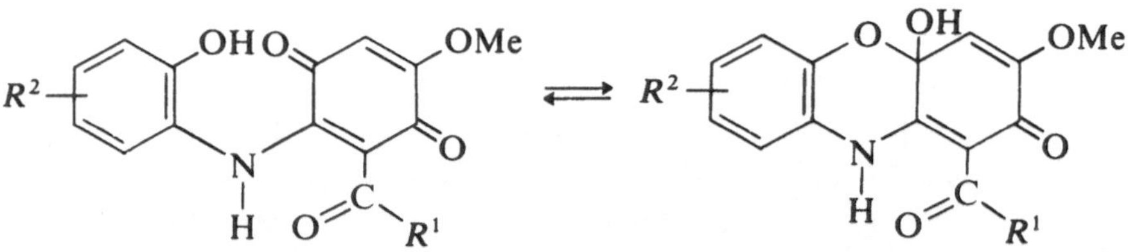

(126A)   **(126B)**

Schäfer and co-workers [429–431] have observed the equilibria **(126A) ⇄ (126B)** in solutions of substituted 2-(2-hydroxyphenylamino)-1,4-benzoquinones. The ring isomer is formed as a result of an intramolecular addition of the phenolic hydroxyl group to the benzoquinone carbonyl group. Previously a similar equilibrium was detected in solutions of 2-(2-hydroxyphenylamino)-1,4-naphthoquinones by the authors [432], who had established the ratio of tautomers by UV spectroscopy. In the spectra of model compounds having a fixed open structure the long wave band appears at 470 nm; however, in the spectra of fixed cyclic model compounds the corresponding absorption was found at 394 nm [430].

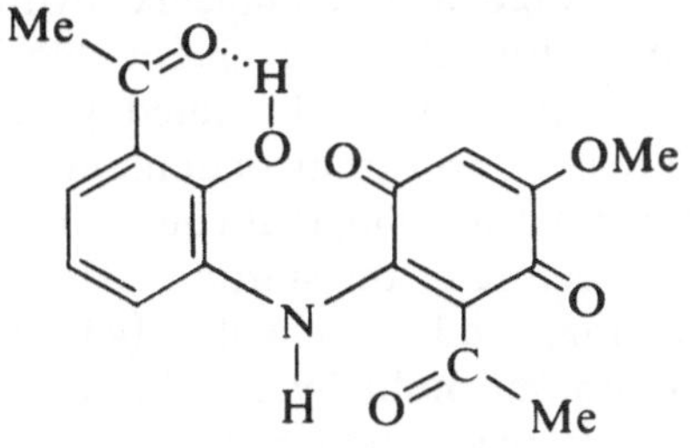

(127)

In an ethanolic solution of 3-acetyl-2-(2-hydroxy-3-acetylphenyl-amino)-5-methoxy-1,4-benzoquinone (126, R^1 = Me, R^2 = 3-MeCO) the cyclic form predominates but in solvents of lower polarity the equilibrium is shifted toward the open form. The authors [430] assume that the open form is stabilized by an intramolecular hydrogen bond between the hydroxyl

Table 23. Isomeric Structures of bis-(Cyclohexanone-2-yl)methanes and Their Derivatives (128)–(131) Obtained in the Solid State [433–437]

R	n	X	Y	Isomeric structure in solid state
(128), H		H_2	H_2	A
Ph		H_2	H_2	A, B[a]
4-MeOC$_6$H$_4$		H_2	H_2	A, B
2-Furyl		H_2	H_2	B[b]
H		O	H_2	B
Ph		O	H_2	B
Ph		O	PhCH	B
Ph		PhCH	H_2	B
4-MeOC$_6$H$_4$		4-MeOC$_6$H$_4$	H_2	B
Ph		PhCH	PhCH	B
(129), H	1			A
H	2			A
Ph	1			B[c]
Ph	2			A
(130)				A
(131)	1			A
	2			B[c]

[a] $K_T = [B]/[A] = 0.25$ (in CDCl$_3$).
[b] $K_T = 0.11$ (in CDCl$_3$).
[c] In solutions equilibrium (A) ⇌ (B) is detected.

group and the 3-acetyl group **(127)**. Both isomers have been isolated: the open quinone as a black powder (after crystallization from benzene), the ring isomer as yellow crystals (from methanol). Obviously, in a proton-donating solvent the cleavage of intramolecular hydrogen bonds proceeds due to the competitive formation of intermolecular hydrogen bonds $MeC=O\cdots H-O$ (solvent), and the liberated phenolic hydroxyl group intramolecularly adds to the $C=O$ group of quinone. UV spectra of solutions of both isomers are identical showing that equilibrium mixtures are formed independently of the structure of the starting compound.

Visotskii, Vershinina, and Tilichenko [433–436] investigated the intramolecular addition of an enolic hydroxyl group to a keto group in the series of bis-(cyclohexanonyl-2)methanes and of some their analogues **(128)–(131)** (see Table 23). In some cases the authors [437] succeeded in obtaining both isomers: **(128A,** R = Ph, 4-MeOC$_6$H$_4$, $X = Y = H_2$**)** and **(128B)**. In solutions equilibria **(128A)** $\rightleftarrows$ **(128B)** are reached only slowly (during 70 h). The isomeric interconversion **(128A)** $\rightleftarrows$ **(128B)** involves an enolization.

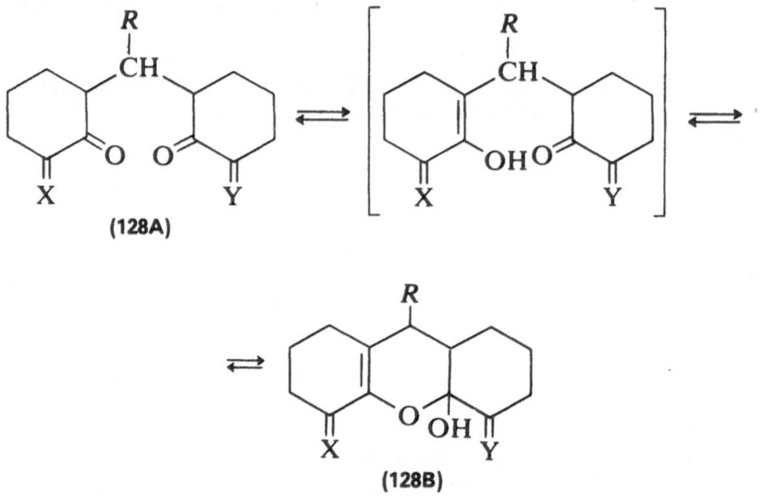

(128A)

(128B)

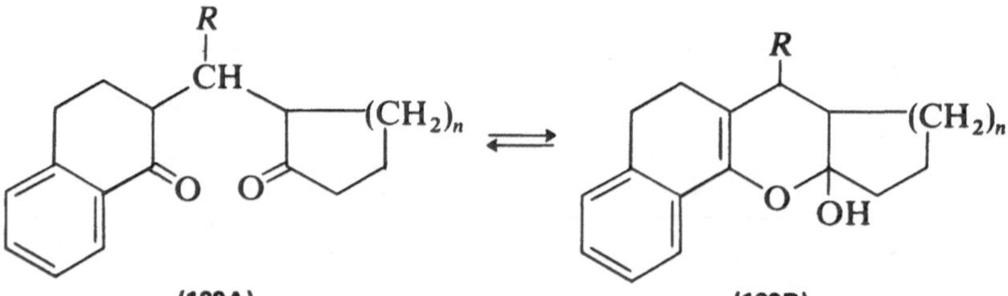

(129A)   (129B)

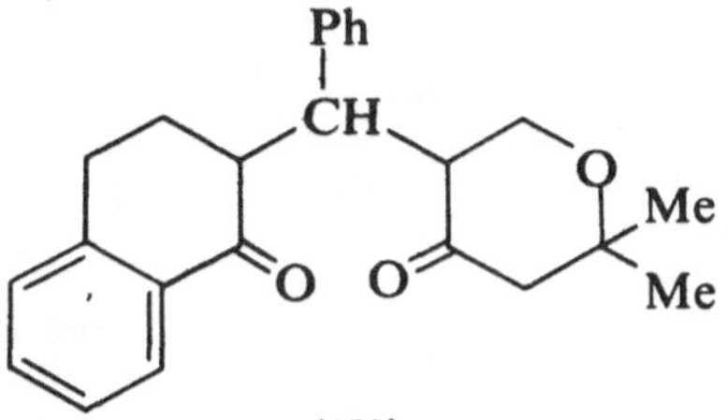

(130)

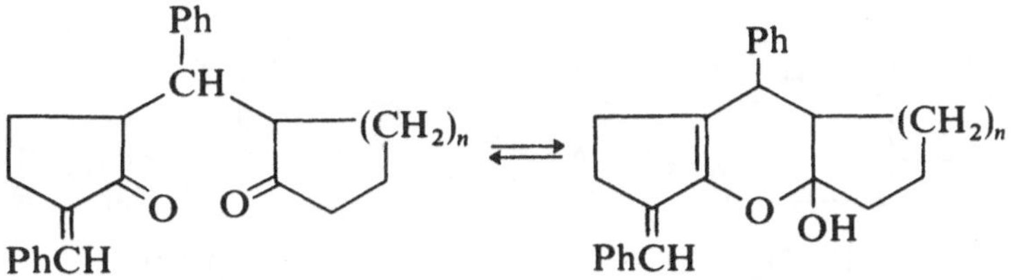

(131A)   **(131B)**

The presence of substituents R at the bridgehead carbon atom stabilizes the cyclic isomer **(128B)** (see Table 23). The presence of one or two arylidene groups in conjugation with the keto groups (X, Y = PhCH) or of a second conjugated keto group in the cyclohexanone ring (X = O) acts in the same direction. In the last case the cyclic isomer is formed even in the absence of a substituent at the bridgehead carbon atom (R = H), which is impossible for the other compounds in this series.

In the series of compounds **(129)** and **(131)** it was determined that contrary to the C=O group of cyclohexanone the keto group of cyclopentanone does not intramolecularly add the hydroxyl group and the cyclic isomers of compounds **(129)** and **(131)**, when n = 1, were not detected. This has been explained assuming [436] that in the five-membered ring the transformation of a planar carbonyl carbon to a tetrahedral configuration is energetically disfavored because of the formation of more eclipsed conformations.

A ring-chain equilibrium **(132A)** ⇌ **(132B)** was observed [438] for 1,5-diketones bearing one keto group in the open chain.

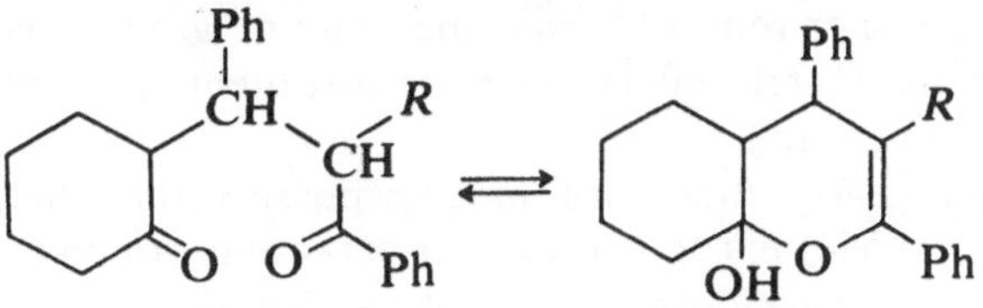

(132A)   **(132B)**

$R = CH_2PO(OR^1)_2$; R^1 = Me, Et.

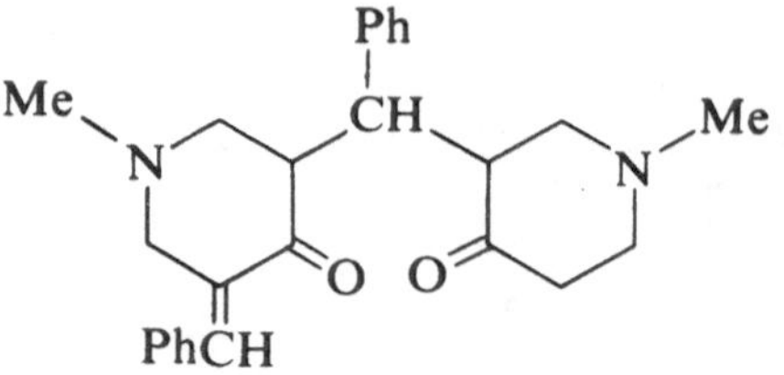

(133A) 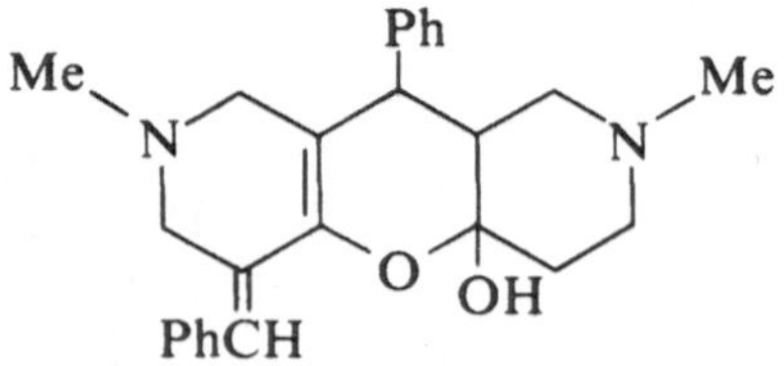   (133B)

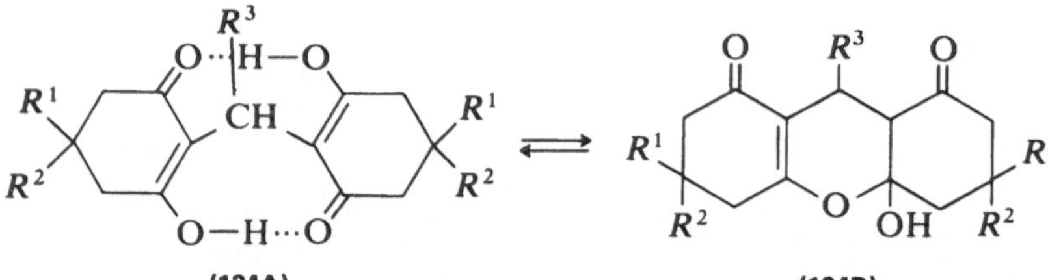

(134A)   (134B)

By a condensation of 1-methyl-4-piperidone with benzaldehyde heterocyclic derivatives of open (133A) and cyclic (133B) structure were obtained [439]; these are analogous to compounds (128).

Bis-(1,3-cyclohexanedione-2-yl)methanes, which can be obtained from condensation reactions of 1,3-cyclohexanediones with aldehydes, can exist in two isomeric forms: the open chelate (134A) stabilized by two intramolecular hydrogen bonds [440–442] and the cyclic hemiketalic decahydro-1,8-xanthenedione (134B) [443].

Gudriniece, Lielbriedis, and Lemba systematically studied the structure and behavior of the large group of bis-(1,3-cyclohexanedione-2-yl)methanes [444–446]. In two cases both isomers were isolated (134A, $R^1 = R^2 = Me$, $R^3 = 4\text{-(ClCH}_2\text{CH}_2)_2\text{NC}_6\text{H}_4$, $4\text{-(NCCH}_2\text{CH}_2)_2\text{NC}_6\text{H}_4$) and (134B) [443, 447, 448]. These compounds were isomerized in both directions: (134A) → (134B) in boiling ethanol or by recrystallization from acetic acid and (134B) → (134A) by slow crystallization from dioxan.

Other compounds of this series were isolated either as a chelate (134A) or a chemiketal (134B). However, no simple correlation is perceptible between the structure of substituents (R^1, R^2, R^3) and the favored isomer in the solid state. This is obviously not surprising since the substituents R^1, R^2, and R^3 are rather removed from the reacting groups. Moreover, it is difficult to estimate steric effects of the substituents in conformationally flexible molecules (134A).

Bis-(dimedone-2-yl)methanes mostly possess the chelate structure (134A, $R^1 = R^2 = Me$) except for the condensation products of dimedone with 2-substituted benzaldehydes existing as hemiketals. The majority of condensation products of unsubstituted 1,3-cyclohexanedione and its 5-methyl and 5-aryl-derivatives with aldehydes are hemiketals (134B).

As a general rule open isomers of *bis*-(1,3-cyclohexane-dione-2-yl)methanes (**134**, R^3 = H) being unsubstituted at the bridgehead carbon are favored. Electron-donating substituents in the aryl group of (**134**, R^3 = aryl) stabilize the open structure (**134A**), electron-withdrawing groups show the opposite effect.

Leaving out unsubstituted derivatives (**134A**, R^3 = H) *bis*-(1,3-cyclohexanedione-2-yl)methanes, irrespective of their isomeric structure in the solid state, slowly give in solution tautomeric equilibria (**134A**) ⇌ (**134B**) which are shifted toward the open form [445]. The methylation of compounds (**134**) with diazomethane led to O-methylderivatives only of the cyclic structure [449].

Diacylformoines (**135**) provide interesting examples of ring–chain tautomerism, an enolic hydroxyl and a keto group participating [450–457]. The enolization can proceed in two directions. The symmetric enediol has an open structure (E-1,2-diacyl-1,2-dihydroxyethylene (**135A**)) stabilized by two intramolecular hydrogen bonds. The asymmetric enediol exists as the cyclic 2,5-dialkyl(or diaryl)-2,4-dihydroxy-3-furanone (**135B**). Both isomers have been identified by IR ((**135A**), $\nu_{C=O}$ 1630–1620 cm^{-1}; (**135B**), $\nu_{C=O}$ 1705–1695 cm^{-1}) and UV spectra [455].

$$\overset{\displaystyle OH}{\underset{\displaystyle |}{R-COCHCOCO-R}}$$

(135)

(135A) ⇌ **(135B)**

R = Me (A)†, *i*-Pr (B), *t*-Bu (B), 4-XC$_6$H$_4$, X = H (B), Me (B), MeO (A), *t*-Bu (B), Me$_2$N (A), Cl (A, B), Br (A, B), NO$_2$ (B), 2,4,6-Me$_3$C$_6$H$_2$ (A), 1-naphthyl (A), 2-naphthyl (A, B)

Both isomeric 4-chloro and 4-bromobenzoyl [455], and 2-naphthoylformoines [457] have been obtained in the solid state. The open isomers (**135A**) crystallize as red crystals from benzene after extended heating of the solution. The cyclic isomers (**135B**) were isolated as yellow crystals from ethanol. Heating of the cyclic (**135B**) results in a thermal isomerization into the open isomer (**135A**).

† In parentheses the isomeric structure of compounds in the solid state is noted.

In other cases either open or ring isomers are formed depending on the structure of the substituent R. The introduction of electron-donating substituents in the aromatic ring of aroylformoines decreases the electrophilicity of the carbonyl carbon and stabilizes the open structure (135A) while electron-withdrawing substituents act in the reverse direction. Branching of the aliphatic substituent R stabilizes the cyclic structure [456]. Mesityloylformoin is not capable of forming the ring isomer because of the steric shielding of the keto group.

In proton acceptor solvents (ethanol, tetrahydrofuran) the equilibrium (135A) ⇌ (135B) is shifted to the cyclic form which is stabilized by intermolecular hydrogen bonds O—H···solvent [453]. In aprotic solvents the equilibrium is displaced in favor of the open form [455, 456].

$$R^1-COCH_2COCOCH_2CO-R^2$$

(136)

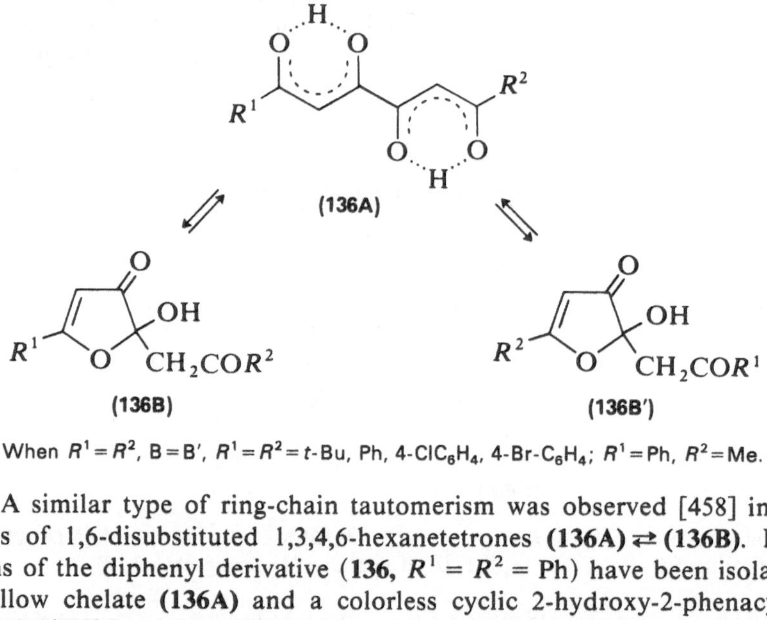

(136A)

(136B)   (136B')

When $R^1 = R^2$, B = B', $R^1 = R^2 = t$-Bu, Ph, 4-ClC$_6$H$_4$, 4-Br-C$_6$H$_4$; R^1 = Ph, R^2 = Me.

A similar type of ring-chain tautomerism was observed [458] in the series of 1,6-disubstituted 1,3,4,6-hexanetetrones (136A) ⇌ (136B). Both forms of the diphenyl derivative (136, $R^1 = R^2$ = Ph) have been isolated: a yellow chelate (136A) and a colorless cyclic 2-hydroxy-2-phenacyl-5-phenyl-3(2H)furanone (136B). At room temperature (136A) slowly transforms into (136B). Both isomers were easily identified by means of their IR ((136A), a broad chelate absorption band at 1600–1490 cm^{-1}; (136B), $\nu_{C=O}$ furanone 1717 cm^{-1}, acyl group $\nu_{C=O}$ 1688 cm^{-1}), UV ((136A), 364 nm; (136B), 320 and 248 nm in methanol) and ^{1}H-NMR spectra (equivalence of methine and substituent R protons in the open isomer). An equilibrium involving 2,2,9,9-tetramethyldecane-3,5,6,8-tetrone (136, $R^1 = R^2 = t$-Bu) in DMSO-d_6 was observed ($K_T = 5.7$, ^{1}H-NMR method). Apart from 16% of the open form two cyclic forms ($[(136B)]/[(136B')] = 6$) have been found in the case of $R^1 \neq R^2$ (136, $R^1 = $ Ph, $R^2 = $ Me). In nonpolar solvents the equilibrium is totally shifted toward the chelate (136A), while in polar solvents the content of (136B) is increased.

It is characteristic for the cases of ring-chain tautomerism discussed so far (134)–(136) that the open form, being stabilized by one or two intramolecular hydrogen bonds, is predominant in apolar solvents and that on going to polar solvents acting as a proton acceptor the equilibrium is displaced in favor of the cyclic form.

An intramolecular reversible addition of gem-diolic hydroxyl groups to aldehydo or keto groups (137) ⇄ (138A) ⇄ (138B) proceeds by the hydration of aliphatic 1,4 or 1,5-dialdehydes [459, 460], phthalic dialdehyde [461], or diketones [462].

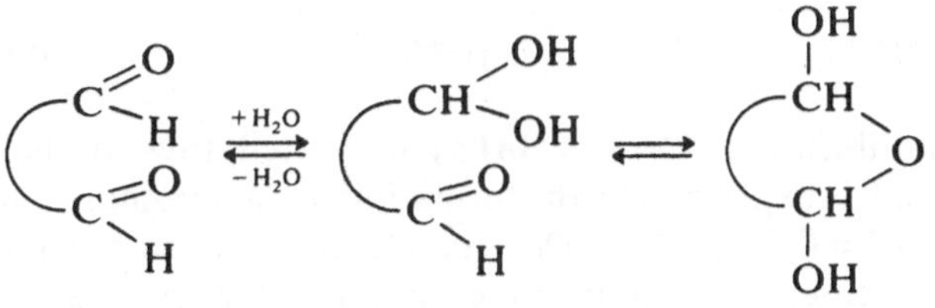

(137)   (138A)   (138B)

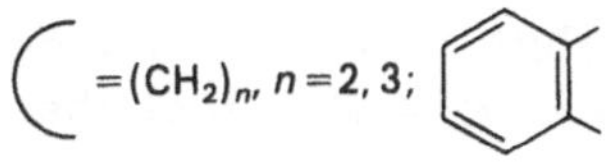

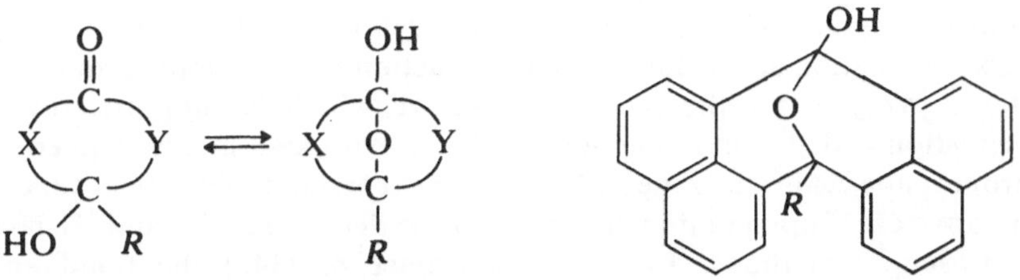

$= (CH_2)_n, n = 2, 3;$

(139A)   (139B)   (140)

A related intramolecular transannular addition yields bicyclic hemiketals (139A) ⇄ (139B). The hemiketals 5-hydroxycyclooctanone (139B, $X = Y = (CH_2)_3$, $R = $ H) [463] and a series of the derivatives of 1,8-(1,8-naphthalyl)naphthalene (140, $R = $ H, Me, OH, NHNH$_2$) [464] are given as examples. A transannular addition is also observed in more complicated systems (erythronolide B, erythromycins [465], a.o. [466]).

2.2.2. Derivatives Containing a Hydroxy Group at Nitrogen (N—OH)

Ring-chain tautomerism of hydroxyderivatives of this type was systematically studied [467–471] in a series of monooximes of 1,3-diketones. In the solutions of these compounds three tautomers were detected: syn and anti-isomers of 3-hydroxyiminoketones **(141A)** and **(141A′)** and 5-hydroxy-Δ^2-isoxazolines **(141B)**:

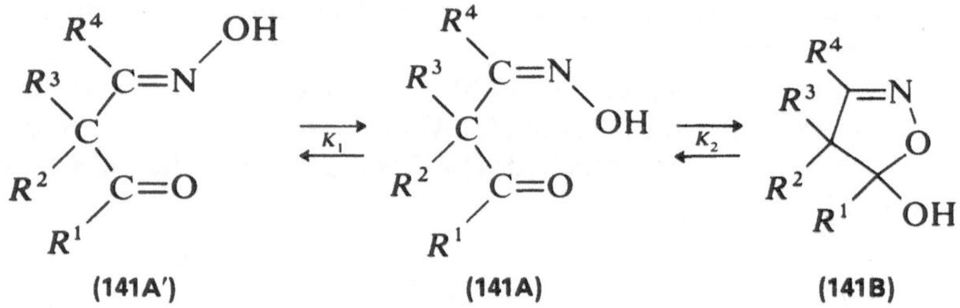

(141A′)   (141A)   (141B)

The equilibrium **(141A′)** $\rightleftarrows$ **(141A)** is shifted toward the syn-isomer ($K_1 > 1$), which is explained by the formation of an intramolecular hydrogen bond NOH$\cdots$O=C in **(141A)**. The ring-chain equilibrium constants K_2 as well as the pertinent thermodynamic parameters were established by ^{1}H-NMR spectroscopy in pyridine-d_5 at different temperatures [471].

Introduction of methyl (or phenyl) groups in positions R^2, R^3, or R^4 of the isoxazoline ring shifts the equilibrium toward the cyclic form. In the doubly substituted **(141**, $R^2 = R^3$ = Me**)** the open form could not be detected.

Two possible reasons for this gem-dimethyl effect were already proposed by Ingold [472]: (i) decrease of the bond angle $Me_2C\lessgtr$ and (ii) restricted rotation as a result of the introduction of two methyl groups. The authors [471] prefer the second alternative which is supported by the observation that in solutions of 5-hydroxy-3-methyl-5-phenyl-4-spiropropane-Δ^2-isoxazoline **(142)** and the corresponding gem-dimethyl derivative **(143)** open tautomers could not be detected by IR and ^{1}H-NMR spectroscopic methods. In the open isomer of **(142)** the bond angle $(CH_2)_2C\lessgtr$ is not reduced but rather is increased by the cyclopropane ring

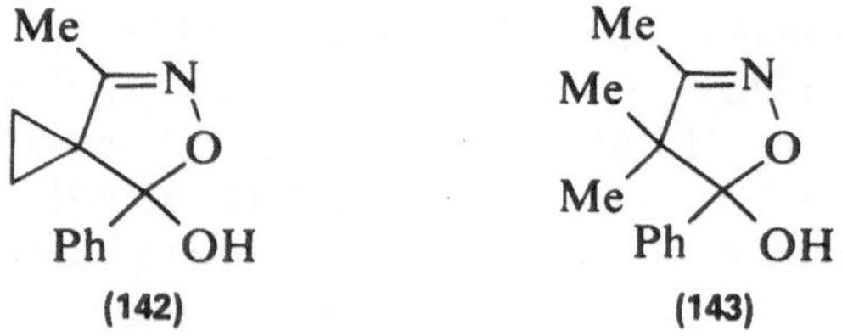

(142)   (143)

Table 24. Ring-Chain Equilibrium Constants of 5-Aryl (or alkyl)-5-hydroxy-Δ^2-isoxazolines (141A, $R^2 = R^3 = R^4 = H$) $\rightleftarrows$ (141B) in Pyridine $-d_5$ Solutions Determined by ^{1}N-NMR at 34°C [471]

$R^1 = 4\text{-XC}_6\text{H}_4$ X	$K_T = [B]/[A]$	R^1	$K_T = [B]/[A]$
Me_2N	0.01	Me	>100
MeO	0.33	Pr	12.8
Me	0.83	Me_2CHCH_2	3.77
F	1.48	Me_3CCH_2	0.97
H	1.68	i-Pr	35.7
Ph	2.54	Ph_2CH	10.1
Cl	2.83	t-Bu	30.3
Br	3.18		
NO_2	40.0		

even compared with the tetrahedral angle $H_2C{\lessgtr}$, which should stabilize the open isomer. The absence of the open form is therefore explained by a restricted free rotation in the open isomer of isoxazoline (142) as compared with the 4,4-unsubstituted derivative.

In the series of 5-hydroxy-5-(4-X-phenyl)-Δ^2-isoxazolines a good linear correlation between the ring-chain equilibrium constants K_2 and the σ^+ coefficients of Brown–Okamoto was detected [471]: $\log K_2/(K_2)_0 = 1.26\,\sigma^+$, $r = 0.993$.

A similar correlation with the inductive and steric coefficients (σ^+ and E_s) of alkyl substituents does not exist for 5-alkyl-5-hydroxy-Δ^2-isoxazolines (141, R^1 = alkyl) (see Table 24), which is explained by the fact that alkyl substituents at the keto group affect the tautomeric equilibrium not only by their electron and steric effects on the reactivity of the keto group but that they also exert a steric influence on the conformational mobility of the whole molecule of the open isomer.

The formation of a five- or six-membered ring by participation of a hydroxyimino group was observed in a series of monooximes of 2-benzoyl [473] and 2-phenacyl-2-aryl-1,3-indanediones [474] which in the solid state are cyclic (144B, $n = 0, 1$). This happens to be the case also for solutions

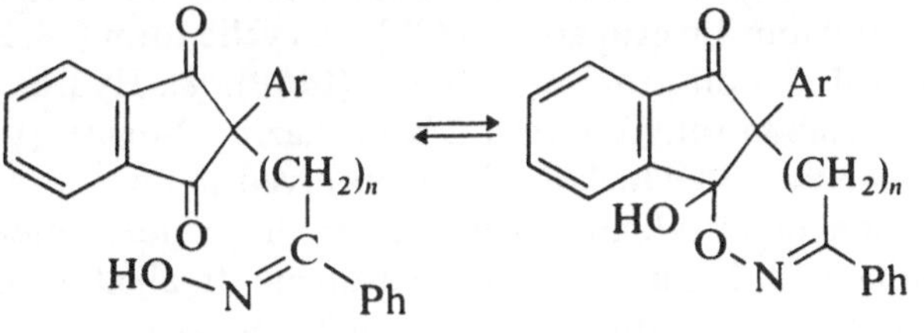

(144A)   (144B)

of monooximes of 2-benzoyl-2-phenyl-1,3-indanedione (**144B**, $n = 0$, Ar = Ph) in acetonitrile. The monooximes of 2-aryl-2-phenacyl-1,3-indanediones, however, form equilibrium mixtures (**144A**, $n = 1$, Ar = Ph, 4-MeOC$_6$H$_4$, 4-Me$_2$NC$_6$H$_4$) $\rightleftarrows$ (**144B**) in dichloroethane.

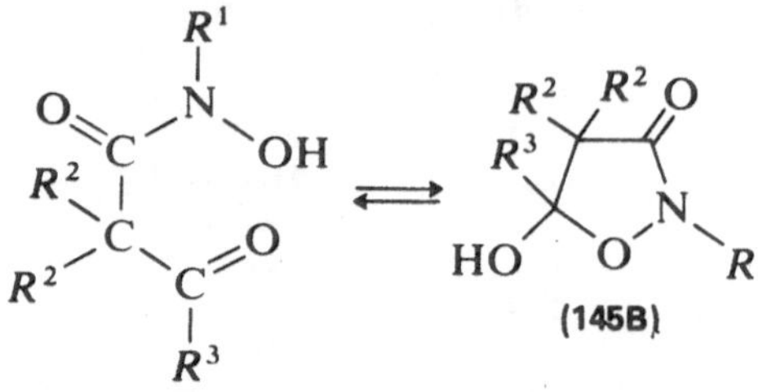

(145A)   (145B)

Equilibria are also obtained in solutions of hydroxamic acids, 5-hydroxy-4,4-diisopropyl-3-isoxazolidinone is an example (**145A**, $R^1 = R^3 =$ H, $R^2 = i$-Pr) $\rightleftarrows$ (**145B**) [475]. Spontaneous isomerization (**145A**, R = Me, aryl, R^2 = H, R^3 = Me) → (**145B**) takes place by the interaction of N-monosubstituted hydroxylamines with diketene [476].

MeCOCH$_2$CH$_2$CH(NO$_2$)$_2$

(146A)

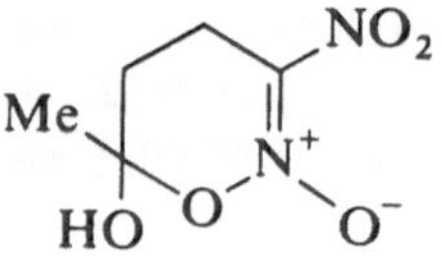

(146B)

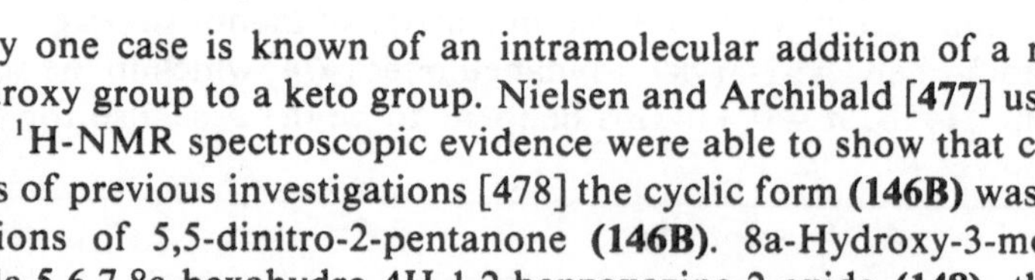

(147)   (148)

Only one case is known of an intramolecular addition of a nitronic acid hydroxy group to a keto group. Nielsen and Archibald [477] using IR, UV, and ^{1}H-NMR spectroscopic evidence were able to show that contrary to results of previous investigations [478] the cyclic form (**146B**) was absent in solutions of 5,5-dinitro-2-pentanone (**146B**). 8a-Hydroxy-3-methyl-4-phenyl-4a,5,6,7,8a-hexahydro-4H-1,2-benzoxazine-2-oxide (**148**), the only cyclic ketonitronic acid which has been observed so far, was obtained [479] by acidification of an alkaline solution of the condensation product of cyclohexanone with 2-nitro-1-phenyl-1-propene. In alkaline medium, compound (**148**) forms the sodium salt of oxonitronic acid (**147**). The structure (**148**) was supported by ^{1}H-NMR spectroscopy [477].

Spectroscopic investigations of other 1-nitro-3(or 4)-oxoalkanes and 1,1-dinitro-3-oxoalkanes confirm their open structure. No data are available on the ring-chain tautomeric equilibrium in solutions of 3 or 4-oxonitronic acids.

2.3. AMINO ALDEHYDES AND KETONES AND RELATED COMPOUNDS

In this section ring-chain isomeric interconversions are described which are brought about by intramolecular addition of N—H groups to a C=O bond. Examples of this kind have already been discussed in the sections on amides and hydrazides of acylcarboxylic acids (see Sec. 2.1.3). The cases treated here comprise amino derivatives of aldehydes and ketones containing primary or secondary amino groups (NH_2 or NHR) located at saturated, ethylene or aromatic carbon atoms. Reactions of covalently hydrated nitrogen-containing heterocycles are an unusual source of aminoaldehydes and aminoketones capable of ring-chain tautomerism. Equilibria are also discussed in which an intramolecular addition of N—H groups of imines, hydroxylamines, hydrazines, triazenes, amidines, ureides, thioureides, guanidines, dithiocarbamines, sulphamides and other functional groups to a C=O bond occurs. Unfortunately, not all these cases have been throroughly investigated.

The closure of a five- or six-membered (rarely a seven-membered) ring, i.e., 1,4 or 1,5 (1,6)-positioned interacting N—H and C=O groups, is a necessary condition for the formation of cyclic isomers. In spite of numerous investigations by different authors it is rather difficult to obtain a general picture of additional structural influences on the relative stability of ring or chain isomers.

2.3.1. Covalent Hydration of Nitrogen-Containing Heterocycles

The tautomeric aminoaldehydes (more rarely aminoketones) *vs.* carbinolamines **(149A)** ⇌ **(149B)** can be formed by covalent hydration of heterocycles containing immonium (*N*-protonated, *N*-alkyl, *N*-aryl) groups. A series of reviews [480–484] have appeared which mainly treat the equilibrium between ionic and covalent forms of the cyclic immonium bases **(149C)** ⇌ **(149B)**.

Beke and Szantay [485, 486], investigating carbinolamines, obtained by the addition of alkali to solutions of 2-substituted isoquinoline or 3,4-dihydroisoquinoline salts, draw the conclusion that a weak basicity of the nitrogen atom of the carbinolamine **(149B)** would favor the formation of the open form **(149A)**.

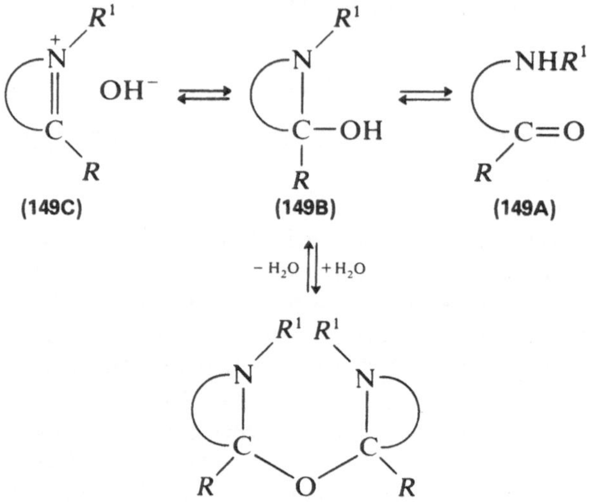

Four cases have been discussed [485]:

1. Strong bases (2-alkyl-3,4-dihydroisoquinolines). In the solid state and in nonpolar solvents they exist as carbinolamines **(149B)**, while in polar solvents only immonium ions **(149C)** are detectable. No evidence was found by chemical or physical methods for the presence of an open form **(149A)**.

2. Moderate bases (2-aryl-3,4-dihydroisoquinolines). In the crystalline state as well as in nonpolar solvents they show structure **(149B)**. In aqueous solution equilibria **(149C)** ⇌ **(149B)** ⇌ **(150)** were observed. The open form **(149A)** is absent.

3. Carbinolamines not capable of forming immonium ions **(149C)** isomerize into the open form **(149A)**. Both forms can be detected in solution and even have been isolated occasionally [486].

4. Compounds showing no pseudobasic properties exist exclusively in the open form **(149A)**. Carbinolamines **(149B)** formed in alkaline solutions of immonium salts of heterocycles are not stable and spontaneously transform into the open isomer **(149A)**.

Hence, it appears that equilibria **(149C)** ⇌ **(149B)** and **(149B)** ⇌ **(149A)** can not be observed simultaneously. Carbinolamines capable of forming the ionic form **(149B)** do not transform into the open isomer **(149A)** and vice versa.

Products of covalent hydration of 1-arylpyridinium salts [487] have the open structure of monoaniles of glutaconic aldehyde. An interesting rearrangement of 3-cyano-1-methylpyridinium iodide into 2-methylamino-3-formylpyridine by the action of alkali involves covalent hydration and opening of the pyridine ring with subsequent cyclization [488].

Many examples of the recyclizations of pyridine derivatives proceeding via covalent hydration, ring opening, and subsequent ring closure have been thoroughly reviewed [489].

Hydration of 2-(2,4-dinitrophenyl)isoquinolinium ion [486] gave both isomers **(151A)** and **(151B)** which were isolated. A study [486] of etherification reaction kinetics (**(151B)** with alcohols in presence of basic catalysts) showed that under the reaction conditions the equilibrium **(151A)** ⇌ **(151B)** took place.

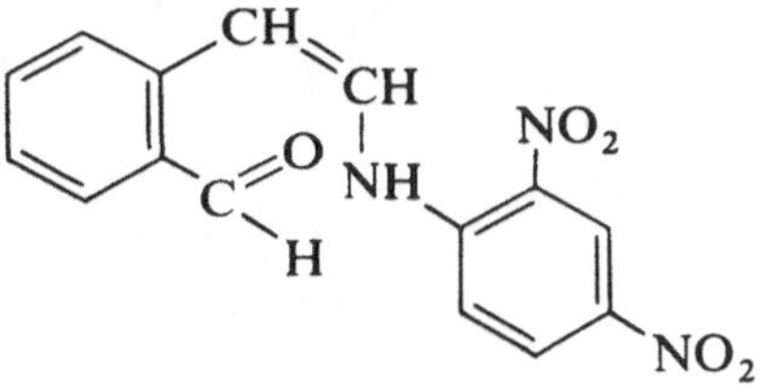

(151A) 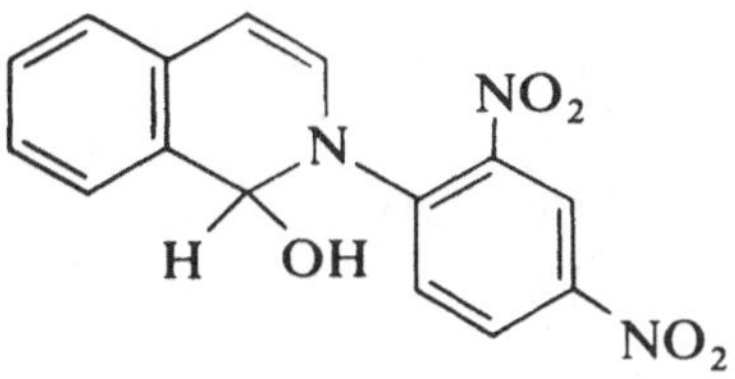   (151B)

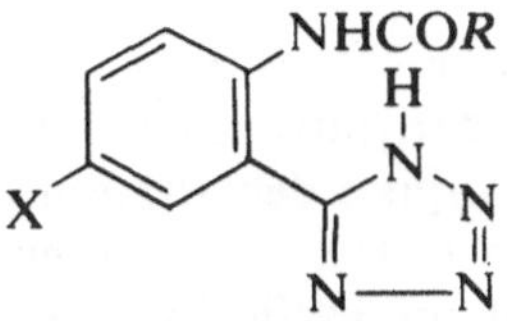

(152A) 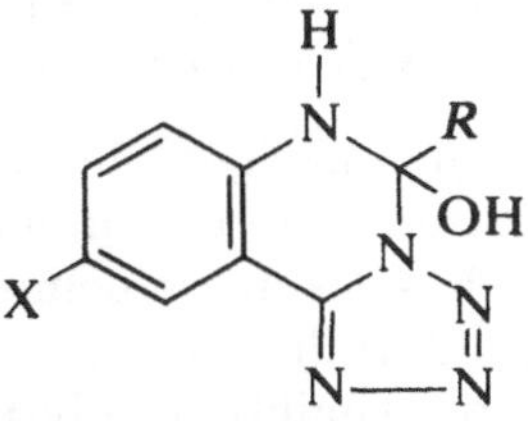   (152B)

In formula **(152)**		Isomeric structure in solid state	References
X	R		
H	H	A, B	490
H	Me	B	491
COOH	Ph	B	492
Br	Me	A	493, 494
Br	Ph	A	493, 494

Postovskii and coauthors [490–494] demonstrated that the covalent hydrates of 5-R-9-X-tetrazolo[1,5-c]quinazolines, depending on the structure of the substituents R and X in the solid state, possess either the open **(152A)** or the cyclic **(152B)** structure. Both isomers of the parent compound

(**152**, X = R = H) were isolated [490]. Using the example of the methyl-derivative (**152**, X = H, R = Me) it was shown [492] that acidification stabilizes the less acidic cyclic form. In pyridine, however, isomerization (**152B**) → (**152A**) proceeds.

Covalent hydrates of pteridines (**153**) [482, 483, 495, 496] as well as of 1,3,6-triazanaphthalenes [497] in acidic medium rapidly transform into open isomers (**154A**). Neutralization leads to the reverse reaction. Evidently, the isomerization (**154B**) → (**154A**) is favored by the addition of a proton to $N_{(1)}$ decreasing the basicity of $N_{(3)}$.

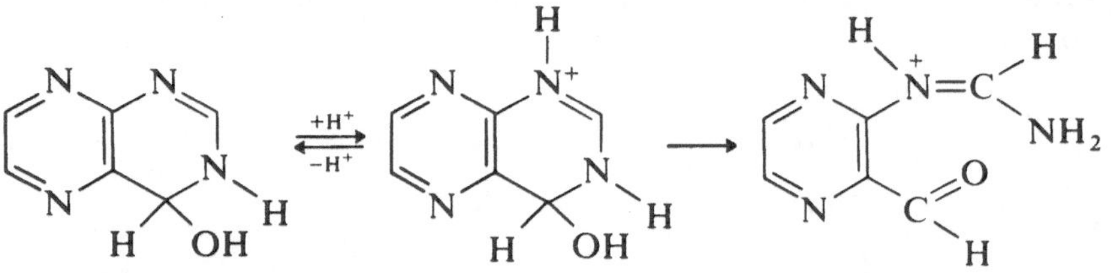

(**153**)   (**154B**)   (**154A**)

2.3.2. Derivatives Containing Amino Groups at sp^3, Aromatic, and sp^2-Carbon Atoms

The high nucleophilicity of a nitrogen atom causes the great stability of cyclic isomers of the aliphatic 4- or 5-aminoderivatives of aldehydes and ketones. Besides, these compounds often undergo a dehydration reaction resulting in cyclic enamines. It is possible to isolate open isomers or to observe the equilibria only if the stability of the cyclic form is lowered because of an unfavorable structure of the linking fragment or of a substituent at the amino group thus lowering the nucleophilicity of the nitrogen atom, e.g., by the introduction of an acyl group.

A ring-chain tautomeric equilibrium was observed [498, 499] in solutions of the *N*-methylamide of 5-acetamino-2-oxovaleric acid (**155A**) ⇌ (**155B**). In chloroform the equilibrium constant is nearly one, in

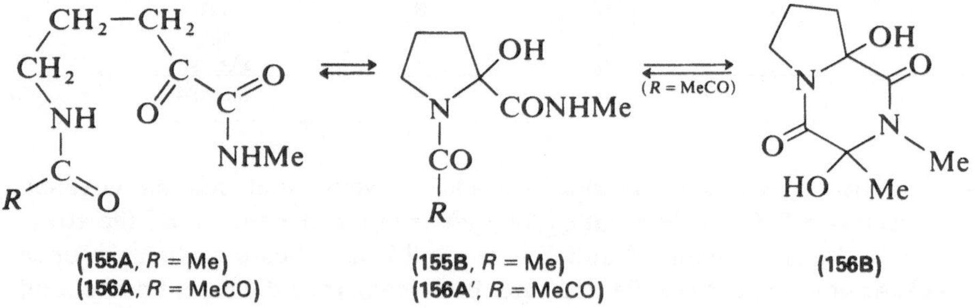

(**155A**, R = Me)   (**155B**, R = Me)   (**156B**)
(**156A**, R = MeCO)   (**156A′**, R = MeCO)

D_2O it is 1.5, and in pyridine the equilibrium is strongly displaced toward the open form (155A). Only the open isomer (155A) was isolated. Surprisingly, the N-methylamide of 2-oxo-5-pyruvoylaminovaleric acid (156A) is stable in chloroform solution [498]. A double cyclization (156A) → (156B) proceeds in a solution of hydrochloric acid. Analogues of (156B) were also obtained [500].

Vanags and Dregeris [501] obtained both isomers of 2-[2-(N-cyclohexyl)aminoethyl]-2-phenyl-1,3-indanedione (157A, R^1 = H, R^2 = cyclo-C_6H_{11}, R^3 = Ph) and (157B). Isomerization with ring closure proceeds in the presence of bases. Perekalin and co-workers [502] obtained cyclic isomers of aminoketones (157B) by the reduction of 2-(2-nitroethyl)-2-substituted 1,3-indanediones. Protonation of the nitrogen atom is accompanied by ring opening (157B) → (158), while deprotonation involves the reverse process [501, 502].

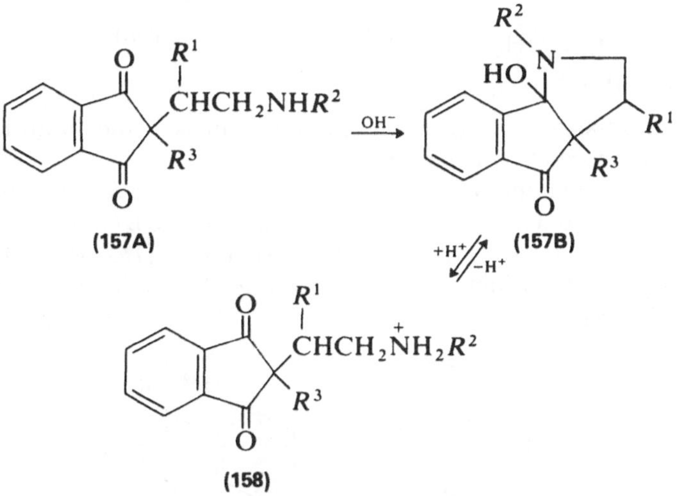

4-Aminomethyl- and 4-(2-amino-2-propyl)-cyclohexanones are not capable of forming carbinolamines [503, 504].

For endo-7-amino-3-bicyclo[3.3.1]nonanone the cyclic structure of 1-hydroxy-2-azaadamantane (159B) has been determined [505]. The homologue endo-7-aminomethyl-3-bicyclo[3.3.1]-nonanone also possesses the ring structure (160B) in the solid state. In solution, however, an equilibrium (160A) ⇌ (160B) containing the open isomer was detected [504, 506, 507]. The absence of a C=O absorption in the IR spectrum of the hydrochloride led to the conclusion that the protonated form is cyclic (161). The influence of protonation of the nitrogen atom on the stability of the cyclic isomer strongly depends on the steric structure of the aminoketone as may be seen from a comparison with the above-mentioned transformation **(157B)** → **(158)**.

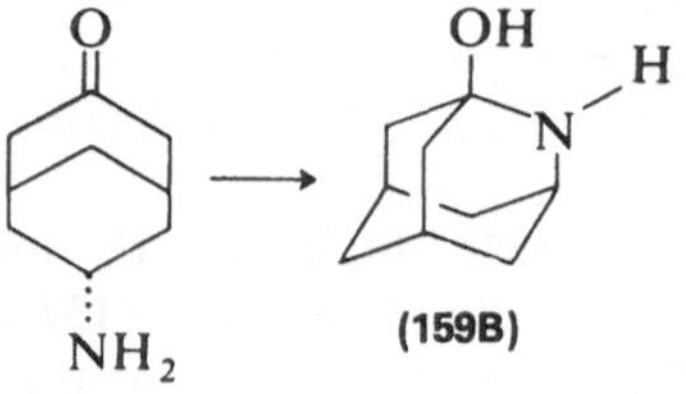

(159A) (159B)

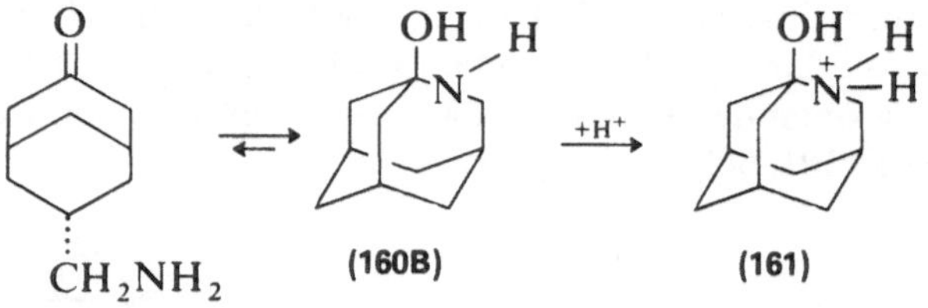

(160A)   (160B)   (161)

The formation of tricyclic carbinolamines from 9-acyl-2-azabicyclo[5.4.0]undecanes was reported [504].

2-Acylbenzylamines having an alkyl substituent at a nitrogen atom exist as stable cyclic isomers of 1-hydroxyisoindolines **(162B)** [508–511]. This is also the case for 2-(2-oxoalkyl or 2-oxoacyl)anilines (**163**, R^1 = alkyl, X = CO, CR_2) [512, 513] which easily undergo ring closure forming 2-hydroxyindolines **(163B)**.

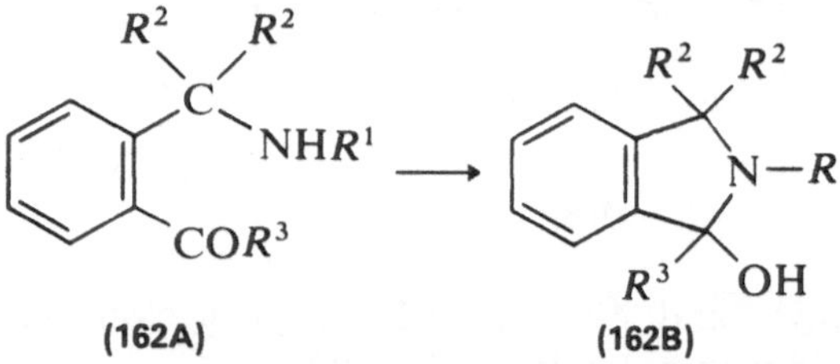

(162A)   (162B)

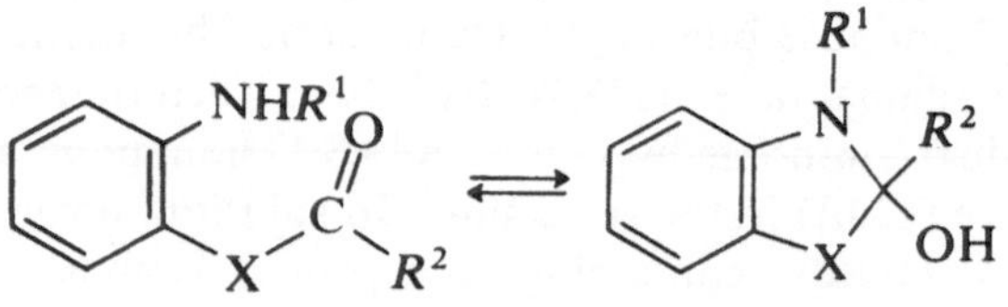

(163A)   (163B)

It is possible to observe the equilibrium **(163A)** ⇌ **(163B)** by lowering the nucleophilicity of the nitrogen atom by means of *N*-acyl substituents.

Attempts to detect tautomeric equilibria in solutions of 4-chloro-6-alkyl-6-hydroxy-5,6-dihydropyrimido[4,5-b] [1,4]-thiazines (166Bb, $R = R^1 = R^2 = R^3 = H$, $R' = Cl$, $R^4 = Me$, CH_2Cl, CH_2COOEt) by means of IR and ^{1}H-NMR spectroscopy failed [533]. Derivatives, however, bearing a methyl group at the nitrogen atom ($R^1 = Me$), form equilibrium mixtures in solutions. In $CDCl_3$ the equilibrium of 6-methylderivatives (166b, $R^1 = R^4 = Me$, $R' = Cl$) is displaced toward the open form, but in the solutions of 6-chloromethyl and 6-ethoxycarbonylmethylderivatives (166b, $R^1 = Me$, $R' = Cl$, $R^4 = CH_2Cl$, CH_2COOEt) the cyclic form prevails. The addition of a proton acceptor solvent, e.g., pyridine-d_5 to the solutions shifts the equilibria toward the cyclic form.

The destabilizing effect of an N-methyl group ($R^1 = Me$) on the ring structure cannot be explained by electronic effects since the methyl group raises the nucleophilicity of the nitrogen atom and, hence, should act in the reverse direction. The experimentally observable effect of the methyl group is explained [533] by a steric interaction of peri-substituents in the cyclic form (166B, $R' = Cl$, $R^1 = Me$) which destabilizes the 1,4-thiazine ring.

Compared with 6-acylalkylthio-5-aminopyrimidines (166b) the equilibrium (166Ad) $\rightleftarrows$ (166Bd) is displaced toward the open form in a higher degree [536]. The introduction of two methyl groups into the quinoxaline ring (166d, $R = Me$) stabilizes the cyclic form. N-Methylaminoquinoxalines (166d, $R^1 = Me$) have exclusively the open structure. This is explained [536] by a steric nonbonded interaction between the substituents $R^1 = Me$ and R^4 or OH of the dihydrothiazine ring which destabilizes the cyclic structure (166Bd). Probably steric interactions of this type take place also in the series (166Bb, $R^1 = Me$). The stabilization of the open form brought about by a phenyl substituent at the keto group ($R^4 = Ph$) is stronger than that of methyl and chloromethyl substituents, respectively. The introduction of a bulky t-butyl group as R^4 hinders the closure of the dihydrothiazine ring. For the tautomeric system (166Ad) $\rightleftarrows$ (166Bd) it was demonstrated that the cyclic form predominates in the solid state and in pyridine solution as compared to a solution in chloroform, probably due to a stabilizing action of intermolecular hydrogen bonds $O-H\cdots N\rightleftharpoons$.

2.3.3. Urea and Thiourea Derivatives

Unkovskii and co-workers [537–539] investigated N-(3-oxo-alkyl)-N'-substituted thioureas (167). In the solid state these compounds possess, with one exception [538], the cyclic structure of 4-hydroxyhexahydropyrimidine-2-thiones (167B), but in solutions they form equilibrium mixtures (167A) $\rightleftarrows$ (167B). Equilibria were reached after 6–15 days. The equilibrium

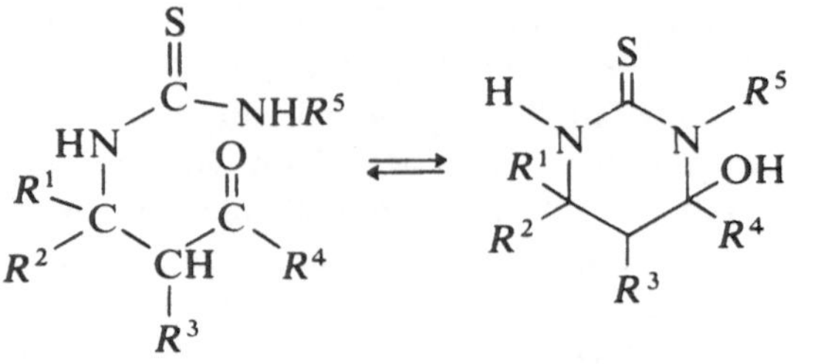

(167A)   (167B)

constants were determined [539] from the intensities of C=O bands in IR spectra of solutions. The introduction of methyl groups as R^1 and R^2 (always R^4 = Me) stabilizes the cyclic form as indicated by the arrow showing the direction in which the stability of the cyclic form (167B) increases:

$$R^1 = \quad H \quad H \quad Me \quad Me \quad Me$$
$$R^2 = \quad H \quad H \quad H \quad H \quad Me$$
$$R^3 = \quad Me \quad H \quad Me \quad H \quad H$$

Increasing the bulk of the substituent at the nitrogen atom (R^5 = Me, Et, Pr) gives rise to an increased rate of ring opening thus shifting the equilibrium toward the open form (167A). One cannot agree with the statement of the authors [539] that increasing electron-donating properties of the substituent (R^5) at the nitrogen atom works in the same direction. Aldehyde derivatives (167, R^4 = H) unlike keto derivatives (R^4 = Me) exist totally in the cyclic form in solution.

A similar phenomenon was also observed in a series of N-(3-oxoalkyl)-S-methylthioureas [540, 541], where the compounds possess a ring structure in the solid state but form ring-chain equilibrium mixtures in solution.

N-(2-Acylphenyl)-ureas and -thioureas [542–545] have a ring structure, being 4-hydroxy-1,2,3,4-tetrahydro-3-quinazolinones (or thiones) (168, X = O, S). Ring-chain equilibria in their solutions have not been detected. N-(Anthraquinone-1-yl)-N'-phenylurea possesses the open structure in the solid state and in solution [307].

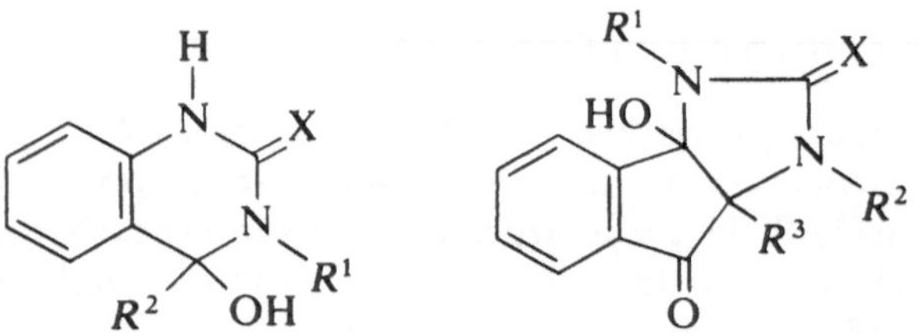

(168)   (169)

2-Ureido [546, 547], 2-thioureido [548, 549] and 2-guanidino-2-substituted 1,3-indanediones [547] are cyclic (**169**, X = O, S, NR).

2.3.4. Dithiocarbamates

Ratios of tautomeric forms of S-(2-oxoalkyl)dithiocarbamates in solution have been determined by IR spectroscopy [550]. The open isomer (**170A**, R^1 = Ph, $R^2 = R^3$ = H, R^4 = t-Bu) was isolated by crystallization from methylene chloride and the cyclic isomer was obtained from methanol-water. The introduction of alkyl groups R^2 and R^3 displaces the equilibrium toward the ring form. Increasing electron-donating properties of the substituents at the keto group (R^4) as well as increasing steric bulk of substituents at the nitrogen atom and keto group (R^1 and R^4 in the compound (**170**)) stabilize the open form [550, 551].

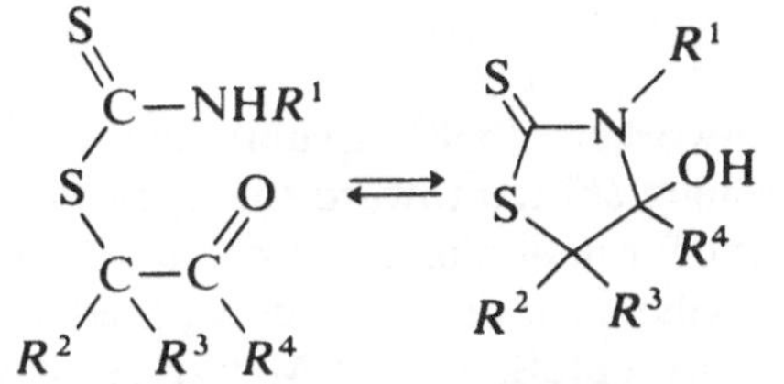

(170A)   (170B)

S-(3-Oxoalkyl)dithiocarbamates [552, 553] show less of a trend to form six-membered 4-hydroxytetrahydro-1,3-thiazine-2-thiones (**171B**). Derivatives containing a gem-dimethyl group (**171B**, $R^2 = R^3$ = Me) are cyclic in the solid state, but in solution the equilibrium is shifted to the open form [553].

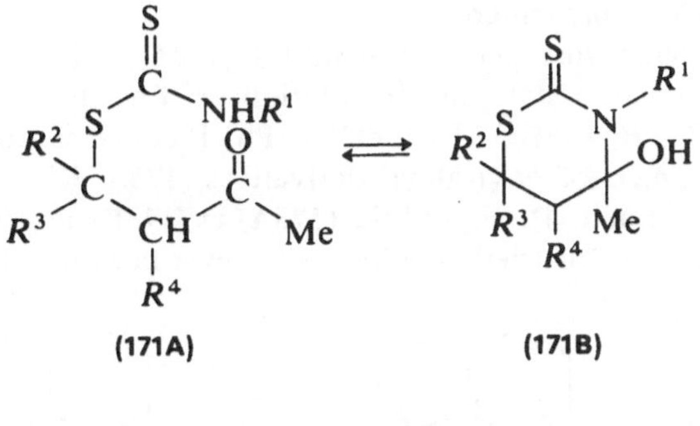

(171A)   (171B)

2.3.5. Sulfonamides and Sulfinamides

Contradictory data are available concerning the structure of 2-acylbenzenesulfonamides. The preparation of 3-alkyl (or phenyl)-3-hydroxy-1,2-

benzisothiazoline-1,1-dioxides (**172B**, R^1 = alkyl, Ph, R^2 = H) from ben-
zisothiazoline-3-one-1,1-dioxide with Grignard reagents has been reported
[554, 555]. However, a repeated investigation [556] accomplished using IR
and ^{1}H-NMR spectroscopic methods failed to support the formation of
compounds (**172B**).

2-Phenyl-3-R-3-hydroxybenzisothiazolines obtained [557] from reactions of 2-phenylbenzisothiazoline-3-one with RMgBr were oxidized by
hydrogen peroxide to their dioxides (**172B**, R^1 = Me, Ph, R^2 = Ph), but no
spectroscopic evidence for the structure (**172B**) was given.

| (172A) | (172B) | (173) |

Hauser and co-workers [558] assume the existence of equilibria
(**172A**) ⇌ (**172B**) strongly shifted toward the open form. The only basis of
such a suggestion is the formation of a cyclic *O*-ethylether (**173**) in a reaction
of *N*-methyl-2-benzoylbenzenesulfonamide with ethanol.

The cyclic structure (**172B**, R^1 = 4-MeC$_6$H$_4$SO$_2$OCH$_2$, R^2 = Ph) was
supported by spectroscopic methods [559].

The structure of the open 2-aroylbenzenesulfonamides (**172A**, R^1 = Ph,
4-ClC$_6$H$_4$, R^2 = Me, PhCH$_2$, Ph) was confirmed by IR spectroscopy [184].
The formation of the cyclic form (**172B**) could not be detected in dioxan
and in the presence of triethylamine or other basic as well as acidic catalysts.
Obviously, the SO$_2$ group decreases the nucleophilicity of the nitrogen atom
to such a degree that contrary to 2-aroylbenzamides the isomerization
(**172A**) → (**172B**) is prevented.

Unlike 2-benzoylbenzenesulfonamides (**172A**), 2-benzoylbenzenesul-
finamides in the solid state and in solution exist as stable cyclic isomers
(**173B**, R^1 = Ph, R^2 = Me, *i*-Pr, PhCH$_2$, Ph) [560]. It is possible only to
obtain open isomers of *N*-(*t*-alkyl) derivatives (**173A**, R^1 = Ph, 4-ClC$_6$H$_4$,
R^2 = *t*-Bu, 1-adamantyl). Equilibria (**173A**) ⇌ (**173B**) in dioxan solution
even in the presence of triethylamine have never been observed.

| (173A) | (173B) | (174) |

During oxidation of 2-benzyl-3-hydroxy-3-phenylbenzisothiazoline-1-oxide (**173B**, R^1 = Ph, R^2 = PhCH$_2$) with hydrogen peroxide ring opening occurs and the sulfonamide (**174**) is formed instead of the expected 3-hydroxybenzisothiazoline-1,1-dioxide (**172B**).

The existence of stable cyclic N-monosubstituted 2-benzoylbenzenesulfinamides (**173B**) as well as the oxidative transformation (**173B**) → (**174**) provide conclusive evidence that the sulfinamide group, due to a higher nucleophilicity of the nitrogen atom of SONHR, is capable of intramolecular nucleophilic addition to a C=O bond but the sulfonamide group is not.

2.3.6. Pyridones

We now discuss more thoroughly investigated examples of the intramolecular addition of an NH group, which is part of a heterocyclic system, to C=O bonds.

The intramolecular nucleophilic addition of lactam N−H groups to a keto group has been demonstrated [561] for 5-carboxy-6-(2'-oxocyclohexyl)methyl-2(1H)-pyridone (**175A**) existing in the cyclic form (**175B**). This phenomenon was studied in detail [562, 563] by means of IR and ^{1}H-NMR methods using 6-(3'-oxoalkyl) and 6-(2'-oxocycloalkyl)methyl-3-hydroxy-2(1H)-pyridones (**176**, R^1 = alkyl, R^2 = alkyl; R^1, R^2 = (CH$_2$)$_n$). In the solid state and in DMSO-d$_6$ solution the open form predominates for ketoderivatives (**176**, R^1 = Me, R^2 = Et; R^1 = Et, R^2 = Me), whereas the

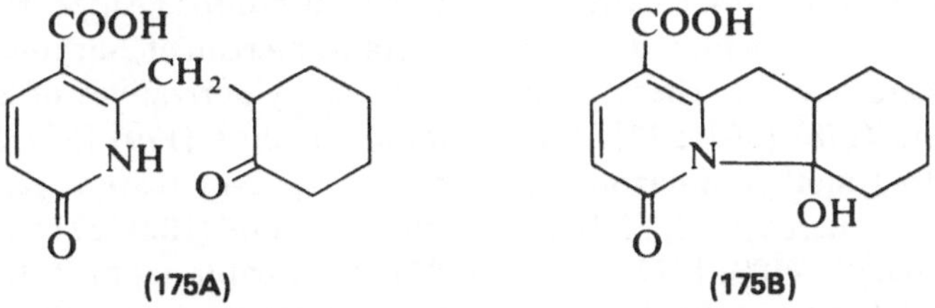

(175A)   (175B)

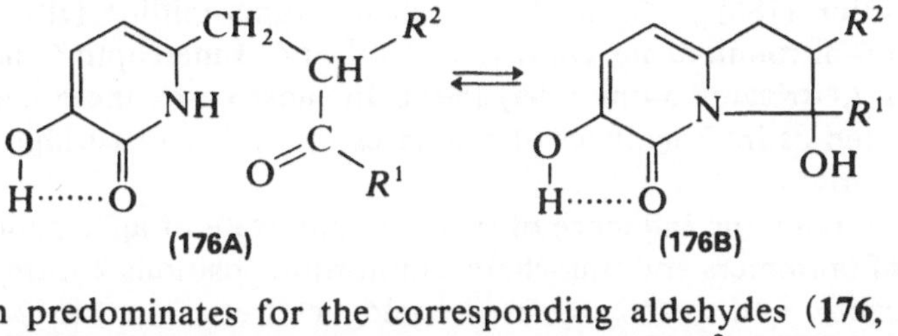

(176A)   ⇌   (176B)

ring form predominates for the corresponding aldehydes (**176**, R^1 = H, R^2 = Me, Et, Pr, i-Pr, Bu). Curiously, aldehydes with R^2 = C$_5$H$_{11}$ and larger alkyl groups are open structured in the solid state but in DMSO-d$_6$ solution the cyclic form predominates. In the series of crystalline cycloalkanones

$(176, R^1, R^2 = (CH_2)_n, n = 3, 4, 5, 6)$ only the derivative with $n = 4$ exists as a cyclic isomer. In DMSO-d_6 solution the tautomeric equilibrium **(176A)** $\rightleftarrows$ **(176B)** ($K_T = 4$) is observed. Higher temperature shifts the equilibrium toward the open form.

2.3.7. 1,3-Diazaheterocycles

An intramolecular addition of an amidine nitrogen atom, part of an imidazoline ring, to a keto group was reported [564] in the case of 2-(2-benzoylphenyl)imidazoline. In the solid state this compound possesses the open structure **(177A**, Ar = Ph), but in solution the equilibrium **(177A)** $\rightleftarrows$ **(177B)** is observed.

An analogous equilibrium is observed also for 2-(2-aroylphenyl)imidazolines substituted in the aroyl group [565, 566].

(177A)   (177B)

Many investigations carried out by several groups (Alper, Kochergin, Singh, a.o.) have considered the ring-chain tautomerism of 2-acylmethylthio-1,3-diazaheterocycles. These comprise S-acylalkylderivatives of 2-mercaptoimidazole **(178)** [567–572], 2-mercaptoimidazoline **(179)** [569, 571–574], 2-mercapto-1-methylimidazolinium **(180)** [575], 3-mercapto-1,2,4-triazole **(181)** [571], 2-mercapto-3,4,5,6-tetrahydropyrimidine **(182)** [571], 2-mercaptobenzimidazole **(183)** [572–574, 576–583], 2-mercaptonaphtho[1,2-d]imidazole **(184)** [572, 584, 585], 8-mercaptopurine **(185)** [572, 586], 8-mercaptotheophilline **(186)** [572, 586, 587], 2-mercaptoperimidine **(187)** [588], 5-mercapto-4-nitroimidazole **(188)** [589, 590] and 3-mercapto-6-methyl-2,5-dihydro-1,2,4-triazine-5-one **(189)** [591]. In most cases these compounds were isolated as intermediates of syntheses of imidazo[2,1-b]thiazoles and their derivatives.

On studying the influence of structure and state of aggregation on the stability of tautomers and ring-chain equilibrium positions 2-formylmethylthioimidazoles and their condensed analogues were found [572, 582, 588] to have a cyclic structure in the solid state as well as in solution. Passing from aldehydes to aliphatic (R^1 = Me) and aromatic (R^1 = Ph) ketones the stability of the open tautomers increases.

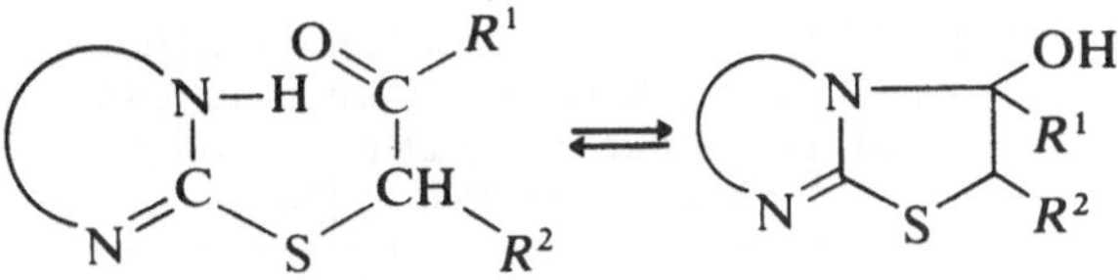

A   B

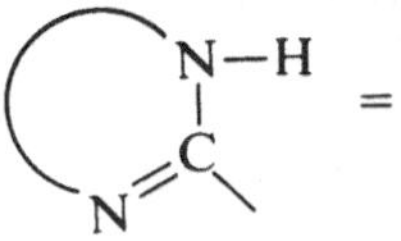

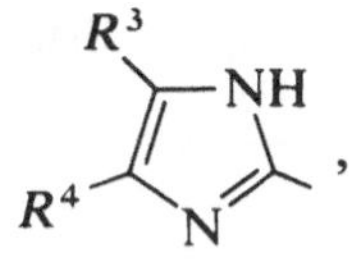

(178) 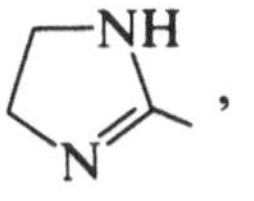   (179)

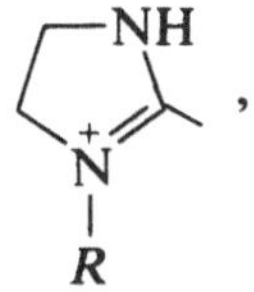

(180)

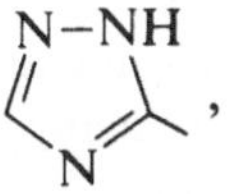

(181) 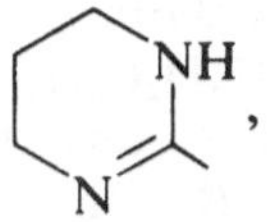   (182) 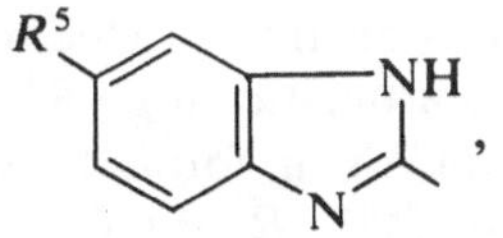   (183)

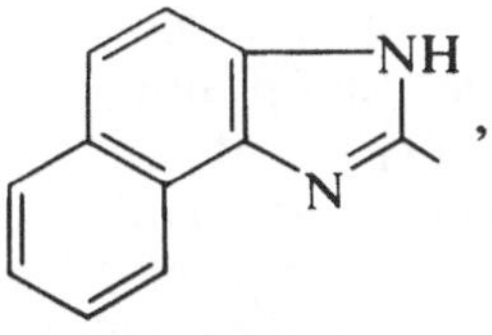

(184)

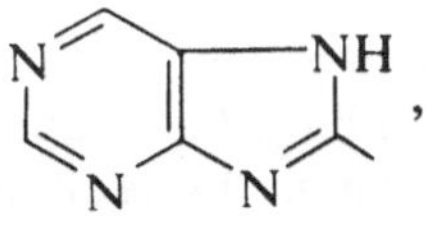

(185) 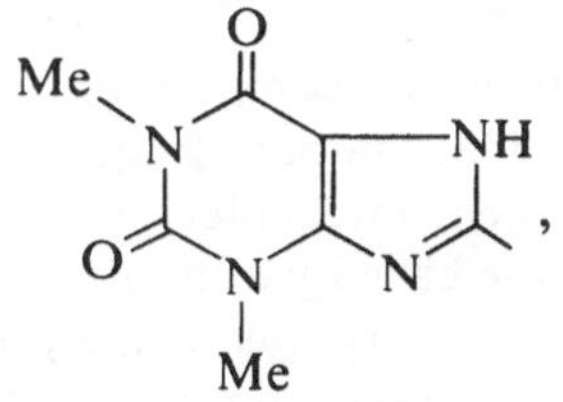   (186) 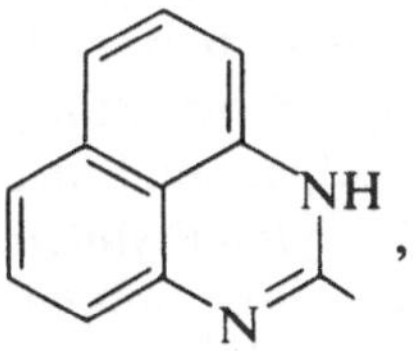   (187)

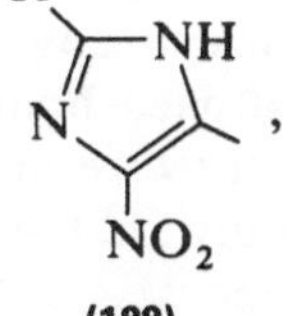

(188) 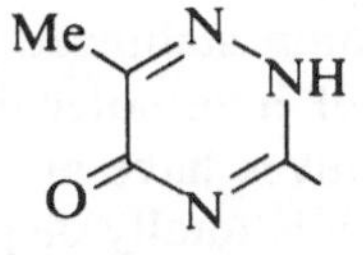   (189)

Varying the structure of substituents R^1 of 2-acylmethylmercaptoben-zimidazoles (**183**, $R^2 = R^5 = H$) it was shown that the open form is stabilized by bulky substituents at keto group (R^1) as well as by decreasing electrophilicity of the keto group brought about by conjugation with aromatic rings, ethoxycarbonyl or cyclopropyl groups. Compounds (**183**, $R^1 = 4\text{-}XC_6H_4$, COOEt, cyclo-C_3H_5, $R^2 = R^5 = H$) in the solid state and in solution exist exclusively in the open form (**183A**).

Table 26. Ring-Chain Equilibrium Constants of
2-Acylmethylmercaptobenzimidazoles (183)
and -perimidines (187) Determined by
^{1}H-NMR in DMSO-d_6 [588]

	$K_T = [B]/[A]$	
R^1	(183A) $\rightleftarrows$ (183B) ($R^2 = R^5 = H$)	(187A) $\rightleftarrows$ (187B) ($R^2 = H$)
Me	0.54	2.0
Et	0.41	1.5
i-Pr	0.11	0.54

Surprisingly, the introduction of electron-withdrawing substituents in the para-position of the aromatic ring ($R^1 = 4$-$NO_2C_6H_4$) does not affect [581] the stability of the open form, though, as was shown using the trifluoromethylderivative (183, $R^1 = CF_3$, $R^2 = R^5 = H$), the presence of the substituent R^1 having a strong -I-effect shifts the equilibrium totally toward the cyclic form. The introduction of a nitro group into the benzimidazole ring (183, $R^5 = NO_2$) decreases the nucleophilicity of the nitrogen atom and displaces the equilibrium toward the open form as compared with the unsubstituted compound.

Alper and Lipschutz [588] comparing equilibrium constants of solutions of 2-acylmethylmercapto-benzimidazoles and -perimidines (see Table 26) revealed the influence of the basicity of a nitrogen atom on the equilibrium position. Higher equilibrium constants, i.e., a greater stability of the cyclic form of 2-acylmethylmercaptoperimidines, were explained by a greater basicity of perimidine ($pK_a = 6.39$) compared with benzimidazole ($pK_a = 5.53$).

In the solid state the ring form is more stabilized [572], which, obviously, is caused by a favorable influence of intermolecular hydrogen bonds. If imidazole acylmethylmercaptoderivatives have an open structure in solid state, then in solution the equilibrium is totally displaced toward the open tautomer. Crystallinic cyclic isomers in solution either possess a cyclic structure or form ring-chain equilibrium mixtures [582]. Dilution of solutions in nonpolar solvents as well as increasing the temperature of the solutions shifts the equilibrium toward the open form [592].

Mass spectroscopic investigations [592] of the ring-chain tautomerism of 2-acylmethylthiobenzimidazoles (183) revealed that in the gaseous phase these compounds exist in the open form. Fragmentation proceeds from the open form of the molecular ion with a predominant bond cleavage in the side chain.

in chloroform to solutions in pyridine or DMSO shifts the equilibrium towards the cyclic form.

2.3.9. Miscellaneous

In addition to examples of a ring-chain tautomerism mentioned above proceeding by a formation of carbinolamines the tautomerism of monoimines of 1,5-diketones [601, 602], O-(3-oxoalkyl)hydroxylamines **(196A)** ⇄ **(196B)** [603, 604], monohydrazones of 1,3-dicarbonylcompounds **(197A)** ⇄ **(197B)** [605–613], 2-(2-formylbenzoyl)-1-hydrazono-1,2-dihydrophthalazine **(198A)** ⇄ **(198B)** [614], N-(2-pyridyl)aminoacetaldehydes **(199A)** ⇄ **(199B)** [615] is reported.

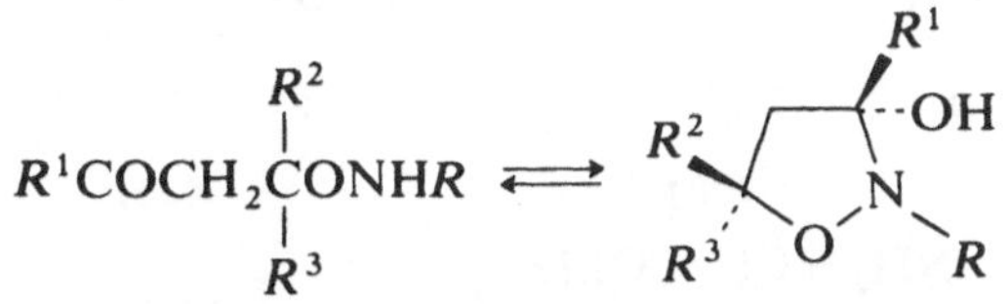

$$R^1COCH_2CONHR$$

(196A) R = H, CONH$_2$ (196B)

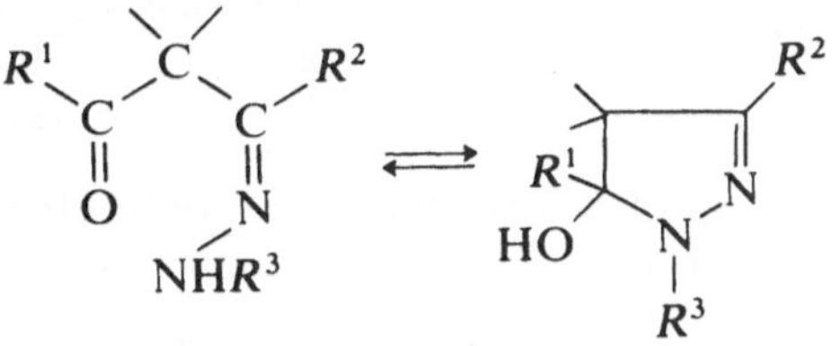

(197A)   (197B)

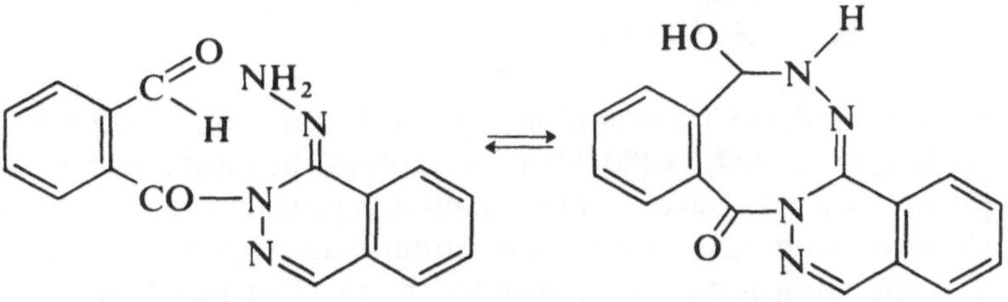

(198A)   (198B)

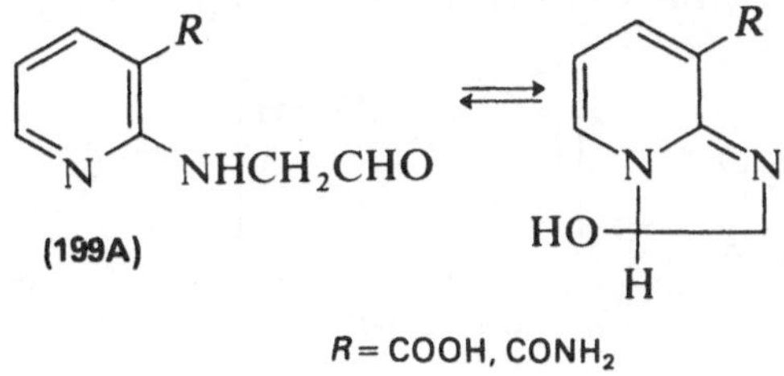

(199A)   (199B)

R = COOH, CONH$_2$

Ring-chain equilibria were repeatedly discussed in connection with investigations of mechanisms of nitrogen-containing heterocycles (see, for example [612, 616]). However, it has rarely been possible to isolate the carbinolamine intermediates.

Investigating the mechanism of Hantzsch's thiazole synthesis from α-haloketone and thioamide intermediates **(200A)** and **(200B)** were isolated [616]. The cyclic structure of 4-hydroxythiazolinium ion **(200B)** is stabilized by the introduction of the substituents R^3, whereas bulky substituents at the nitrogen atom (R^1) and at the keto group (R^4) stabilize the open structure.

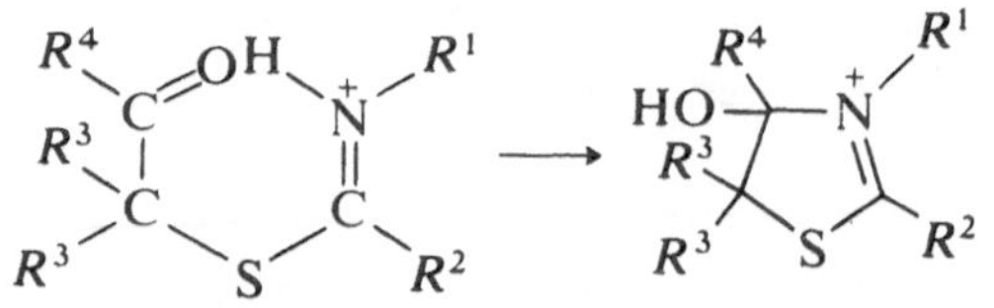

(200A)   **(200B)**

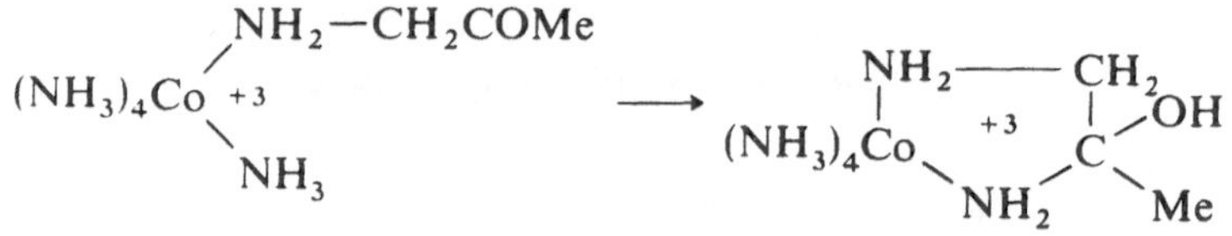

(201A)   **(201B)**

An interesting case of a formation of a carbinolamine structure was recently detected [617] for a cobalt(III) amine complex **(201A)** → **(201B)**.

2.4. INTRAMOLECULAR MIGRATION OF O-, N-, AND S-ACYL GROUPS

The intramolecular migration of acyl functions between two nucleophilic groups **(202)** ⇌ **(204)** is an example of chemical transformations which proceed via heterocycles of low stability (cyclic tetrahedral intermediates). The close relationship of these reactions with ring-chain tautomerism under consideration is best demonstrated by the fact that both reactions proceed by an intramolecular nucleophilic addition to an acyl carbonyl group $(R-CO-O, R-CO-N, R-CO-S)$ [618-623].

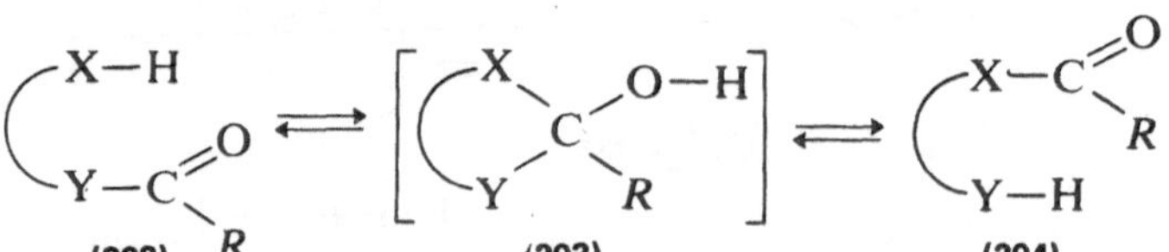

(202)   **(203)**   **(204)**

X, Y = O, NR, S

Migration of acyl groups is observed in acylated aminophenols, aminoalcohols, polyalcohols, polyphenols, diamines, thioglycols, amino-thiophenols, aminothiols, etc. Not only acyl but also other groups accessible to nucleophilic attack are subjects of migration between two nucleophilic centers (OH, NH, SH).

Many examples using kinetic methods show [620] that acyl group migration ($O \rightarrow N$, $N \rightarrow O$, $O \rightarrow O'$, $N \rightarrow N'$, $S \rightarrow O$, $S \rightarrow N$, $N \rightarrow S$ e.o.) proceeds along the general mechanism via cyclic tetrahedral intermediates. Following an intramolecular nucleophilic addition of the X-H group (X = O, NR, S) to the acyl C=O group the cyclic tetrahedral intermediate (203) depending on the pH value of the medium opens along the most easily polarizable bond (C-X or C-Y). Cyclic intermediates (203) rarely are isolated because they are of low stability [624].

Both isomers (205A) and (205B) of a rearrangement of 2-benzoyloxy-benzyl alcohol have been isolated [625], and their interconversions in both directions were accomplished.

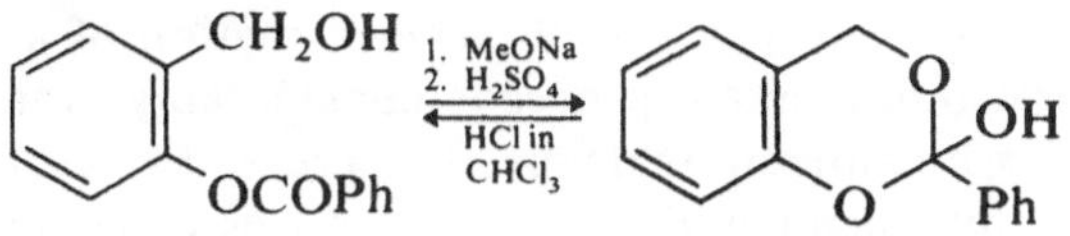

(205A)   (205B)

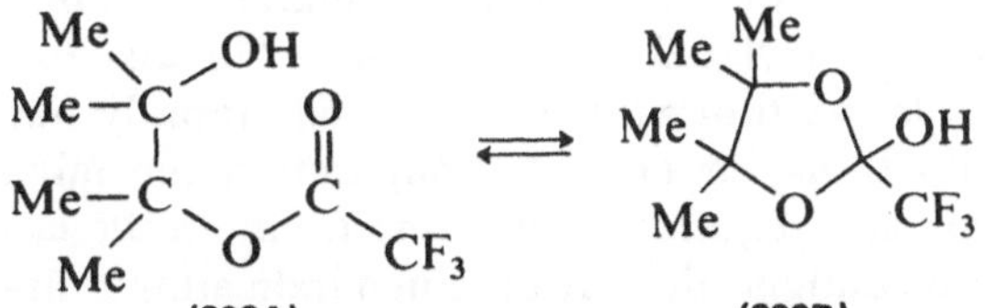

(206A)   (206B)

The ring-chain equilibrium (206A) ⇌ (206B) has been observed in so-lutions of pinacol trifluoroacetate using ^{1}H-NMR spectroscopy [626]. In this case the cyclic form is stabilized by two gem-dimethyl groups and a strong electron-withdrawing substituent (CF_3) at the keto group. Analogous

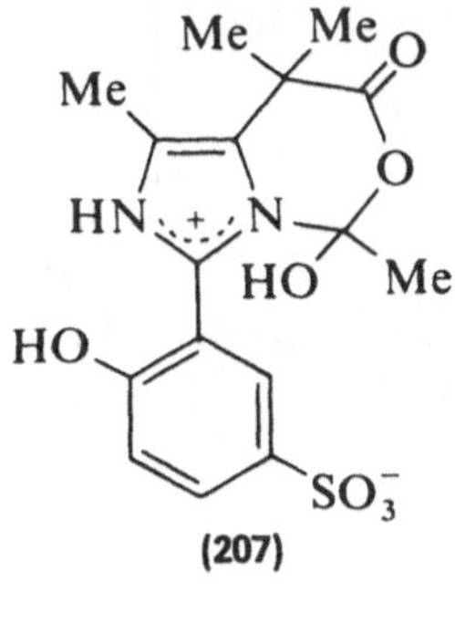

(207)

structural factors, i.e., a protonated imidazole as an electron-withdrawing group and a gem-dimethyl group stabilize the tetrahedral intermediate (207) which was isolated also [627, 628].

The acidity of the medium exerts a great influence on the rates and direction of acyl group migration. Thus, an O → N migration of acyl groups proceeds only in alkaline solution while in acidic medium the reverse N → O migration occurs. This is explained by the greater basicity of the nitrogen atom compared with that of the oxygen atom.

The formation of the cyclic tetrahedral intermediate (203) is supported by experiments concerning the influence of the steric proximity of both nucleophilic centers on the migration rate.

In a series of S-acetylmercaptoalkylamines of general formula $MeCOS(CH_2)_n NH_2$ it has been demonstrated [629] that the transacetylation rate of S-acetylmercaptoethylamine ($n = 2$) is higher by two orders than that of 3-S-acetylmercaptopropylamine ($n = 3$). If the chain between the nucleophilic centers amounts to more than three carbon atoms, the transacetylation does not take place at all. The formation of a five-membered cyclic tetrahedral intermediate proceeds considerably more rapidly [630] than that of a six-membered analogue. The greatest rate of transacylation including cyclic five-membered intermediates was observed in a series of acetylthioalkanols [631] and similar compounds [620].

Differing rates of acyl group migration are observed for diastereoisomers of acylated aminoalcohols. For cis- or threo-forms of 1,2-aminoalcohols the transacylation proceeds rapidly but for analogous compounds with trans- or erythro-configuration the migration proceeds extremely slowly and requires severe conditions. In the last case products with the reversed configuration are obtained indicating a different migration mechanism (intramolecular electrophilic addition to a carbonyl group) [620, 632–634]:

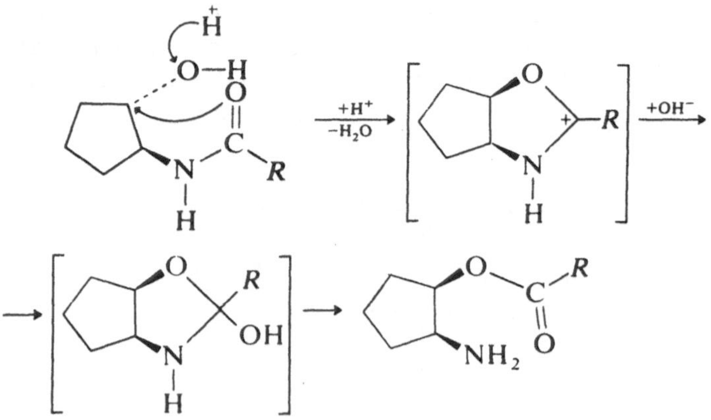

The migration rate increases with rising acyl group electrophilicity. For this reason an acetyl group migrates more rapidly than a benzoyl group. Thus, for example, the rate of S-acetylmercaptoethylamine transformation into the corresponding thiol is higher by an order of magnitude than that of the rearrangement of S-benzoylmercaptoethylamine [635]. The acceleration of the migration of a trichloroacetyl group is even greater.

Though structural factors favoring the formation of the cyclic tetrahedral intermediate (203) in general affect positively the migration of acyl groups, the total rate of the process depends as well on the ring opening (203) → (204). For example, it was determined that for S → N migration the rate is greater than O → N. However, in the case of O-acylaminoalcohols the formation of an intermediate cyclic compound proceeds more rapidly by two orders of magnitude than in the case of S-acylaminothiols. This is caused by a greater electrophilicity of the −CO−O− group carbon atom compared to that of the −CO−S− group. Considering this one should expect that the rate of migration of an acyl group from O → N could be greater than that of the migration S → N. Actually the subsequent opening of a hydroxythiazoline ring, leading to the formation of N-acylaminothiol, in alkaline medium proceeds significantly more rapidly than hydroxyoxazolidine ring opening, and as a result the rate of S → N transacylation appears to be greater [620, 636].

Minkin, Olekhnovich, and Zhdanov [637–639] detected and thoroughly investigated the migration of acyl and nitroaryl groups in the systems (208) ⇌ (210), (211) ⇌ (213), and similar ones [638, 640–642] proceeding through bipolar cyclic intermediates (209) or (212).

Contrary to the prototropic migration processes for acyl groups discussed so far, which proceed with a simultaneous proton migration, these transformations relate to acylotropic or arylotropic (carbonotropic) rearrangements.

Electronic and steric effects of substituents and the structure of molecules favoring or disfavoring the formation of cyclic intermediates and thus controlling the migration rate have the same influence as those in the prototropic ring-chain equilibrium systems already discussed.

Thus, for example, increasing the electron-withdrawing properties of the substituent R in (208) accelerates the migration of acyl groups, electron-withdrawing properties of the substituents R^1 and R^3 on the other hand retard it. The migration process is accelerated as well by a steric interaction of the reacting groups, e.g., by passing from (208) to (211).

The introduction of three electron-withdrawing substituents in the migrating aryl group (211, R = 2,4,6-trinitrophenyl), the introduction of methyl groups in positions 3 and 7 of tropolone ring (211, $R^1 = R^2$ = Me), sterically favoring the approaching of interacting groups, and also increasing

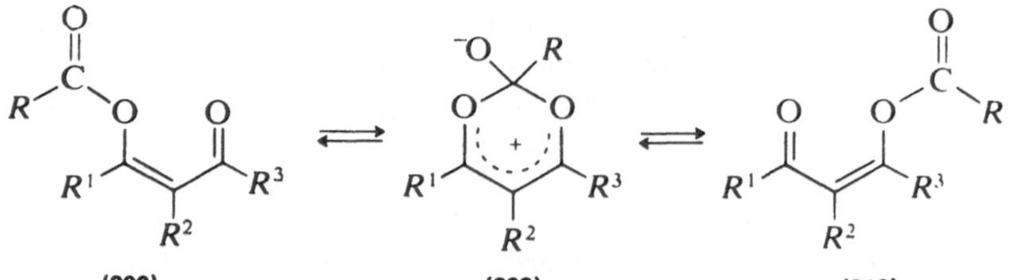

(208)   (209)   (210)

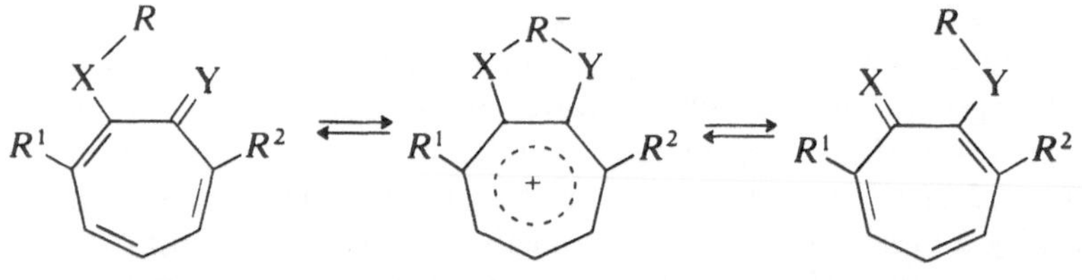

(211)   (212)   (213)

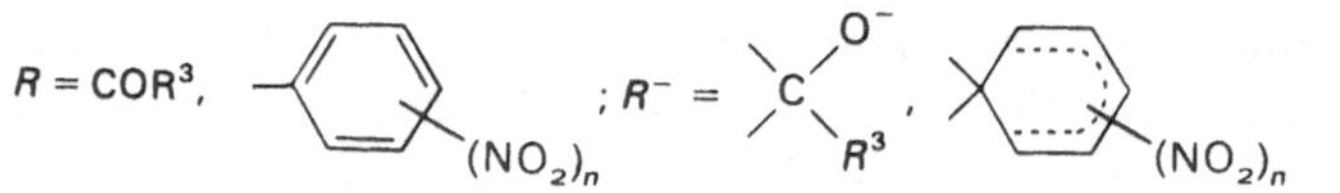

$R = COR^3$, [benzene ring](NO_2)_n ; $R^- = $ [structure] , [ring](NO_2)_n

X, Y = O, NR^4, S $n = 1, 2, 3$

the nucleophilicity of the centers X and Y in the series O < NR' < S stabilize the bipolar intermediate **(212)**. It does not matter if centers X and Y differ greatly in nucleophilicity as indicated by arylthiotropones **(211, X = O, Y = S, $R^1 = R^2 = H$)** where the equilibrium is totally displaced toward **(213)**.

It is not within the scope of this book to discuss in detail intramolecular migrations of acyl and similar groups (see reviews [618–624]). We only want to show that structural factors, governing intramolecular additions of XH groups to C=O bonds in ring-chain tautomeric systems, also rule intramolecular acyl migrations and similar reactions which proceed via cyclic tetrahedral intermediates.

2.5. OXA-, AZA-, AND THIACYCLOLS

The insertion of a hydroxy-, amino- or mercapto-carboxylic acid moiety into the ring of a lactone, or a cyclic peptide or depsipeptide, represents a particular case of intramolecular transacylation. These reactions proceed according to the scheme given below, "cyclols" **(215)**, i.e.: oxacyclols (X = O), azacyclols (X = NR) and thiacyclols (X = S) being intermediates of the reactions. Finally peptide **(216, X = NH)**, depsipeptide **(216, X = O)** or thiadepsipeptide **(216, X = S)** macrocycles (see Table 28) are formed.

Wrinch [646-648] was the first to discuss the possibilities of the formation of cyclols from peptide molecules. The scheme of interconversions **(214)** → **(216)** was proposed by Shemyakin [649, 650]. Hofman's synthesis of ergotamine [651-654] was based on this principle.

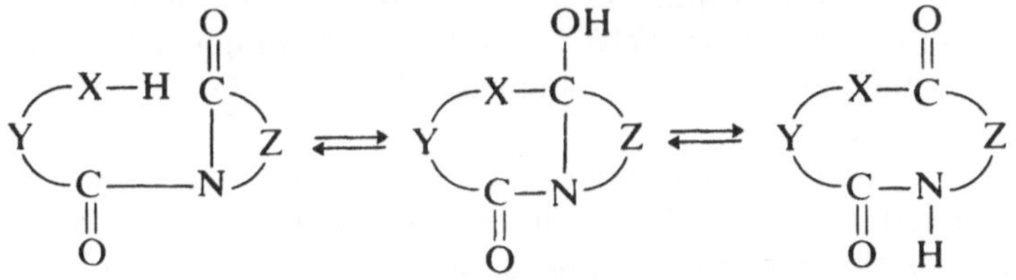

(214)   **(215)**   **(216)**

X = O, NR, S

Shemyakin, Antonov, Shkrob [649, 650, 655-658] investigating these reactions systematically showed that the reaction sequence **(214)** → **(216)** represents a new way of inserting hydroxy and amino acids into peptide chains or rings. The formation of stable cyclols **(215)** is a particular case of these reactions requiring definite structural and steric conditions.

The investigation of N-hydroxyacyllactams (**214**, X = O) showed [643, 644, 656, 659-661] that depending on the size of the cycle (Z) and on the structure of hydroxyacylchain (Y) either a spontaneous isomerization **(214)** → **(216)** proceeds or oxacyclols (**215**, X = O) are isolated, not being capable of isomerization into macrocycles **(216)**.

Table 28. Isomeric Structures of
N-Hydroxyacyllactams (**214**, X=O,
Z=$(CH_2)_n$) [643-645]

Y	n	Structure
CH_2	3	**(214)**
MeCH<	3	**(214)**
CH_2CH_2	3	**(214)**
CH_2	4	**(214)** ⇌ **(215)**
MeCH<	4	**(214)** ⇌ **(215)**
$Me_2C<$	4	**(215)**
CH_2CH_2	4	**(216)**
CH_2	5	**(214)** ⇌ **(215)**
MeCH<	5	**(215)**
CH_2CH_2	5	**(216)**
CH_2	11	**(216)**
o-C_6H_4	3	**(214)**
o-C_6H_4	4	**(215)**
o-C_6H_4	5	**(215)**

N-Glycolyl- and *N*-(3-hydroxypropionyl)butyrolactams (**214**, X = O, Y = (CH$_2$) and (CH$_2$)$_2$, Z = (CH$_2$)$_3$) do not isomerize into cyclols (**215**). For *N*-glycolyl- and *N*-(2-hydroxypropionyl)-valerolactams (**214**, X = O, Y = CH$_2$, CHMe, Z = (CH$_2$)$_4$) a tautomeric equilibrium (**214**) $\rightleftarrows$ (**215**) is observed in solution. If macrocycles containing eleven or more atoms are formed, cyclodepsipeptides (**216**, X = O) emerge in high yields. Sometimes even ten-membered rings are obtained (**216**, X = O, Y = (CH$_2$)$_2$, Z = (CH$_2$)$_4$) (see Table 28).

Unlike *N*-salicyloylbutyrolactam, *N*-salicyloylvalero- and -caprolactams form stable cyclols. The latter are not capable of isomerization into the corresponding cyclodepsipeptides, the formation of which is prevented by the presence of a benzene ring which flattens the macrocycle (**216**) thus favoring the intramolecular transannular addition reaction between the amide N—H group and the ester C=O group.

In solutions of *N*-(α-hydroxyacyl)diketopiperazines an equilibrium (**214**) $\rightleftarrows$ (**215**) is observed [662]. On heating the equilibrium shifts toward *N*-hydroxyacyllactams (**214**). The rate of the equilibration in solutions (dioxan, tetrahydrofuran) appreciably increases in the presence of trace amounts of water.

Isomeric interconversions (**214**) $\rightleftarrows$ (**215**) $\rightleftarrows$ (**216**) were observed also in a series of azacyclols. However, azacyclols (**215**, X = NR) have only rarely been isolated since in most cases they dehydrate yielding cyclic acylamidines (**217**). Alternatively, they may isomerize into cyclodipeptides (**216**, X = NH) [656, 657, 663–665].

(**215**, X = NH)   (**217**)

The isomerization of azacyclols into cyclopeptides (**215**, X = NH) → (**216**) proceeds more rapidly than the analogous conversion of oxacyclols [656, 665]. It has been assumed that cyclopeptides are less prone to a reverse transannular reaction between two amide groups (**216**, X = NH) → (**215**) than cyclodepsipeptides in a similar reaction between amide and ester groups (**216**, X = O) → (**215**).

An azacyclol was isolated from *N*-(2-methylaminobenzoyl)butyrolactam (**215**, X = NMe, Y = *o*-C$_6$H$_4$, Z = (CH$_2$)$_3$) [666].

Italian scientists [667] obtained azacyclols (**218**) substituted at the nitrogen atom and therefore not capable of dehydration by cyclization

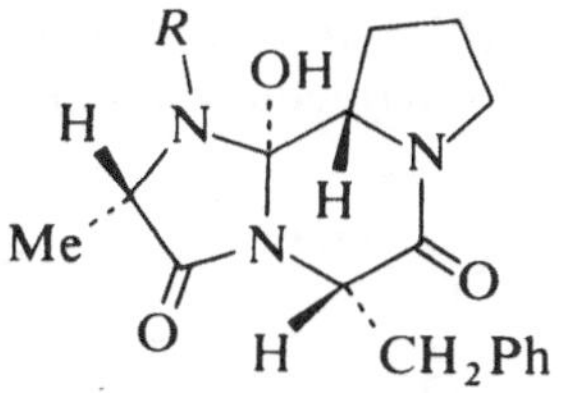

(218) R = PhCH$_2$OCO, 4-BrC$_6$H$_4$OCO

of N-benzyloxycarbonyl- and N-4-bromobenzyloxycarbonyl-L-alanyl-L-phenylalanyl-L-proline-4-nitrophenyl esters in aqueous alkaline solution. Other azacyclols [668, 669] as well as oxacyclol [670] of tripeptide structure have also been obtained by them.

A stable azacyclol, 7-hydroxy-2-oxo-1-azabicyclo[5.3.0]decane **(220)**, is formed by a transannular addition of the amide N—H group to the keto group in 1-azacyclodecane-2,7-dione **(219)** [671].

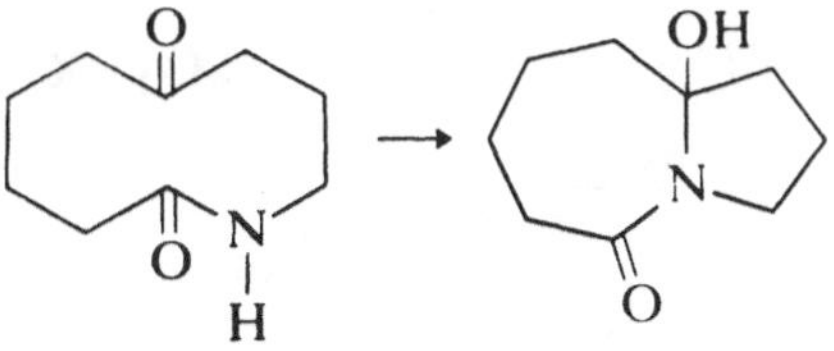

(219)   **(220)**

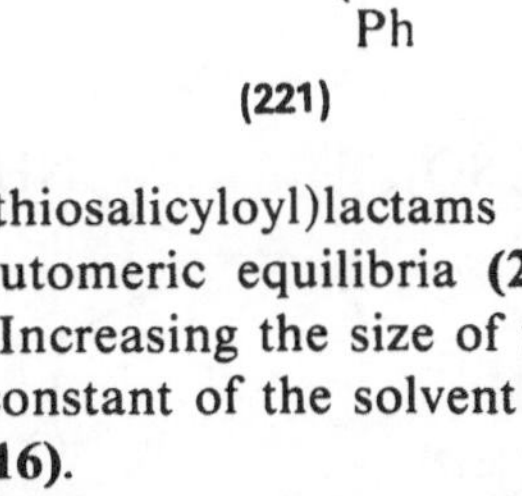

(221)

In solutions of N-(thiosalicyloyl)lactams **(214**, X = S Y = o-C$_6$H$_4$, Z = (CH$_2$)$_n$, n = 3–5), tautomeric equilibria **(214)** ⇌ **(215)** ⇌ **(216)** have been detected [672–674]. Increasing the size of the lactam ring as well as increasing the dielectric constant of the solvent displaces the equilibrium toward the macrocycle **(216)**.

A stable thiacyclol **(221)** was obtained [675] by the cyclization of [(RS)-2-tritylthiopropionyl]-L-phenylalanyl-L-proline.

Additional analogous formations of heterocycles via intermediate cyclols are known. The isomerization of 3-(N-ethylaminocarbonylmethyl)-

1,3-benzoxazine-2,4-dione **(222)** into 1-salicyloyl-3-ethylhydantoin **(224)** proceeds most likely via an intermediate cyclol **(223)** [676].

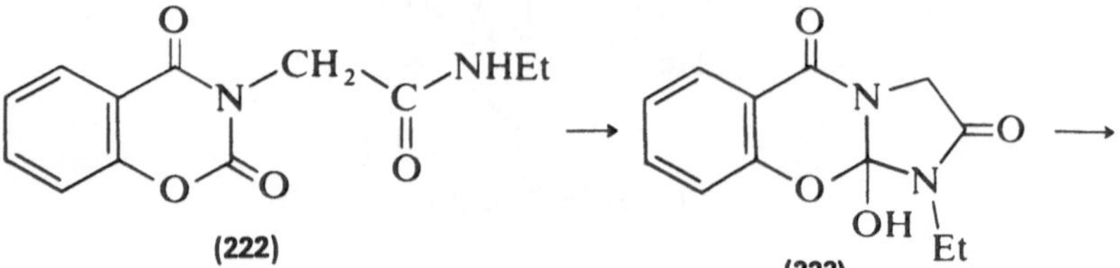

(222)   **(223)**

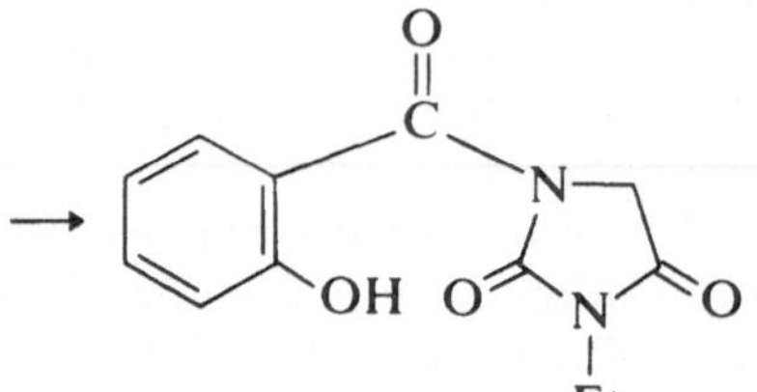

(224)

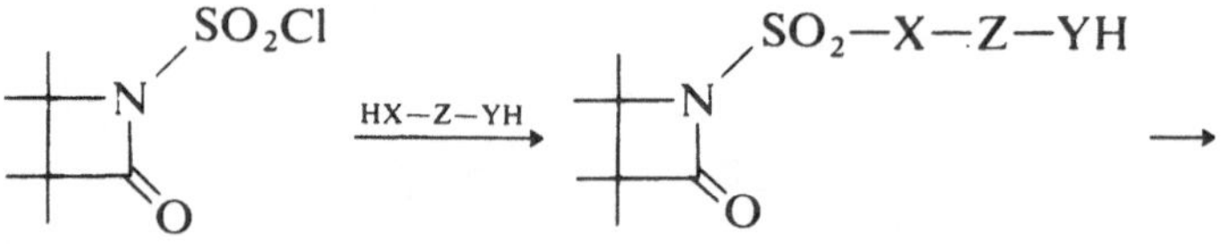

(225)   **(226)**

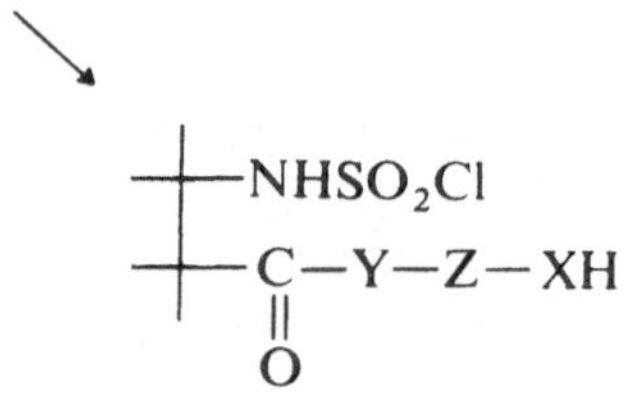

(228)

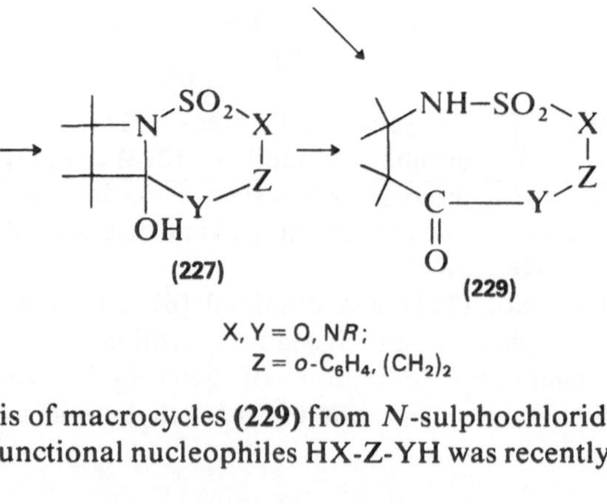

(227)   **(229)**

X, Y = O, NR;
Z = o-C$_6$H$_4$, (CH$_2$)$_2$

A synthesis of macrocycles **(229)** from *N*-sulphochlorides of β-lactams **(225)** and bis-functional nucleophiles HX-Z-YH was recently reported [677].

The mechanism was thought to proceed via intermediate cyclols (227). However, an alternative mechanism as well is under discussion, based on the assumption of an initial attack of the nucleophile at the carbonyl group followed by opening of the lactam ring and repeated cyclization (225) → (228) → (229).

Hesse and co-workers proposed [678–684] a convenient method for the synthesis of macrocyclic polyaminolactams (231) by an alkaline isomerization of N-3-aminopropyllactams (230, $m = 0$, $n = 6, 7, 9, 11$). The KAPA reagent (1,3-propanediamine/potassium 3-aminopropylamide) was used as a catalyst for the isomerization.

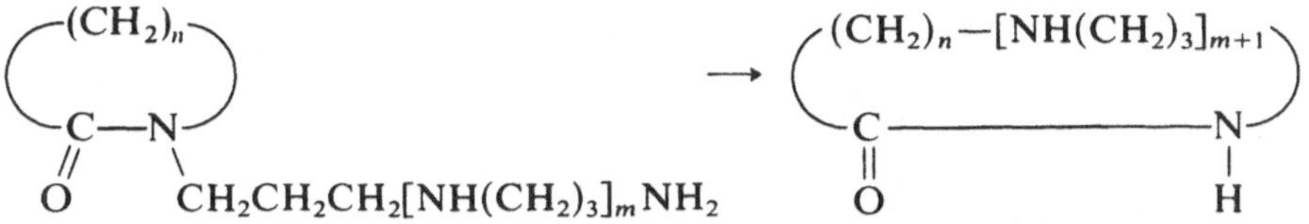

The reaction works also when the chain at the lactam nitrogen contains several aminopropyl residues (230, $n = 11$, $m = 1, 2, 4, 9$). Using this route [679] even a 53-membered polyaminolactam was obtained. The authors suppose [683] that the introduction of aminopropyl residues into the ring proceeds stepwise via intermediate cyclol anions:

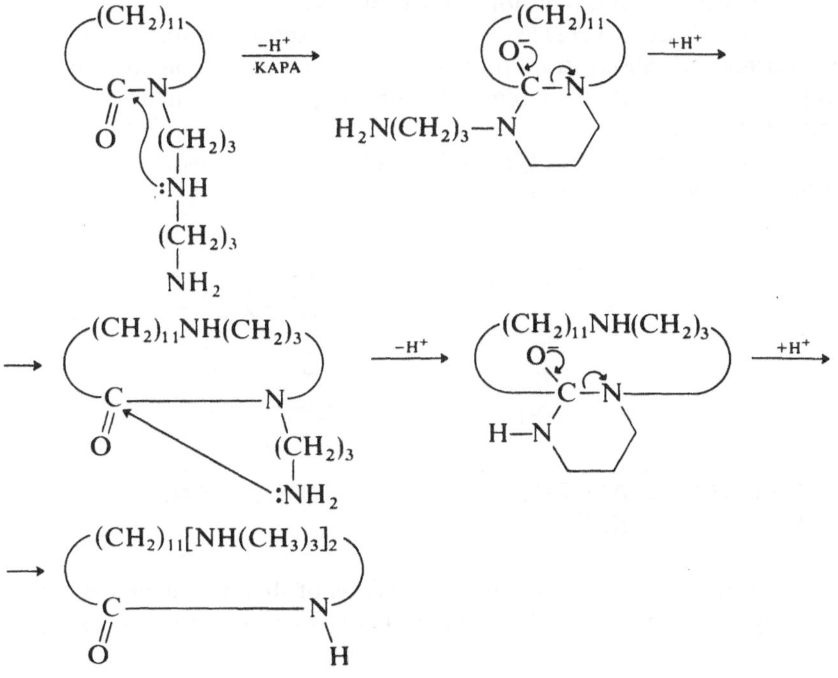

In analogy to a zip-fastener this reaction is called a "zip-reaction."

Reimschuessel [685] proposed a cyclol mechanism for an isomerizing polymerization of β-carboxy- and β-carboxymethyllactams, e.g., for the polymerization of β-carboxymethyl caprolactam:

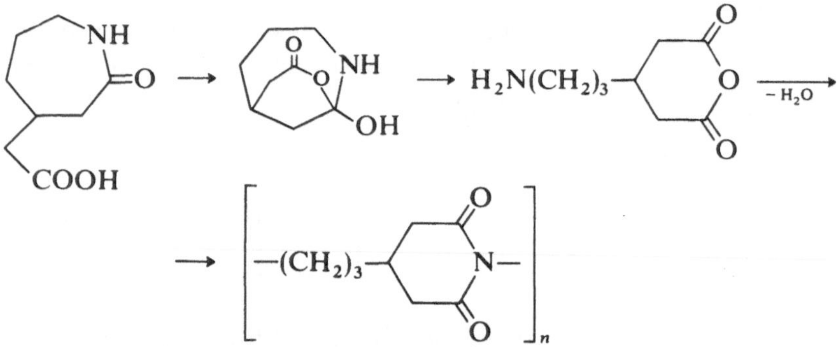

The examples reported so far convincingly demonstrate that cyclols are of great importance as intermediates in the formation of many macrocyclic systems, particularly of cyclopeptides and cyclodepsipeptides. Unfortunately, the information available from the literature does not allow a more detailed discussion of the influence of molecular structure on the relative stabilities of cyclols (215) and their isomers (214) and (216), probably with the exception of oxacyclols (see Table 28).

The transformation (216) → (215) is a transannular isomerization. Useful information about transannular interaction and reactions of type (232) → (233) and other similar reactions has been given in reviews by Leonard [686, 687].

The transformations of type (232) → (233) not being isomerizations will not be discussed in detail here.

2.6. MERCAPTO ALDEHYDES AND KETONES AND RELATED COMPOUNDS

The ring-chain isomeric interconversions of this type have been comparatively rarely investigated. As generally known sulfur possesses a higher nucleophilicity than oxygen. Thus, if the structure of the linking chain does not hinder intramolecular addition, ring isomers (234B) predominate. They are often subjected to dehydration reactions or to more complicated transformations.

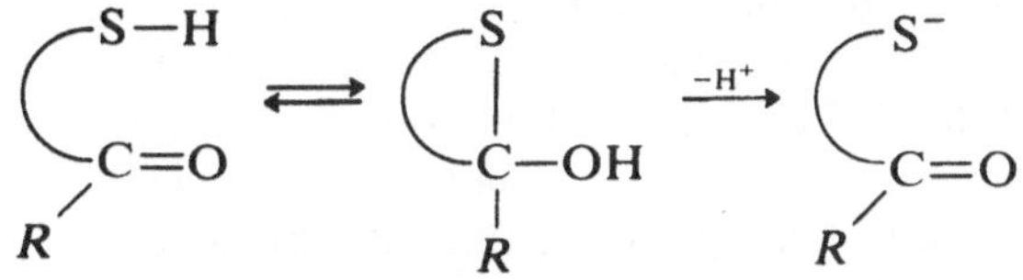

(234A)   (234B)   (235)

Since an SH group has a greater acidity than an OH group [688], anions can be expected to possess the open structure (235) as discussed below.

N-(2-Oxo-1,1-dimethylalkyl)- and N-(2-oxo-1,1-diphenyl)alkyl-dithiocarbaminic acids (236A, R^1 = Me, Ph, R^2 = Me, Ph) in the solid state and in solution are cyclic 5-hydroxythiazolidine-2-thiones (236B), but the anion possesses the open structure (237).

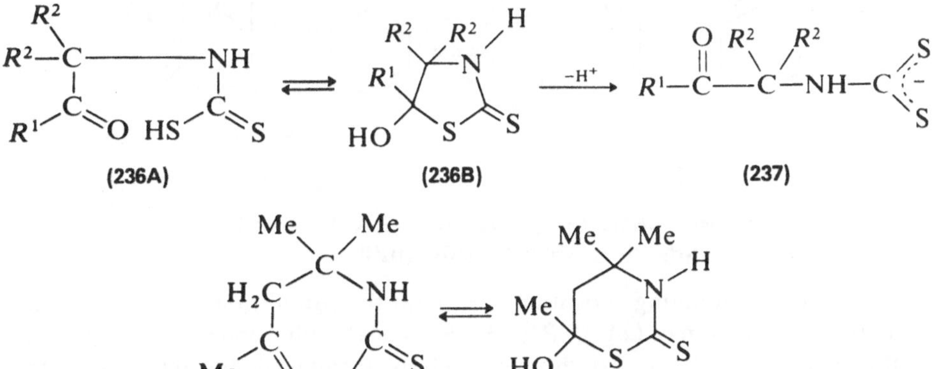

(236A)   (236B)   (237)

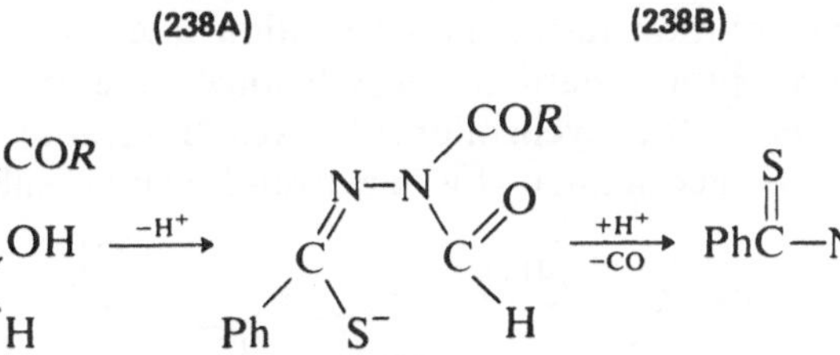

(238A)   (238B)

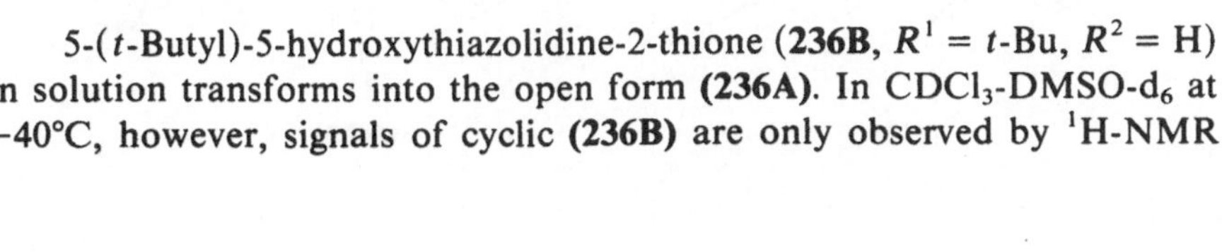

(239)   (240)   (241)

5-(t-Butyl)-5-hydroxythiazolidine-2-thione (236B, R^1 = t-Bu, R^2 = H) in solution transforms into the open form (236A). In CDCl$_3$-DMSO-d$_6$ at −40°C, however, signals of cyclic (236B) are only observed by ^{1}H-NMR spectroscopy. Heating the solution above 0°C results in ring opening **(236B)** → **(236A)**. Analogous reactions were observed in solutions of 5-hydroxy-5-phenylthiazolidine-2-thione **(236B,** R^1 = Ph, R^2 = H) and 6-hydroxy-4,4,6-trimethyl-1,3-thiazine-2-thione **(238B)** [689, 690].

The products of the covalent hydration of 3-acyl-5-phenyl-1,4,5-thiadiazolium salts are cyclic 3-acyl-2-hydroxy-5-phenyl-1,4,5-thiadiazolines **(239)** [691, 692]. The anion **(240)** has the open structure, and *N*-acyl-*N'*-thiobenzoylhydrazines **(241)** have been isolated by acidification of alkaline solutions. Reactions of compounds **(239)** with diazomethane gave S-methyl-derivatives of open structure while with diazoethane a mixture of S- and O-derivatives was formed.

In acidic aqueous solution α-thiosemicarbazones, dithiocarboxyhydrazones, and thiobenzoylhydrazones of isatin are in equilibrium with 9b-hydroxy-1,3,4-thiadiazino[5,6-b]indoles **(242A)** ⇄ **(242B)** [693, 694]. Increasing medium acidity displaces the equilibrium toward the cyclic **(242B)** which apparently is more basic.

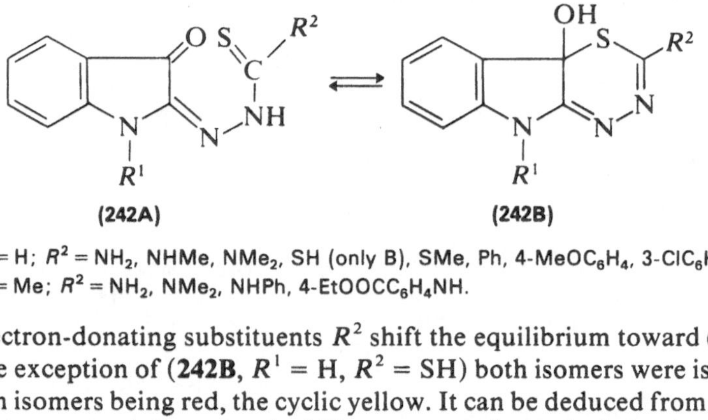

(242A)   **(242B)**

R^1 = H; R^2 = NH$_2$, NHMe, NMe$_2$, SH (only B), SMe, Ph, 4-MeOC$_6$H$_4$, 3-ClC$_6$H$_4$.
R^2 = Me; R^2 = NH$_2$, NMe$_2$, NHPh, 4-EtOOCC$_6$H$_4$NH.

Electron-donating substituents R^2 shift the equilibrium toward **(242B)**. With the exception of **(242B,** R^1 = H, R^2 = SH) both isomers were isolated, the open isomers being red, the cyclic yellow. It can be deduced from kinetic measurements carried out by the UV spectroscopic method that increasing the acidity of the solution raises the cyclization rate. In the rate-limiting step of cyclization a proton participates. Obviously, the protonation of the keto group precedes the cyclization. In acidic solution the reaction **(242A)** → **(242B)** is accelerated by electron-donating substituents R^2.

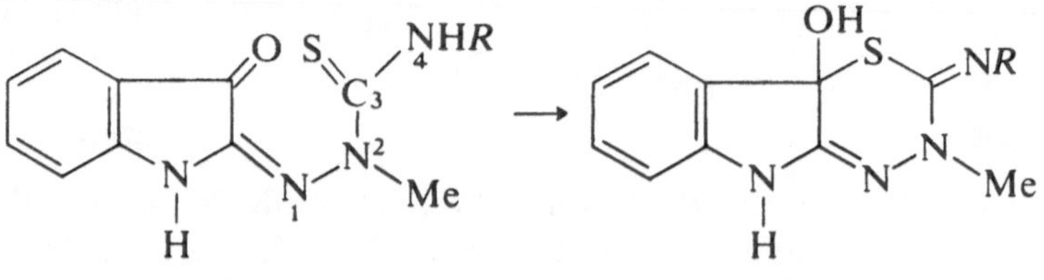

(243A)   **(243B)**

However, under neutral conditions the influence of substituents R^2 on the rate of cyclization could not be evaluated.

The introduction of a methyl group at $N_{(2)}$ of isatin α-thiosemicarbazone and its 4-methylderivative (**243A**, R = H, Me) significantly accelerates cyclization (**243A**, R = H, Me) $\rightarrow$ (**243B**) compared with $N_{(2)}$ unsubstituted derivatives (**242A**, R^1 = H, R^2 = NH_2, NHMe).

2.7. INTRAMOLECULAR ADDITION OF C—H GROUPS

Intramolecular addition of C—H groups to a C=O bond occurs rather frequently. However, these as well as the reverse reactions generally are practical only in the presence of alkali, acids, or other catalysts. The need to use severe reaction conditions often results in a dehydration of the products of intramolecular addition yielding unsaturated compounds.

An important condition for the above-mentioned reactions is the presence of electron-withdrawing substituents at a carbon atom, which increase the acidity of the C—H bond, thus favoring either the formation of a carbanion (basic catalysis) or an electrophilic substitution of hydrogen atom (acidic catalysis).

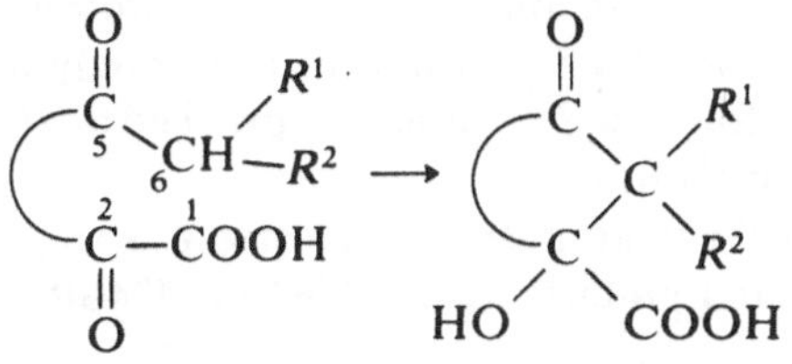

(244A)   (244B)

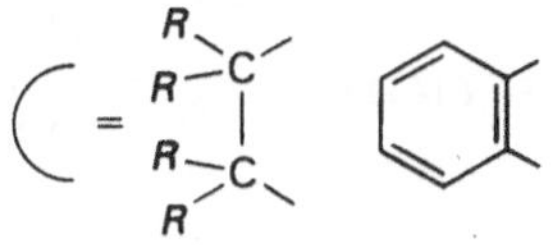

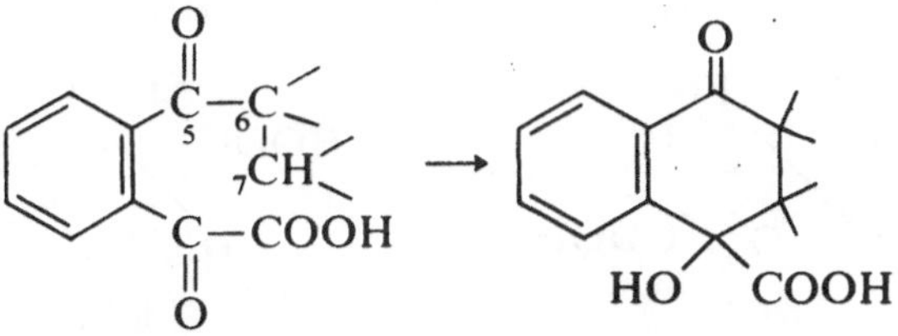

(245A)   (245B)

Shemyakin and Shchukina [695] summarized a large amount of the information concerning the formation of hydroxycarboxylic acids in cyclization reactions (**244A**) $\rightarrow$ (**244B**) and (**245A**) $\rightarrow$ (**245B**). They came to the conclusion that a comparatively important influence on both reactions is exerted by the carboxylic group increasing the electrophilicity of the adjacent keto group.

In **(244A)** the keto group in position 5 as well as substituents R^1 and R^2 (Cl, OH, $\overset{+}{N}C_5H_5$ a.o.) [695–697] possessing a -I-effect raise the C—H acidity and favor cyclization. The same is true for substituents at the atoms $C_{(6)}$ and $C_{(7)}$ in **(245)**. Hydroxycarboxylic acids **(245B)** under conditions of their formation are often subjected to dehydration or subsequent oxidative decarboxylation to quinones.

Transformations of this kind are important for a discussion of a mechanism of oxidative-hydrolytic transformations of quinones and particularly of the Hooker-reaction [697–699]. Unfortunately, no spectroscopic investigation of this type of equilibrium has been carried out. This is obviously caused by difficulties which result from the severe reaction conditions necessary to carry out an intramolecular addition of C—H groups to the C=O bond.

Unlike the nucleophilic addition of O—H, N—H, and S—H groups to the C=O bond which is due to a participation of *p*-electrons of these heteroatoms and a final migration of the proton, a nucleophilic addition of a C—H group cannot take place before the proton has been removed forming a carbanion in alkaline solution. In strongly acidic media, where the keto group is protonated, reactions can proceed as electrophilic substitution of the hydrogen atom (C—H).

Thorpe [700–705] quantitatively investigated ring-chain equilibria of 2-oxo-3,3-disubstituted glutaric acids **(246A)** ⇌ **(246B)**.

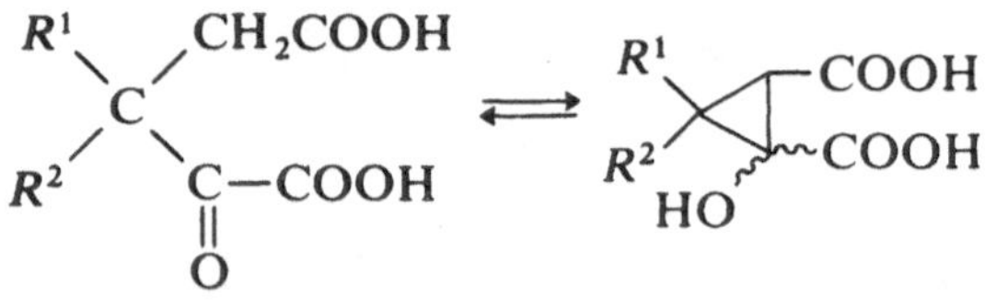

(246A)   (246B)

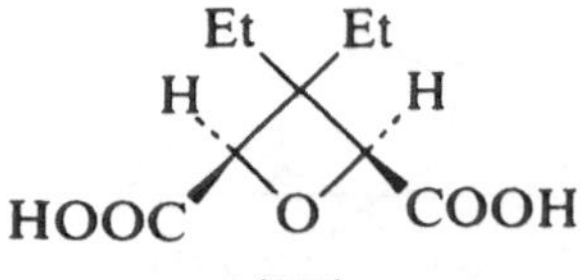

(247) 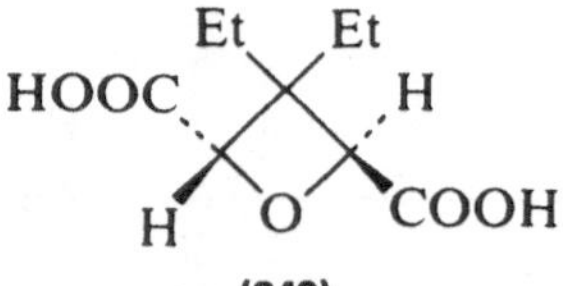   (248)

These studies gave rise to the very useful hypothesis [472] of the gem-dimethyl or gem-dialkyl effect which has been called the Thorpe-Ingold effect. This effect was discussed earlier in this book and used repeatedly to explain the stabilizing influence of the gem-dialkyl substituents in a chain on cyclic tautomers as well as the accelerating influence on the cyclization reactions. The effect was experimentally supported by many cyclization reactions [47-55, 706-708] and was theoretically substantiated [52-54].

$$\text{OH}$$
$$\bigg| $$

CH—COOH

CH—COOH
$$\bigg|$$
OH

(249)

Curiously, the structures of the compounds **(246A)** and **(246B)**, the investigation of which initiated the conceptualization of the Thorpe-Ingold effect, are incorrect. After it was stated that the first experiments gave no strict evidence [48] a reinvestigation [709] using ^{1}H-NMR spectroscopy revealed that the compounds described by Thorpe as **(246A, $R^1 = R^2 =$ Et)** and **(246B)** were actually oxethanes **(247)** and **(248)**. According to other work [710] the compound **(246B, R^1, $R^2 = (CH_2)_5$)** really possesses structure **(249)**.

REFERENCES

1. P. R. Jones, *Chem. Rev.* **1963**, *63*, 461.
2. V. M. Andreev, G. P. Kugatova-Schemyakina, and S. A. Kazaryan, *Uspekhi Khimii* **1968**, *37*, 559.
3. R. E. Valters, *Uspekhi Khimii* **1973**, *42*, 1060.
4. D. J. Chadwick and J. D. Dunitz, *J. Chem. Soc., Perkin Trans 2* **1979**, 276.
5. R. P. Bell, B. G. Cox, and B. A. Timini, *J. Chem. Soc., Sect. B* **1971**, 2247.
6. R. P. Bell and B. G. Cox, *Chem. Soc., Perkin Trans. 2* **1975**, 1349.
7. M. M. Shemyakin, D. N. Shigorin, L. A. Shchukina, and E. P. Semkin, *Izv. Akad. Nauk SSSR, Otd. Khim. Nauk* **1959**, 695.
8. E. Bernatek, *Acta Chem. Scand.* **1960**, *14*, 785.
9. R. E. Lutz, C. T. Clark, and J. P. Feifer *J. Org. Chem.* **1960**, *25*, 346.
10. R. E. Lutz and H. Moncure, Jr., *J. Org. Chem.* **1961**, *26*, 746.
11. D. H. Kim and D. N. Harpp, *Chem. Ind.* **1965**, 183.
12. J. Kagan, *J. Org. Chem.* **1967**, *32*, 4060.
13. A. Winston, J. C. Sharp, K. E. Atkins, and D. E. Battin, *J. Org. Chem.* **1967**, *32*, 2166.
14. B. Paul and W. Korytnyk, *Chem. Ind.* **1967**, 230.
15. S. Seltzer and K. D. Stevens, *J. Org. Chem.* **1968**, *33*, 2708.
16. N. P. Buu-Hoï, M. Dufour, and P. Jacquignon, *Bull. Soc. Chim. Fr.* **1970**, 137.
17. N. P. Buu-Hoï, M. Dufour, and P. Jacquignon, *Bull. Soc. Chim. Fr.* **1971**, 2999.
18. B. Paul and W. Korytnyk, *J. Heterocycl. Chem.* **1976**, *13*, 701.
19. C. Pascual, D. Wegmann, U. Graf, R. Scheffold, P. F. Sommer, and V. Simon, *Helv. Chim. Acta* **1964**, *47*, 213.

20. R. Scheffold and P. Dubs, *Helv. Chim. Acta* **1967**, *50*, 798.
21. H. des Abbayes, *Bull. Soc. Chim. Fr.* **1970**, 3671.
22. K. Bowden and M. P. Henry, *J. Chem. Soc., Perkin Trans. 2* **1972**, 206.
23. R. P. Bell and A. D. Covington, *J. Chem. Soc., Perkin Trans. 2* **1975**, 1343.
24. R. P. Bell, D. W. Earls, and J. B. Henshall, *J. Chem. Soc., Perkin Trans. 2* **1976**, 39.
25. R. P. Bell, *The Proton in Chemistry*, 2nd edition, Chapman and Hall, London, 1973.
26. K. Bowden and G. R. Taylor, *J. Chem. Soc., Sect. B* **1971**, 1390.
27. K. Bowden and M. P. Henry, *J. Chem. Soc., Perkin Trans. 2* **1972**, 201.
28. M. S. Newman and C. Courduvelis, *J. Org. Chem.* **1965**, *30*, 1795.
29. P. R. Jones and P. J. Desio, *J. Org. Chem.* **1965**, *30*, 4293.
30. N. M. Tsybina, T. V. Protopopova, S. G. Rosenberg, and A. I. Talygina, *Zh. Org. Khim.* **1971**, *7*, 253.
31. M. V. Bhatt and K. M. Kamath, *Tetrahedron Lett.* **1966**, 3885.
32. M. V. Bhatt and K. M. Kamath, *J. Chem. Soc., Sect. B* **1968**, 1036.
33. W. Flitsch, *Chem. Ber.* **1970**, *103*, 3205.
34. A. Winston, J. P. M. Bederka, W. G. Isner, P. C. Juliano, and J. C. Sharp, *J. Org. Chem.* **1965**, *30*, 2784.
35. H. Schroeter, R. Appel, R. Brammer, and G. O. Schenck, *Justus Liebigs Ann. Chem.* **1966**, *697*, 42.
36. K. Bowden and G. R. Taylor, *J. Chem. Soc., Sect. B* **1971**, 1395.
37. L. Christiaens and M. Renson, *Bull. Soc. Roy. Sci. Liege* **1972**, *41*, 139.
38. K. Bowden and A. M. Last, *J. Chem. Soc., Perkin Trans. 2* **1973**, 1144.
39. R. E. Valters, V. R. Zinkovska, A. V. Burkevica, and S. P. Valtere, *Latv. PSR Zinat. Akad. Vestis, Kim. Ser.* **1974**, 118.
40. P. T. Lansbury and J. F. Bieron, *J. Org. Chem.* **1963**, *28*, 3564.
41. D. S. Erley, W. J. Potts, P. R. Jones, and P. J. Desio, *Chem. Ind.* **1964**, 1915.
42. J. Finkelstein, T. Williams, V. Toome, and S. Traiman, *J. Org. Chem.* **1967**, *32*, 3229.
43. M. Kuchar and B. Kakac, *Collect. Czech. Chem. Commun.* **1971**, *36*, 2298.
44. H. des Abbayes and C. Neveu, *Compt. Rend. Acad. Sci., Ser. C* **1974**, *278*, 805.
45. H. des Abbayes, F. Salmon-Legagneur, and C. Neveu, *Compt. Rend. Acad. Sci., Ser. C* **1971**, *273*, 302.
46. H. des Abbayes, F. Salmon-Legagneur, and C. Neveu, *Compt. Rend. Acad. Sci., Ser. C* **1972**, *274*, 1950.
47. E. L. Eliel, *Stereochemistry of Carbon Compounds*, McGraw-Hill Book Company, New York, 1962, Chapter 7.
48. G. S. Hammond, In: Steric Effects in Organic Chemistry, Ed. M. S. Newman, John Wiley and Sons, New York, 1956, Chapter 9.
49. B. Capon and S. T. McManus, *Neighbouring Group Participation*, Vol. I, Plenum Press, New York, 1976, p. 58.
50. J. Hine, *Structural Effects on Equilibria in Organic Chemistry*, Wiley Interscience, New York, 1975, p. 284.
51. R. E. Valters, *Uspekhi Khimii* **1982**, *51*, 1374.
52. N. L. Allinger and V. Zalkov, *J. Org. Chem.* **1960**, *25*, 701.
53. C. Danforth, A. W. Nicholson, J. C. James, and G. M. Loudon, *J. Am. Chem. Soc.* **1976**, *98*, 4275.
54. R. E. Vinans and Ch. F. Wilcox, Jr., *J. Am. Chem. Soc.* **1976**, *98*, 4281.
55. S. Milstien and L. A. Cohen, *Proc. Nat. Acad. Sci. USA* **1970**, *67*, 1143; S. Milstien and L. A. Cohen, *J. Am. Chem. Soc.* **1972**, *94*, 9158.
56. F. C. Baddar and A. Habashi, *J. Chem. Soc.* **1959**, 4119.
57. Yu. A. Pentin, I. S. Trubnikov, R. B. Teplinskaya, N. P. Shusherina, and R. Ya. Levina, *Zh. Obshch. Khim.* **1962**, *32*, 1927.

58. G. Leclerc, C.-G. Vermuth, and J. Schreiber, *Bull. Soc. Chim. Fr.* **1967**, 1302.
59. R. E. Valters and A. E. Kipina, *Latv. PSR Zinat. Akad. Vestis, Kim. Ser.* **1969**, 470.
60. A. N. Kost, M. A. Yurovskaya, and M. T. Nguen, *Khim. Geterotsikl. Soedin.* **1975**, 659.
61. I. S. Trubnikov, *Zh. Org. Khim.* **1965**, *1*, 1526.
62. J. F. Grove and H. A. Willis, *J. Chem. Soc.* **1951**, 877.
63. D. D. Wheeler, D. C. Young, and D. S. Erley, *J. Org. Chem.* **1957**, *22*, 547.
64. M. Renson, *Bull. Soc. Chim. Belg.* **1961**, *70*, 77.
65. H. Sterk, *Monatsh. Chem.* **1968**, *99*, 1764.
66. P. R. Jones and S. L. Congdon, *J. Am. Chem. Soc.* **1959**, *81*, 4291.
67. N. R. Bruvele and E. J. Gudriniece, *Latv. PSR Zinat. Akad. Vestis, Kim. Ser.* **1970**, 198.
68. R. E. Valters, *Latv. PSR Zinat. Akad. Vestis, Kim. Ser.* **1969**, 91.
69. A. Aebi, E. Gyurech-Vago, E. Hofstetter, and P. Waser, *Pharm. Acta Helv.* **1963**, *48*, 407.
70. R. E. Valters, *Latv. PSR Zinat. Akad. Vestis, Kim. Ser.* **1969**, 699.
71. R. E. Valters, A. E. Kipina, and S. P. Valtere, *Latv. PSR Zinat. Akad. Vestis, Kim. Ser.* **1970**, 206.
72. N. R. Bruvele and E. J. Gudriniece, *Latv. PSR Zinat. Akad. Vestis, Kim. Ser.* **1970**, 368.
73. R. E. Valters and A. E. Kipina, *Latv. PSR Zinat. Akad. Vestis, Kim. Ser.* **1971**, 200.
74. A. Hantzsch and A. Schwiete, *Ber. Dtsch. Chem. Ges.* **1916**, *49*, 213.
75. A. Hantzsch, *J. prakt. Chem.* **1927**, *117*, 151.
76. A. H. Zicmanis and A. K. Arens, *Latv. PSR Zinat. Akad. Vestis, Kim. Ser.* **1970**, 489.
77. V. V. Korshak, S. V. Vinogradova, G. N. Melekhina, S. N. Salazkin, L. I. Komarova, P. V. Petrovskii, and P. O. Okulevich, *Izv. Akad. Nauk SSSR, Ser. Khim.* **1976**, 368.
78. R. E. Valters and A. E. Bace, *Latv. PSR Zinat. Akad. Vestis, Kim. Ser.* **1971**, 335.
79. R. E. Valters, A. E. Bace, and S. P. Valtere, *Latv. PSR Zinat. Akad. Vestis, Kim. Ser.* **1972**, 61.
80. W. Graf, E. Girod, E. Schmid, and W. G. Stoll, *Helv. Chim. Acta* **1959**, *42*, 1085.
81. L. Christiaens and M. Renson, *Bull. Soc. Chim. Belg.* **1969**, *78*, 359.
82. M. S. Newman and Ch. W. Muth, *J. Am. Chem. Soc.* **1951**, *73*, 4627.
83. J. H. P. Tyman and A. A. Najam, *Spectrochim. Acta* **1977**, *33A*, 479.
84. V. N. Eraksina, T. M. Ivanova, T. A. Babushkina, A. M. Vasil'ev, and N. N. Suvorov, *Khim. Geterotsikl. Soedin.* **1979**, 916.
85. F. H. Pinkerton and S. F. Thames, *J. Organomet. Chem.* **1970**, *24*, 623.
86. S. Gronowitz, B. Gestblom, and B. Mathiasson, *Ark. Kemi* **1963**, *20*, 407.
87. R. E. Valters and J. R. Mednis, *Zh. Org. Khim.* **1974**, *10*, 1248.
88. J. R. Mednis and R. E. Valters, *Latv. PSR Zinat. Akad. Vestis, Kim. Ser.* **1976**, 88.
89. M. V. Gorelik, M. V. Kazankov, and M. I. Bernadskii, *Zh. Org. Khim.* **1976**, *12*, 2041.
90. M. V. Gorelik, M. V. Kazankov, and M. I. Bernadskii, *Zh. Org. Khim.* **1978**, *14*, 1535.
91. N. Hellström, *Nature* **1960**, *187*, N4732, 146.
92. G. F. Muzychenko, L. A. Badovskaya, V. G. Kul'nevich, and A. I. Suprunova, *Zh. Org. Khim.* **1971**, *7*, 1594.
93. F. Farina and M. V. Martin, *Ann. quim.* **1971**, *77*, 315.
94. W. J. Conradie, C. F. Garbers, and P. S. Steyn, *J. Chem. Soc.* **1964**, 594.
95. E. I. Vinogradova and M. M. Shemyakin, *Zh. Obshch. Khim.* **1946**, *16*, 709.
96. D. T. Mowry, *J. Am. Chem. Soc.* **1950**, *72*, 2535.
97. E. Kuh and R. L. Sheppard, *J. Am. Chem. Soc.* **1953**, *75*, 4597.
98. S. Kovac, E. Solcaniova, E. Beska, and P. Rapos, *J. Chem. Soc., Perkin Trans. 2* **1973**, 105.
99. H. H. Wassermann and F. M. Precopio, *J. Am. Chem. Soc.* **1952**, *74*, 326.
100. E. A. Shaw, *J. Am. Chem. Soc.* **1946**, *68*, 2510.
101. N. L. Wendler and H. L. Slates, *J. Org. Chem.* **1967**, *32*, 849.
102. N. Sugiyama, T. Gasha, H. Kataoka, and Ch. Kashima, *Bull. Chem. Soc. Jpn.* **1968**, *41*, 971.

103. N. Sugiyama, H. Kataoka, Ch. Kashima, and K. Yamada, *Bull. Chem. Soc. Jpn.* **1969**, *42*, 1098.

104. R. A. Raphael, *J. Chem. Soc.* **1947**, 805.

105. J. H. Ford, A. R. Johnson, and J. W. Hinman, *J. Am. Chem. Soc.* **1950**, *72*, 4529.

106. R. E. Lutz, P. S. Bailey, C. K. Dien, and J. W. Rinker, *J. Am. Chem. Soc.* **1953**, *75*, 5039.

107. C. L. Browne and R. E. Lutz, *J. Org. Chem.* **1953**, *18*, 1638.

108. M. Semonsky, E. Rockova, V. Zikan, B. Kakac, and V. Jelinek, *Coll. Czech. Chem. Commun.* **1963**, *28*, 377.

109. G. Rio and J.-C. Hardy, *Bull. Soc. Chim. Fr.* **1970**, 3572.

110. I. A. Kuzovnikova, L. A. Badovskaya, Ya. I. Tur'yan, and V. G. Kul'nevich, *Khim. Geterotsikl. Soedin.* **1974**, 737.

111. M. Kuchar, *Coll. Czech. Chem. Commun.* **1968**, *33*, 880.

112. M. Kuchar and B. Kakac, *Coll. Czech. Chem. Commun.* **1969**, *34*, 3343.

113. J. Fowler and S. Seltzer, *J. Org. Chem.* **1970**, *35*, 3529.

114. S. Seltzer, *J. Org. Chem.* **1981**, *46*, 2643.

115. Yu. A. Pentin, I. S. Trubnikov, N. P. Shusherina, and R. Ya. Levina, *Zh. Obshch. Khim.* **1961**, *31*, 2092.

116. Yu. A. Pentin, I. S. Trubnikov, R. B. Teplinskaya, N. P. Shusherina, and R. Ya. Levina, *Dokl. Akad. Nauk SSSR* **1961**, *139*, 1121.

117. I. S. Trubnikov, R. B. Teplinskaya, Yu. A. Pentin, N. P. Shusherina, and R. Ya. Levina, *Zh. Obshch. Khim.* **1963**, *33*, 1210.

118. H. Meerwein, *J. Prakt. Chem.* **1927**, *116*, 229.

119. M. Renson and L. Christiaens, *Bull. Soc. Chim. Belg.* **1962**, *71*, 379.

120. M. Renson and L. Christiaens, *Bull. Soc. Chim. Belg.* **1962**, *71*, 394.

121. R. G. Hiskey and M. A. Harpold, *J. Org. Chem.* **1967**, *32*, 1986.

122. M. Avram, D. Constantinescu, I. G. Dinulescu, O. Constantinescusimon, G. D. Mateescu, and C. D. Nenitzescu, *Rev. Roum. Chim.* **1970**, *15*, 1097.

123. V. M. Rodionov and A. M. Fedorova, *Izv. Akad. Nauk SSSR, Otd. Khim. Nauk,* **1950**, 247.

124. D. V. Nightingale, W. S. Wagner, and R. H. Wise, *J. Am. Chem. Soc.* **1953**, *75*, 4701.

125. P. R. Jones and A. A. Lavigne, *J. Org. Chem.* **1960**, *25*, 2020.

126. V. Balasubramaniyan, *Chem. Rev.* **1966**, *66*, 567.

127. H. E. French and J. E. Kircher, *J. Am. Chem. Soc.* **1944**, *66*, 298.

128. R. E. Valters and V. R. Zinkovska, *Khim. Geterotsikl. Soedin.* **1973**, 1127.

129. P. Lingens and B. Sprössler, *Justus Liebigs Ann. Chem.* **1967**, *702*, 169.

130. P. F. Wegfahrt and H. Rapoport, *J. Org. Chem.* **1969**, *34*, 3035.

131. N. P. Buu-Hoï and Ch. K. Lin, *Compt. Rend. Acad. Sci., Ser. C* **1939**, *209*, 221.

132. M. V. Bhatt, S. H. El Ashry, and M. Balakrishnan, *Proc. Indian. Acad. Sci.* **1979**, *A88*, 421.

133. M. V. Bhatt and S. H. El Ashry, *Indian J. Chem.* **1980**, *19B*, 487.

134. H. J. Hediger, *Infrarotspektroskopie, Grundlagen, Anwendungen, Interpretation,* Frankfurt am Main, 1971, p. 86.

135. W. I. Elliot and I. Fried, *J. Org. Chem.* **1978**, *43*, 2708.

136. M. S. Newman and Zia ud Din, *J. Org. Chem.* **1971**, *36*, 2740.

137. M. V. Bhatt, S. H. El Ashry, and V. Somayaji, *Indian J. Chem.* **1980**, *19B*, 473.

138. H. G. Kuivila, *J. Org. Chem.* **1960**, *25*, 284.

139. G. H. Schmid and L. S. J. Weiler, *Can. J. Chem.* **1965**, *43*, 1242.

140. M. S. Newman, N. Gill, and B. Darre, *J. Org. Chem.* **1966**, *31*, 2713.

141. B. M. Sheiman, L. Ya. Denisova, S. F. Dimova, A. A. Shereshevskii, I. M. Kustanovich, and V. M. Berezovskii, *Khim. Geterotsikl. Soedin.* **1971**, 190.

142. J. Cason and E. J. Reist, *J. Org. Chem.* **1958**, *23*, 1492.

143. J. Cason and E. J. Reist, *J. Org. Chem.* **1958**, *23*, 1668.

144. Y. Kubota and T. Tasuno, *Chem. Pharm. Bull. Tokyo* **1971**, *19*, 1226.

145. P. M. Pojer, E. Ritchie, and W. C. Taylor, *Austr. J. Chem.* **1967**, *21*, 1375.
146. R. E. Valters and S. P. Valtere, *Latv. PSR Zinat. Akad. Vestis, Kim. Ser.* **1969**, 704.
147. R. E. Valters and S. P. Valtere, *Latv. PSR Zinat. Akad. Vestis, Kim. Ser.* **1971**, 208.
148. W. L. F. Armarego and S. C. Sharma, *J. Chem. Soc., Sect. C* **1970**, 1600.
149. W. L. F. Armarego, B. A. Milloy, and S. C. Sharma, *J. Chem. Soc., Perkin Trans. I* **1972**, 2485.
150. M. S. Newman and C. Courduvelis, *J. Am. Chem. Soc.* **1966**, *88*, 781.
151. M. V. Bhatt, K. M. Kamath, and M. Ravindranathan, *J. Chem. Soc., Sect. C* **1971**, 1772.
152. M. V. Bhatt, K. M. Kamath, and M. Ravindranathan, *J. Chem. Soc., Sect. C* **1971**, 3344.
153. R. E. Lutz, *J. Am. Chem. Soc.* **1930**, *52*, 3405.
154. R. E. Lutz and R. J. Taylor, *J. Am. Chem. Soc.* **1933**, *55*, 1168.
155. R. E. Lutz and R. J. Taylor, *J. Am. Chem. Soc.* **1933**, *55*, 1593.
156. M. S. Newman and C. Courduvelis, *J. Am. Chem. Soc.* **1964**, *86*, 2442.
157. M. S. Newman, In: *Steric Effects in Organic Chemistry*, Ed. M. S. Newman, John Wiley and Sons, New York, 1956, Chapter 4.
158. R. E. Valters, A. E. Bace, A. V. Burkevica, and R. B. Kampare, *Zh. Org. Khim.* **1976**, *12*, 173.
159. H. Glinka and A. Fabrycy, *Roczn. Chem.* **1970**, *44*, 1703.
160. G. J. Duburs and G. J. Vanags, *Latv. PSR Zinat. Akad. Vestis, Kim. Ser.* **1961**, 235.
161. D. W. H. Macdowell, R. A. Jourdenais, R. W. Naylor, and J. C. Wisowaty, *J. Org. Chem.* **1972**, *37*, 4406.
162. R. E. Valters, *Latv. PSR Zinat. Akad. Vestis, Kim. Ser.* **1970**, 339.
163. R. E. Valters and S. P. Valtere, *Latv. PSR Zinat. Akad. Vestis, Kim. Ser.* **1971**, 213.
164. G. A. Karlivans and R. E. Valters, *Zh. Org. Khim.* **1982**, *18*, 2226.
165. J. Cason and E. J. Reist, *J. Org. Chem.* **1958**, *23*, 1675.
166. J. M. Holland and D. W. Jones, *J. Chem. Soc., Sect. C* **1970**, 530.
167. J. M. Holland and D. W. Jones, *J. Chem. Soc., Sect. C* **1970**, 536.
168. V. R. Zinkovska and R. E. Valters, *Latv. PSR Zinat. Akad. Vestis, Kim. Ser.* **1974**, 207.
169. R. E. Valters, V. R. Zinkovska, and A. E. Bace, *Latv. PSR Zinat. Akad. Vestis, Kim. Ser.* **1973**, 316
170. R. E. Valters and V. P. Ciekure, *Khim. Geterotsikl. Soedin.* **1972**, 502.
171. V. R. Zinkovska and R. E. Valters, *Latv. PSR Zinat. Akad. Vestis, Kim. Ser.* **1976**, 65.
172. G. H. Schmid, *Can. J. Chem.* **1966**, *44*, 2917.
173. E. Ott, *Justus Liebigs Ann. Chem.* **1912**, *392*, 245.
174. R. E. Lutz and R. J. Taylor, *J. Am. Chem. Soc.* **1933**, *55*, 1585.
175. E. Ott, In: *Organic Syntheses, Collective Vol. 2*, Ed. A. Blatt, Wiley, New York, 1946.
176. M. V. Bhatt and S. H. El Ashry, *Proc. Indian. Acad. Sci.* **1980**, *89*, 7.
177. N. Dawies, A. H. Hambly, and G. S. C. Semmens, *J. Chem. Soc.* **1933**, 1309.
178. Ch. Rüchardt and S. Rochlitz, *Justus Liebigs Ann. Chem.* **1974**, 15.
179. W. Lonsky and W. Mayer, *Chem. Ber.* **1975**, *108*, 1593.
180. H. Brinkmann and Ch. Rüchardt, *Tetrahedron Lett.* **1972**, 5221.
181. Ch. Rüchardt and H. Brinkman, *Chem. Ber.* **1975**, *108*, 3224.
182. J. F. King, B. L. Huston, A. Hawson, J. Komery, D. M. Deaken, and D. R. K. Harding, *Can. J. Chem.* **1971**, *49*, 933.
183. D. E. Balode and R. E. Valters, *Zh. Org. Khim.* **1979**, *15*, 878.
184. D. E. Balode and R. E. Valters, *Latv. PSR Zinat. Akad. Vestis, Kim. Ser.* **1979**, 588.
185. J. F. King, A. Hawson, B. L. Huston, L. J. Danks, and J. Komery, *Can. J. Chem.* **1971**, *49*, 943.
186. D. Sh. Rozina, L. T. Nesterenko, and Yu. I. Vainshtein, *Zh. Obshch. Khim.* **1958**, *28*, 2878.
187. V. N. Klyuev, A. B. Korzhenevskii, and B. D. Berezin, *Izv. VUZ SSSR, Khim. i Khim. Tekhnol.* **1978**, *21*, 31.

188. V. N. Klyuev, A. B. Korzhenevskii, and B. D. Berezin, *Izv. VUZ SSSR, Khim. i Khim. Tekhnol.* **1978**, *21*, 189.

189. O. Keller and V. Prelog, *Helv. Chim. Acta* **1971**, *54*, 2572.

190. R. B. Kampare, R. E. Valters, E. E. Liepins, and G. A. Karlivans, *Latv. PSR Zinat. Akad. Vestis, Kim. Ser.* **1981**, 244.

191. M. V. Bhatt and M. Ravindranathan, *J. Chem. Soc., Perkin Trans 2* **1973**, 1160.

192. L. Yu. Yuzefovich, B. M. Sheiman, T. M. Filippova, and V. G. Mairanovskii, *Khim. Geterotsikl. Soedin.* **1978**, 758.

193. G. A. Karlivans, R. E. Valters, and S. P. Valtere, *Zh. Org. Khim.* **1977**, *13*, 805.

194. G. A. Karlivans, V. P. Ciekure, and R. E. Valters, *Latv. PSR Zinat. Akad. Vestis, Kim. Ser.* **1981**, 244.

195. R. E. Valters, S. P. Valtere, and A. E. Kipina, *Zh. Org. Khim.* **1968**, *4*, 445.

196. R. E. Valters and S. P. Valtere, In: *Biological Active Compounds*, Nauka, Leningrad, 1968, p. 218 (In Russian).

197. N. H. Cromwell and K. E. Cook, *J. Am. Chem. Soc.* **1958**, *80*, 4573.

198. R. E. Valters, S. P. Valtere, and A. E. Kipina, In: *Biological Active Compounds*, Nauka, Leningrad, 1968, p. 213 (In Russian).

199. R. Chiron and Y. Graff, *Bull. Soc. Chim. Fr.* **1970**, 575.

200. R. Chiron and Y. Graff, *Bull. Soc. Chim. Fr.* **1971**, 2145.

201. R. E. Valters and G. A. Karlivans, *Latv. PSR Zinat. Akad. Vestis, Kim. Ser.* **1974**, 705.

202. R. Ramachandran, Y. Graff, and R. Chiron, *Bull. Soc. Chim. Fr.* **1972**, 1031.

203. R. Chiron, Y. Graff, and R. Ramachandran, *Bull. Soc. Chim. Fr.* **1972**, 3396.

204. A. S. Wexler, *Appl. Spectr. Rev.* **1967**, *1*, 29.

205. R. E. Valters and S. P. Valtere, *Latv. PSR Zinat. Akad. Vestis, Kim. Ser.* **1969**, 753.

206. M. Ahmed and J. M. Vernon, *J. Chem. Soc., Perkin Trans. 1* **1975**, 2048.

207. R. E. Valters and S. P. Valtere, *Khim. Geterotsikl. Soedin.* **1972**, 1577.

208. S. M. Abdel Rahman, M. F. Ismail, and Z. M. Ismail, *Rev. Roum. Chim.* **1976**, *21*, 889.

209. M. Ueda, A. Kimura, M. Ishimori, and Y. Imai, *J. Chem. Soc. Jpn., Chem. and Ind. Chem.* **1976**, 1502.

210. A. M. Islam, I. B. Hannout, N. M. Taha, and A. A. El-Magharaby, *Indian. J. Chem.* **1977**, *15B*, 58.

211. A. Marsili and V. Scartoni, *Tetrahedron Lett.* **1968**, 2511.

212. A. Marsili and V. Scartoni, *Gazz. Chim. Ital.* **1972**, *102*, 507.

213. R. E. Valters, A. E. Bace, S. P. Valtere, and R. B. Kampare, *Latv. PSR Zinat. Akad. Vestis, Kim. Ser.* **1983**, 111.

214. R. E. Valters and V. R. Zinkovska, *Latv. PSR Zinat. Akad. Vestis, Kim. Ser.* **1978**, 310.

215. S. P. Valtere, Z. E. Zarina, G. A. Karlivans, and R. E. Valters, *Latv. PSR Zinat. Akad. Vestis, Kim. Ser.* **1978**, 575.

216. B. A. Kakac, K. Mnoucek, M. Semonsky, V. Zikan, and A. Cerny, *Collect. Czech. Chem. Commun.* **1968**, *33*, 1256.

217. R. E. Valters, *Latv. PSR Zinat. Akad. Vestis, Kim. Ser.* **1970**, 223.

218. J. B. Jones and J. M. Young, *Can. J. Chem.* **1966**, *44*, 1059.

219. H. Sterk, *Monatsh. Chem.* **1968**, *99*, 1770.

220. M. Winn and H. E. Zaugg, *J. Org. Chem.* **1968**, *33*, 3779.

221. Y. Gouriou, C. Fayat, and A. Foucoud, *Bull. Soc. Chim. Fr.* **1970**, 2293.

222. R. E. Valters, A. E. Bace, and R. B. Kampare, *Latv. PSR Zinat. Akad. Vestis, Kim. Ser.* **1983**, 234.

223. L. Wolf, *Justus Liebigs Ann. Chem.* **1885**, *229*, 249.

224. *Beilsteins Handbuch der Organischen Chemie*, 4th edition, Verlag von Julius Springer, Berlin, 1921, Volume 3, p. 676.

225. R. Lukes and V. Prelog, *Collect. Czech. Chem. Commun.* **1929**, *1*, 282.

226. R. Lukes and V. Prelog, *Collect. Czech. Chem. Commun.* **1929**, *1*, 334.

227. R. Lukes and V. Prelog, *Collect. Czech. Chem. Commun.* **1929**, *1*, 617.

228. R. Lukes and V. Prelog, *Chem. listy* **1931**, *25*, 76; 101.

229. E. Walton, *J. Chem. Soc.* **1940**, 438.

230. R. Lukes and Z. Linhartova, *Collect. Czech. Chem. Commun.* **1960**, *25*, 502.

231. W. Flitsch, R. Heidhues, H. Peters, E. Gerstmann, V. v. Weissenborn, H.-D. Bartfeld, B. Müter, and K. Gurke, *Forschungsber. des Landes Nordrhein-Westfalen*, Nr. 2220, Westdeutscher Verlag, Opladen, 1972.

232. B. M. Sheiman, L. Ya. Denisova, S. F. Dymova, and V. M. Berezovskii, *Khim. Geterotsikl. Soedin.* **1975**, 787.

233. R. Chiron and Y. Graff, *Bull. Soc. Chim. Fr.* **1967**, 3715.

234. K. E. Schulte and J. Reisch, *Arch. Pharmaz.* **1959**, *292*, 125.

235. J. B. Jones and J. M. Young, *J. Med. Chem.* **1968**, *11*, 1176.

236. W. Flitsch and V. v. Weissenborn, *Chem. Ber.* **1966**, *99*, 3444.

237. A. Gossauer, W. Hirsch, and R. Kutschan, *Angew. Chem. Int. Ed. Engl.* **1976**, *15*, 626.

238. I. A. Strakova, A. J. Strakov, E. J. Gudriniece, and N. I. Sikht, *Khim. Geterotsikl. Soedin.* **1974**, 1256.

239. N. M. Cybina, B. I. Bryantsev, N. A. Loshakova, T. V. Protopopova, S. G. Rozenberg, and A. P. Skoldinov, *Zh. Org. Khim.* **1973**, *9*, 496.

240. P. de Mayo and S. T. Reid, *Chem. Ind.* **1962**, 1576.

241. A. J. McAlees and R. McCrindle, *J. Chem. Soc., Sect. C* **1969**, 2425.

242. O. H. Wheeler and O. Rosado, In: *The Chemistry of Amides*, Interscience Publishers, London, 1970, 335.

243. J. C. Hubert, W. N. Speckamp, and H. O. Huisman, *Tetrahedron Lett.* **1972**, 4493.

244. J. C. Hubert, J. B. P. A. Wijnberg, and W. N. Speckamp, *Tetrahedron* **1975**, *31*, 1437.

245. J. Hubert, *Cyclic Imides in the Synthesis of Azasteroids and Alkaloids. The Sodium Borohydride Reduction of Cyclic Imides*, Rotterdam, 1974.

246. H. des Abbayes, *Compt. Rend. Acad. Sci., Ser. C* **1968**, *267*, 983.

247. H. des Abbayes, *Bull. Soc. Chim. Fr.* **1970**, 3661.

248. B. M. Sheiman, L. Ya. Denisova, S. F. Dymova, and V. M. Berezovskii, *Khim. Geterotsikl. Soedin.* **1973**, 22.

249. B. M. Sheiman, L. Yu. Yuzefovich, T. M. Filippova, V. G. Mairanovskii, and V. M. Berezovskii, *Khim. Geterotsikl. Soedin.* **1977**, 634.

250. L. Yu. Yuzefovich, B. M. Sheiman, V. G. Mairanovskii, and T. M. Filippova, *Khim. Geterotsikl. Soedin.* **1979**, 616.

251. R. E. Valters, *Ring-Chain Isomerism of Keto, Imino, and Cyano Carboxamides, Synopsis of Thesis of Doctoral Dissertation*, Academy of Sciences of Latvian SSR, Riga, 1975 (In Russian).

252. W. L. Meyer and N. G. Schnautz, *J. Org. Chem.* **1962**, *27*, 2011.

253. O. R. Rodig and N. J. Johnson, *J. Org. Chem.* **1969**, *34*, 1942.

254. A. Bowers, *J. Org. Chem.* **1961**, *26*, 2043.

255. W. Nagata, S. Hirai, H. Itazaki, and K. Takeda, *J. Org. Chem.* **1961**, *26*, 2413.

256. W. Nagata, S. Hirai, H. Itazaki, and K. Takeda, *Justus Liebigs Ann. Chem.* **1961**, *641*, 184.

257. A. Mondon, G. Aumann, and E. Oelrich, *Chem. Ber.* **1972**, *105*, 2025.

258. E. W. Warnhoff, W. T. Tai, and Y. C. Toong, *Can. J. Chem.* **1978**, *56*, 93.

259. S. Mincev and B. V. Aleksiev, *Compt. Rend. Acad. Bulg. Sci.* **1976**, *29*, 363.

260. S. Gabriel, *Ber. Dtsch. Chem. Ges.* **1885**, *18*, 1251.

261. S. Gabriel, *Ber. Dtsch. Chem. Ges.* **1885**, *18*, 2433.

262. S. Gabriel, *Ber. Dtsch. Chem. Ges.* **1885**, *18*, 2451.

263. A. Ruhemann, *Ber. Dtsch. Chem. Ges.* **1891**, *24*, 3964.

264. C. Graebe and F. Ullmann, *Justus Liebigs Ann. Chem.* **1896**, *291*, 8.

265. F. Sachs and A. Ludwig, *Ber. Dtsch. Chem. Ges.* **1904**, *37*, 385.

266. H. Meyer, *Monatsh. Chem.* **1907**, *28*, 1211.

267. S. Wawzonek, H. A. Laitinen, and S. J. Kwiatkowski, *J. Am. Chem. Soc.* **1944**, *66*, 830.

268. Z.-I. Horii, Ch. Iwata, and Y. Tamura, *J. Org. Chem.* **1961**, *26*, 2273.

269. V. A. Usov and J. F. Freimanis, *Khim. Geterotsikl. Soedin.* **1969**, 640.

270. Y. Kubota and T. Tasuno, *Chem. Pharm. Bull. Tokyo* **1971**, *19*, 1226.

271. C. Broquet and J. P. Genet, *Compt. Rend. Acad. Sci., Ser. C* **1967**, *265*, 117.

272. H.-R. Müller and M. Seefelder, *Justus Liebigs Ann. Chem.* **1969**, *728*, 88.

273. H. Fritz and S. Schenk, *Justus Liebigs Ann. Chem.* **1975**, 255.

274. L. Barsky, H. W. Gschwend, J. McKenna, and H. R. Rodriguez, *J. Org. Chem.* **1976**, *41*, 3651.

275. E. Breuer and S. Zbaida, *Tetrahedron* **1975**, *31*, 499.

276. M. Sekiya and Y. Terao, *Chem. Pharm. Bull. Tokyo* **1970**, *18*, 947.

277. A. M. Islam and Y. M. El-Gharby, *Egypt. J. Chem.* **1973**, *16*, 1.

278. F. Micheel and W. Flitsch, *Chem. Ber.* **1961**, *94*, 1749.

279. W. Flitsch, *Justus Liebigs Ann. Chem.* **1965**, *684*, 141.

280. M. Sekiya and Y. Terao, *J. Pharmac. Soc. Jpn.* **1968**, *88*, 1085.

281. L. Sh. Arsenievich and D. B. Stefanovich, *Glasn. Khem. Drushtva* **1970**, *35*, 209.

282. W. S. Ang and B. Halton, *Austr. J. Chem.* **1971**, *24*, 851.

283. D. Ben Ishai and Z. Inbal, *J. Org. Chem.* **1973**, *38*, 2251.

284. Y. Kanaoka and Y. Hatanaka, *Chem. Pharm. Bull. Tokyo* **1974**, *22*, 2205.

285. W. Metlesics, T. Anton, and L. H. Sternbach, *J. Org. Chem.* **1967**, *32*, 2185.

286. P. Aeberli and W. J. Houlihan, *J. Org. Chem.* **1969**, *34*, 165.

287. H. Bredereck and H. W. Vollmann, *Chem. Ber.* **1972**, *105*, 2271.

288. H.-J. W. Vollmann, K. Bredereck, and H. Bredereck, *Chem. Ber.* **1972**, *105*, 2933.

289. T. W. M. Spence and G. Tennant, *J. Chem. Soc., Perkin Trans 1* **1972**, 835.

290. G. A. Karlivans, R. E. Valters, R. B. Kampare, and V. P. Ciekure, *Khim. Geterotsikl. Soedin.* **1979**, 780.

291. A. R. Katritzky, S. Rahimj-rastgoo, and N. K. Ponkshe, *Synthesis*, **1981**, 127.

292. S. Petersen and H. Heitzer, *Justus Liebigs Ann. Chem.* **1978**, 283.

293. G. A. Karlivans and R. E. Valters, *Khim. Geterotsikl. Soedin.* **1980**, 335.

294. R. E. Valters and G. A. Karlivans, *Latv. PSR Zinat. Akad. Vestis, Kim. Ser.* **1976**, 61.

295. R. E. Valters and G. A. Karlivans, *Khim. Geterotsikl. Soedin.* **1976**, 1207.

296. G. A. Karlivans, R. E. Valters, and V. P. Ciekure, *Khim. Geterotsikl. Soedin.* **1977**, 763.

297. O. S. Anisimova, Yu. N. Sheinker, and R. E. Valters, *Khim. Geterotsikl. Soedin.* **1982**, 666.

298. G. A. Karlivans and R. E. Valters, *Zh. Org. Khim.* **1978**, *14*, 890.

299. E. Regel and K.-H. Büchel, *Justus Liebigs Ann. Chem.* **1977**, 145.

300. V. Scartoni, I. Morelli, A. Marsili, and S. Catalano, *J. Chem. Soc., Perkin Trans. 1* **1977**, 2332.

301. R. E. Valters, A. E. Bace, and S. P. Valtere, *Khim. Geterotsikl. Soedin.* **1973**, 1124.

302. K. Schenker, *Helv. Chim. Acta* **1968**, *51*, 413.

303. J. L. Moniot, D. M. Hindenlang, and M. Shamma, *J. Org. Chem.* **1979**, *44*, 4343.

304. J. R. Mednis and R. E. Valters, *Latv. PSR Zinat. Akad. Vestis, Kim. Ser.* **1976**, 441.

305. G. J. Duburs and G. J. Vanags, *Latv. PSR Zinat. Akad. Vestis, Kim. Ser.* **1962**, 125.

306. J. R. Mednis, R. E. Valters, V. E. Kampars, and O. J. Neilands, *Khim. Geterotsikl. Soedin.* **1977**, 1411.

307. J. R. Mednis and R. E. Valters, *Latv. PSR Zinat. Akad. Vestis, Kim. Ser.* **1977**, 74.

308. E. G. Howard, R. V. Lindsey, Jr., and C. W. Theobald, *J. Am. Chem. Soc.* **1959**, *81*, 4355.

309. M. Semonsky, E. Rockova, V. Zikan, B. Kakac, and V. Jelinek, *Collect. Czech. Chem. Commun.* **1968**, *33*, 2698.

310. G. B. Quistad and D. A. Lightner, *Tetrahedron Lett.* **1971**, 4417.

311. P. C. Jocelyn and A. Queen, *J. Chem. Soc.* **1957**, 4437.

312. M. Semonsky, V. Zikan, H. Skvorova, and B. Kakac, *Collect. Czech. Chem. Commun.* **1968**, *33*, 2690.

313. K. Yamada, T. Kato, and Y. A. Hirata, *J. Chem. Soc., Chem. Commun.* **1969**, 1474.

314. I. K. Kalnina and E. J. Gudriniece, *Latv. PSR Zinat. Akad. Vestis, Kim. Ser.* **1972**, 110.

315. A. Queen and A. Reipas, *J. Chem. Soc., Sect. C* **1967**, 245.

316. H. Junek, B. Hornisher, and H. Hamböck, *Monatsh. Chem.* **1969**, *100*, 503.

317. H. McKennis, E. R. Bowman, L. D. Quin, and R. C. Denney, *J. Chem. Soc., Perkin Trans. I* **1973**, 2046.

318. M. Semonsky, A. Cerny, B. Kakac, and A. Subrt, *Collect. Czech. Chem. Commun.* **1963**, *28*, 3278.

319. W. I. Awad, F. G. Baddar, M. A. Omara, and S. M. A. R. Omran, *J. Chem. Soc.* **1965**, 2040.

320. K. Yakushyin, M. Kozuka, T. Morishita, and H. Furukawa, *Chem. Pharm. Bull. Tokyo* **1981**, *29*, 2420.

321. V. Zikan, B. Kakac, J. Holubek, H. Vesela, and M. Semonsky, *Collect. Czech. Chem. Commun.* **1976**, *41*, 3113.

322. R. Lukes and J. Gorocholinskij, *Collect. Czech. Chem. Commun.* **1936**, *8*, 223.

323. R. Lukes and K. Blaha, *Chem. Listy* **1952**, *46*, 726.

324. R. Lukes, A. Fabryova, S. Dolezal, and L. Novotny, *Collect. Czech., Chem. Commun.* **1960**, *25*, 1063.

325. E. J. Cragoe, Jr. and A. M. Pietruszkiewicz, *J. Org. Chem.* **1957**, *22*, 1338.

326. E. Winterfeldt, *Chem. Ber.* **1964**, *97*, 2436.

327. E. Tagmann, E. Sury, and K. Hoffmann, *Helv. Chim. Acta* **1954**, *37*, 185.

328. A. Warshawsky and D. Ben-Ishai, *J. Heterocycl. Chem.* **1970**, *7*, 917.

329. T. A. Favorskaya, N. Yu. Baron, and S. I. Yakimovich, *Zh. Org. Khim.* **1969**, *5*, 1187.

330. R. E. Valters and V. R. Balina, *Latv. PSR Zinat. Akad. Vestis, Kim. Ser.* **1971**, 741.

331. L. Kronberg and B. Danielson, *Acta Pharm. Suec.* **1971**, *8*, 373.

332. S. Gabriel, *Ber. Dtsch. Chem. Ges.* **1887**, *20*, 2863.

333. D. E. Horning, G. Laccasse, and J. M. Muchovski, *Can. J. Chem.* **1971**, *49*, 2785.

334. R. E. Valters and V. R. Zinkovska, *Khim. Geterotsikl. Soedin.* **1972**, 1707.

335. D. W. Jones, *J. Chem. Soc., Sect. C* **1969**, 1729.

336. W. E. Kreighbaum, W. F. Kavanaugh, and W. T. Comer, *J. Heterocycl. Chem.* **1973**, *10*, 317.

337. W. E. Kreighbaum, W. F. Kavanaugh, W. T. Comer, and D. Deitchman, *J. Med. Chem.* **1972**, *15*, 1131.

338. R. Nowicki and A. Fabrycy, *Khim. Geterotsikl. Soedin.* **1976**, 1103.

339. A. Warshawsky and D. Ben-Ishai, *J. Heterocycl. Chem.* **1969**, *6*, 681.

340. D. P. Langlois and H. Wolf, *J. Am. Chem. Soc.* **1948**, *70*, 2624.

341. M. S. Newman, S. S. Gupte, and S. K. Sankarrapa, *J. Org. Chem.* **1970**, *35*, 2757.

342. K. Bowden and G. R. Taylor, *J. Chem. Soc., Sect. B* **1971**, 145.

343. K. Bowden and G. R. Taylor, *J. Chem. Soc., Sect. B* **1971**, 149.

344. V. V. Korshak, S. V. Vinogradova, S. N. Salazkin, and A. A. Kul'kov, *Zh. Org. Khim.* **1973**, *9*, 640.

345. M. V. Bhatt, K. S. Rao, and G. V. Rao, *J. Org. Chem.* **1977**, *42*, 2697.

346. E. Beska, P. Rapos, and P. Winternitz, *J. Chem. Soc., Sect. C* **1969**, 728.

347. I. K. Kalnina, E. J. Gudriniece, and E. E. Liepins, *Latv. PSR Zinat. Akad. Vestis, Kim. Ser.* **1971**, 103.

348. K. Bowden and M. P. Henry, *J. Chem. Soc., Sect. B* **1971**, 156.

349. M. S. Newman and S. Mladenovic, *J. Am. Chem. Soc.* **1966**, *88*, 4523.

350. M. S. Newman and S. S. Gupte, *J. Org. Chem.* **1970**, *35*, 4176.

351. K. Bowden and A. M. Last, *J. Chem. Soc., Chem. Commun.* **1970**, 1315.

352. K. Bowden and A. M. Last, *J. Chem. Soc., Perkin Trans. 2* **1973**, 345.

353. K. Bowden and A. M. Last, *J. Chem. Soc., Perkin Trans. 2* **1973**, 351.
354. K. Bowden and A. M. Last, *J. Chem. Soc., Perkin Trans. 2* **1973**, 358.
355. K. Bowden and F. A. El-Kaissi, *J. Chem. Soc., Perkin Trans. 2* **1977**, 526.
356. K. Bowden and F. A. El-Kaissi, *J. Chem. Soc., Perkin Trans. 2* **1977**, 1927.
357. D. R. Buckle, B. C. C. Cantello, N. J. Morgan, H. Smith, and B. A. Spicer, *J. Med. Chem.* **1975**, *18*, 733.
358. M. S. Newman and K. Naiki, *J. Org. Chem.* **1962**, *27*, 863.
359. M. S. Newman and L. K. Lala, Tetrahedron Lett. **1967**, 3267.
360. M. S. Newman and C. Courduvelis, *J. Am. Chem. Soc.* **1964**, *86*, 1893.
361. M. S. Newman and S. Hishida, *J. Am. Chem. Soc.* **1962**, *84*, 3582.
362. M. L. Bender, J. A. Reinstein, M. S. Silver, and R. Mikulak, *J. Am. Chem. Soc.* **1965**, *87*, 4545.
363. K. Bowden and G. R. Taylor, *J. Chem. Soc., Chem. Commun.* **1967**, 1112.
364. M. S. Newman and A. L. Leegwater, *J. Am. Chem. Soc.* **1968**, *90*, 4410.
365. H. D. Burrows and R. M. Topping, *J. Chem. Soc., Chem. Commun.* **1969**, 904; *J. Chem. Soc., Sect. B* **1970**, 1323.
366. K. C. Kemp and M. L. Mieth, *J. Chem. Soc., Chem. Commun.* **1969**, 1260.
367. M. V. Bhatt, G. V. Rao, and K. S. Rao, *J. Org. Chem.* **1979**, *44*, 984.
368. B. Capon, *Quart. Rev. Chem. Soc.* **1964**, *18*, 45.
369. H. Schmid, M. Hochweber, and H. v. Halban, *Helv. Chim. Acta* **1948**, *31*, 354.
370. M. S. Newman and L. K. Lala, *J. Org. Chem.* **1967**, *32*, 3225.
371. S. Wawzonek and J. H. Fossum, *J. Electrochem. Soc.* **1949**, *96*, 234.
372. E. Schmitz and I. Eichorn, In: *The Chemistry of Ether Linkage*, Interscience Publishers, London, 1967, p. 309.
373. W. Pigman and H. S. Isbell, *Adv. Carbohydr. Chem.* **1968**, *23*, 11.
374. H. S. Isbell and W. Pigman, *Adv. Carbohydr. Chem.* **1969**, *24*, 13.
375. B. Capon, *Chem. Rev.* **1969**, *69*, 407.
376. J. F. Stoddart, *Stereochemistry of Carbohydrates*, Wiley-Interscience, New York, 1971, Chapter 5.
377. B. Capon and R. B. Walker, *J. Chem. Soc., Perkin Trans 2* **1974**, 1600.
378. P. W. Wertz, J. C. Garver, and L. Anderson, *J. Am. Chem. Soc.* **1981**, *103*, 3916.
379. I. S. Trubnikov and Yu. A. Pentin, *Zh. Obshch. Khim.* **1962**, *32*, 3590.
380. H. Kosmol, K. Kieslich, and H. Gibian, *Justus Liebigs Ann. Chem.* **1968**, *711*, 42.
381. H. J. J. Loozen, E. F. Godefroi, and J. S. M. M. Besters, *J. Org. Chem.* **1975**, *40*, 892.
382. Ch. D. Hurd and W. H. Saunders, Jr., *J. Am. Chem. Soc.* **1952**, *74*, 5324.
383. L. Cottier and G. Descotes, *Bull. Soc. Chim. Fr.* **1971**, 4557.
384. W. Lütke, *Chem. Ber.* **1950**, *83*, 571.
385. Yu. A. Pentin and I. S. Trubnikov, *Dokl. Akad. Nauk SSSR* **1962**, *146*, 107.
386. H. Sterk, *Monatsh. Chem.* **1968**, *99*, 2107.
387. J. E. Whitting and J. T. Edward, *Can. J. Chem.* **1971**, *49*, 3799.
388. I. D. Kalikhman, O. B. Bannikova, V. N. Elokhina, A. S. Nakhmanovich, and M. G. Voronkov, *Khim. Geterotsikl. Soedin.* **1982**, 473.
389. L. Cottier and G. Descotes, *Bull. Soc. Chim. Fr.* **1973**, 2451.
390. M. Cazaux and B. de Jeso, *Compt. Rend. Acad. Sci., Ser. C* **1980**, *290*, 49.
391. J. T. Edward, M. Kaufman, R. K. Wojtowski, D. M. S. Wheeler, and T. M. Barrett, *Can. J. Chem.* **1973**, *51*, 1610.
392. J. A. Zalikowski, K. E. Gilbert, and W. T. Borden, *J. Org. Chem.* **1980**, *45*, 346.
393. R. E. Lutz, J. A. Freek, and A. S. Murphey, *J. Am. Chem. Soc.* **1948**, *70*, 2015.
394. R. E. Lutz and R. H. Jordan, *J. Am. Chem. Soc.* **1949**, *71*, 996.
395. R. E. Lutz and R. S. Murphey, *J. Am. Chem. Soc.* **1949**, *71*, 478.
396. N. H. Cromwell and K.-Ch. Tsou, *J. Am. Chem. Soc.* **1949**, *71*, 993.

397. R. E. Lutz and J. W. Baker, *J. Org. Chem.* **1956**, *21*, 49.
398. B. M. Mikhailov and A. N. Makarova, *Zh. Obshch. Khim.* **1957**, *27*, 2526.
399. R. E. Lutz and C. E. Griffin, *J. Org. Chem.* **1960**, *25*, 928.
400. Ch. Herzig and J. Gasteiger, *Chem. Ber.* **1981**, *114*, 2348.
401. E. Bauer and H. Berg, *Roczn. Chem.* **1961**, *35*, 329.
402. H.-P. Rettig and H. Berg, *Z. Phys. Chem. (Leipzig)* **1963**, *222*, 193.
403. J. H. Day and A. Joachim, *J. Org. Chem.* **1965**, *30*, 4107.
404. K. D. McMurtrey and G. D. Daves, Jr., *J. Org. Chem.* **1970**, *35*, 4252.
405. J. Harron, R. A. McClelland, Ch. Thankachan, and Th. T. Tidwell, *J. Org. Chem.* **1981**, *46*, 903.
406. L. A. Pavlova, *Zh. Obshch. Khim.* **1959**, *29*, 1588.
407. L. A. Pavlova and I. V. Samartseva, *Zh. Org. Khim.* **1966**, *2*, 1712.
408. V. S. Sorokina and L. A. Pavlova, *Zh. Org. Khim.* **1973**, *9*, 1970.
409. J. G. Smith and R. T. Wikman, *Tetrahedron* **1974**, *30*, 2603.
410. D. A. Oparin, T. G. Melent'eva, and L. A. Pavlova, *Zh. Org. Khim.* **1980**, *16*, 1530.
411. E. D. Venus-Danilova, L. A. Pavlova, and A. Fabrycy, *Vestn. Leningr. Univ.* **1956**, N16 (3), 117.
412. R. Howe, B. C. Rao, and H. Heineker, *J. Chem. Soc., Sect. C* **1967**, 2510.
413. M. C. Sacquet, B. Graffe, and P. Maitte, *Tetrahedron Lett.* **1972**, 4453.
414. C. Perrot and E. Cerrutti, *Bull. Soc. Chim. Fr.* **1974**, 2591.
415. W. Treibs and R. Schöllner, *Chem. Ber.* **1961**, *94*, 2983.
416. M. Moreau, R. Quagliaro, R. Longeray, and J. Dreux, *Bull. Soc. Chim. Fr.* **1969**, 1362.
417. A. Rykowski, P. Nantka-Namirski, *Polish J. Pharmacol. Pharm.* **1973**, *25*, 455.
418. T. Tanaka and I. Iijima, *Tetrahedron* **1973**, *29*, 1285.
419. H. Obara and J. Onodera, *Bull. Chem. Soc. Jpn.* **1968**, *41*, 2798.
420. H. Obara and J. Onodera, *Bull. Chem. Soc. Jpn.* **1969**, *42*, 3345.
421. J. Borbely, V. Szabo, and P. Sohar, *Tetrahedron* **1981**, *37*, 2307.
422. E. Morera and G. Ortar, *Tetrahedron Lett.* **1981**, *22*, 1273.
423. P. J. Wittek and T. M. Harris, *J. Am. Chem. Soc.* **1973**, *95*, 6865.
424. I. K. Korobitsyna, Chen-li In, and Yu. K. Yur'ev, *Zh. Obshch. Khim.* **1961**, *31*, 2548.
425. A. Schönberg and E. Singer, *Tetrahedron* **1978**, *34*, 1285.
426. A. Schönberg, E. Singer, G. A. Hoyer, and D. Rosenberg, *Chem. Ber.* **1977**, *110*, 3954.
427. H. I. Roth and W. Kok, *Arch. Pharmaz.* **1976**, *309*, 81.
428. J.-P. Poupelin, G. Saint-Ruf, J.-C. Perche, J. C. Roussey, B. Laude, G. Narcisse, F. Bakri-Logeais, and F. Hubert, *Eur. J. Med. Chem.* **1980**, *15*, 253.
429. W. Schäfer and H. Schlude, *Tetrahedron Lett.* **1968**, 2161.
430. W. Schäfer and H. Schlude, *Tetrahedron* **1971**, *27*, 4721.
431. W. Schäfer, I. Geyer, and H. Schlude, *Tetrahedron* **1972**, *28*, 3811.
432. A. Butenandt, E. Biekert, and W. Schäfer, *Justus Liebigs Ann. Chem.* **1960**, *632*, 143.
433. V. I. Visotskii and M. N. Tilichenko, *Khim. Geterotsikl. Soedin.* **1971**, 299.
434. V. I. Visotskii, N. V. Vershinina, M. N. Tilichenko, V. V. Isakov, and T. M. Belokon', *Zh. Org. Khim.* **1973**, *9*, 2427.
435. V. I. Visotskii, N. V. Vershinina, and M. N. Tilichenko, *Khim. Geterotsikl. Soedin.* **1974**, 746.
436. N. V. Vershinina, V. I. Visotskii, L. M. Eremeeva, V. A. Kaminskii, and M. N. Tilichenko, *Khim. Geterotsikl. Soedin.* **1977**, 1315.
437. V. I. Visotskii, N. V. Vershinina, and M. N. Tilichenko, *Khim. Geterotsikl. Soedin.* **1975**, 898.
438. V. I. Visotskii, G. V. Pavel', K. G. Chupranova, V. A. Shchukin, and M. N. Tilichenko, *Zh. Obshch. Khim.* **1979**, *49*, 1952.
439. G. L. Lyle, J. J. Dziark, J. Connor, and C. S. Huber, *Tetrahedron* **1973**, *29*, 4039.

440. D. F. Martin, M. Shamma, and W. C. Fernelius, *J. Am. Chem. Soc.* **1958**, *80*, 5851.

441. S. N. Ananchenko, I. V. Berezin, and I. V. Torgov, *Izv. Akad. Nauk SSSR, Otd. Khim. Nauk* **1960**, 1644.

442. G. V. Kondrat'eva, G. A. Kogan, and S. I. Zav'yalov, *Izv. Akad. Nauk SSSR, Otd. Khim. Nauk* **1962**, 1441.

443. I. E. Lielbriedis and E. J. Gudriniece, *Latv. PSR Zinat. Akad. Vestis, Kim. Ser.* **1966**, 684.

444. J. K. Lemba and I. E. Lielbriedis, *Latv. PSR Zinat. Akad. Vestis, Kim. Ser.* **1973**, 598.

445. J. K. Lemba and I. E. Lielbriedis, *Latv. PSR Zinat. Akad. Vestis, Kim. Ser.* **1975**, 589.

446. J. K. Lemba, E. V. Blums, and I. E. Lielbriedis, *Latv. PSR Zinat. Akad. Vestis, Kim. Ser.* **1976**, 207.

447. I. E. Lielbriedis and E. J. Gudriniece, *Latv. PSR Zinat. Akad. Vestis, Kim. Ser.* **1965**, 738.

448. I. E. Lielbriedis and J. K. Lemba, *Latv. PSR Zinat. Akad. Vestis, Kim. Ser.* **1975**, 57.

449. J. K. Lemba and I. E. Lielbriedis, *Khim. Geterotsikl. Soedin.* **1975**, 324.

450. R. Goto, Y. Miyagi, and M. Inokawa, *Bull. Chem. Soc. Jpn.* **1963**, *36*, 147.

451. Y. Miyagi and R. Goto, *Bull. Chem. Soc. Jpn.* **1963**, *36*, 650.

452. Y. Miyagi and R. Goto, *Bull. Chem. Soc. Jpn.* **1963**, *36*, 961.

453. Y. Miyagi, *Bull. Chem. Soc. Jpn.* **1964**, *37*, 12.

454. S. Kimura, Y. Miyagi, and R. Goto, *Bull. Chem. Soc. Jpn.* **1966**, *39*, 1333.

455. Y. Miyagi, S. Kimura, and R. Goto, *Bull. Chem. Soc. Jpn.* **1968**, *41*, 2927.

456. L. Horner and F. Maurer, *Chem. Ber.* **1968**, *101*, 1783.

457. L. Horner and F. Maurer, *Justus Liebigs Ann. Chem.* **1970**, *736*, 145.

458. M. Poje and K. Balenovic, *J. Heterocycl. Chem.* **1979**, *16*, 417.

459. P. M. Hardy, A. C. Nicholls, and H. N. Rydon, *J. Chem. Soc., Perkin Trans 2* **1972**, 2270.

460. E. B. Whipple and M. Ruta, *J. Org. Chem.* **1974**, *39*, 1666.

461. R. S. McDonald and E. V. Martin, *Can. J. Chem.* **1979**, *57*, 506.

462. O. H. Mattsson and C. A. Wachtmeister, *Tetrahedron Lett.* **1967**, 1855.

463. A. C. Cope, M. A. McKervey, and N. M. Weinshenker, *J. Org. Chem.* **1969**, *34*, 2229.

464. W. C. Agosta, *J. Am. Chem. Soc.* **1967**, *89*, 3505.

465. R. S. Egan, J. R. Martin, Th. J. Perun, and L. A. Mitscher, *J. Am. Chem. Soc.* **1975**, *97*, 4578.

466. K. Dornberger, H. Thrum, and G. Engelhardt, *Tetrahedron Lett.* **1976**, 4469.

467. J. Castells and A. Colombo, *J. Chem. Soc., Chem. Commun.* **1969**, 1062.

468. R. Escale, F. Petrus, and J. Verducci, *Bull. Soc. Chim. Fr.* **1974**, 725.

469. R. Escale and J. Verducci, *Bull. Soc. Chim. Fr.* **1974**, 1203.

470. R. Jacquier, F. Petrus, J. Verducci, and Y. Vidal, *Tetrahedron Lett.* **1974**, 387.

*471. R. Escale, R. Jacquier, B. Ly, F. Petrus, and J. Verducci, *Tetrahedron* **1976**, *32*, 1369.

472. C. K. Ingold, *J. Chem. Soc.* **1921**, *119*, 305.

*473. L. S. Geita, I. E. Dalberga, and A. K. Grinvalde, *Latv. PSR Zinat. Akad. Vestis, Kim. Ser.* **1976**, 704.

474. E. J. Ozola and A. K. Arens, *Latv. PSR Zinat. Akad. Vestis, Kim. Ser.* **1970**, 457.

475. J. A. Schutyser and F. C. de Schryver, *Chem. Ind.* **1972**, 465.

476. J. Perronet, P. Girault, and J. P. Demoute, *J. Heterocycl. Chem.* 1980, *17*, 727.

477. A. T. Nielsen and T. G. Archibald, *J. Org. Chem.* **1969**, *34*, 1470.

478. H. Schechter, D. E. Ley, and L. Zeldin, *J. Am. Chem. Soc.* **1952**, *74*, 3664.

479. E. B. Hodge and R. Abbott, *J. Org. Chem.* **1962**, *27*, 2254.

480. A. Albert, In: *Physical Methods in Heterocyclic Chemistry*, Ed. A. R. Katritzky, Academic Press, New York, 1963, Chapter 1.

481. A. Albert, *Angew. Chem.* **1967**, *79*, 913.

482. A. Albert and W. L. F. Armarego, *Adv. Heterocycl. Chem.* **1965**, *4*, 1.

483. D. D. Perrin, *Adv. Heterocycl. Chem.* **1965**, *4*, 43.

* See also additions in proof on page 168.

484. A. Albert, *Adv. Heterocycl. Chem.* **1976**, *20*, 117.
485. D. Beke, *Adv. Heterocycl. Chem.* **1963**, *1*, 167.
*486. D. Beke and C. Szantay, *Justus Liebigs Ann. Chem.* **1961**, *640*, 127.
487. N. E. Grigor'eva, M. M. Fetisova, and G. A. Fetisov, *Zh. Org. Khim.* **1963**, *3*, 2167.
488. J. H. Blanch and K. Fretheim, *J. Chem. Soc., Sect. C* **1971**, 1892.
489. A. N. Kost, S. P. Gromov, and R. S. Sagitullin, *Tetrahedron* **1981**, *37*, 3423.
490. N. N. Vereshchagina, I. Ya. Postovskii, and S. L. Mertsalov, *Khim. Geterotsikl. Soedin.* **1967**, 1096.
491. I. Ya. Postovskii and N. N. Vereshchagina, *Khim. Geterotsikl. Soedin,* **1967**, 944.
492. B. V. Golomolzin and I. Ya. Postovskii, *Khim. Geterotsikl. Soedin.* **1970**, 855.
493. B. V. Golomolzin and I. Ya. Postovskii, *Khim. Geterotsikl. Soedin.* **1971**, 133.
494. I. Ya. Postovskii and B. V. Golomolzin, *Khim. Geterotsikl. Soedin.* **1970**, 100.
495. Y. Inoue and D. D. Perrin, *J. Chem. Soc.* **1962**, 2648.
496. D. D. Perrin, *J. Chem. Soc.* **1962**, 645.
497. W. L. F. Armarego, *J. Chem. Soc.* **1962**, 4094.
498. E. Öhler and U. Schmidt, *Chem. Ber.* **1975**, *108*, 2907.
499. H. Poisel and U. Schmidt, *Chem. Ber.* **1975**, *108*, 2917.
*500. J. Häusler and U. Schmidt, *Chem. Ber.* **1974**, *107*, 2804.
501. G. J. Vanags and J. J. Dregeris, *Latv. PSR Zinat. Akad. Vestis, Kim. Ser.* **1964**, 559.
502. E. M. Danilova, V. V. Perekalin, and T. Ya. Paperno, *Zh. Org. Khim.* **1967**, *3*, 1860.
503. T. A. Wnuk and P. Kovacic, *J. Am. Chem. Soc.* **1975**, *97*, 5807.
504. D. Thon and W. Schneider, *Chem. Ber.* **1976**, *109*, 2743.
505. H. Stetter, P. Tacke, and J. Gärtner, *Chem. Ber.* **1964**, *97*, 3480.
506. S. J. Padegimas and P. Kovacic, *J. Org. Chem.* **1972**, *37*, 2672.
507. H. Quast and P. Eckert, *Justus Liebigs Ann. Chem.* **1974**, 1727.
508. W. Theilacker and H. Kalenda, *Justus Liebigs Ann. Chem.* **1953**, *584*, 87.
509. L. A. Pavlova and I. V. Samartseva, *Zh. Org. Khim.* **1968**, *4*, 2235.
510. J. D. White and N. E. Mann, *Adv. Heterocycl. Chem.* **1969**, *10*, 113.
511. I. V. Samartseva and L. A. Pavlova, *Zh. Org. Khim.* **1972**, *8*, 1964.
512. E. Braudeau, S. David, and J. Fischer, *Tetrahedron* **1974**, *30*, 1445.
513. T. H. C. Bristov, H. E. Toster, and M. Hooper, *J. Chem. Soc., Chem. Commun.* **1974**, 677.
514. O. Buchardt, J. Becher, and C. Lohse, *Acta Chem. Scand.* **1966**, *20*, 2467.
515. O. Buchardt, B. Jensen, and I. K. Larsen, *Acta Chem. Scand.* **1967**, *21*, 1841.
516. J. Streith, H. K. Darrah, and M. Weil, *Tetrahedron Lett.* **1966**, 5555.
517. K. Takayama, K. Harano, and T. Taguchi, *J. Pharm. Soc. Jpn.* **1974**, *94*, 548.
518. O. Buchardt, A. M. Duffield, and C. Djerassi, *Acta Chem. Scand.* **1968**, *22*, 2329.
519. C. W. Rees and C. R. Sabet, *J. Chem. Soc.* **1965**, 870.
520. G. Bianchi, A. Gamba-Invernizzi, and R. Gandolfi, *J. Chem. Soc., Perkin Trans. 1* **1974**, 1757.
521. T. Eicher and J. L. Weber, *Tetrahedron Lett.* **1973**, 1541.
522. N. Chaterjie, R. Shapiro, Sih-Gwan Quo, and R. A. Stephani, *Tetrahedron. Lett.* **1975**, 2535.
523. M. Weigele, J. P. Tengi, S. de Bernardo, R. Czajkovski, and W. Leimgruber, *J. Org. Chem.* **1976**, *41*, 388.
524. T. S. Safonova and L. G. Levkovskaya, *Khim. Geterotsikl. Soedin.* **1968**, 997.
525. L. G. Levkovskaya and T. S. Safonova, *Khim. Geterotsikl. Soedin.* **1969**, 970.
526. T. S. Safonova and L. G. Levkovskaya, *Khim. Geterotsikl. Soedin.* **1970**, 1096.
527. T. S. Safonova and L. G. Levkovskaya, *Khim. Geterotsikl. Soedin.* **1971**, 78.
528. T. S. Safonova, L. G. Levkovskaya, V. V. Makeeva, and T. F. Vlasova, *Khim. Geterotsikl. Soedin.* **1973**, 1262.

* See also additions in proof on page 168.

529. M. P. Nemeryuck and T. S. Safonova, *Khim. Geterotsikl. Soedin.* **1967**, 486.

530. T. S. Safonova and M. P. Nemeryuck, *Khim. Geterotsikl. Soedin.* **1968**, 735.

531. T. S. Safonova, M. P. Nemeryuck, and G. P. Syrova, *Khim. Geterotsikl. Soedin.* **1970**, 1423.

532. M. P. Nemeryuck and T. S. Safonova, *Khim. Geterotsikl. Soedin.* **1971**, 73.

533. T. S. Safonova, J. N. Sheinker, M. P. Nemeryuck, and G. P. Syrova, *Tetrahedron* **1971**, *27*, 5455.

534. T. S. Safonova and L. A. Mishkina, *Khim. Geterotsikl. Soedin.* **1970**, 1092.

535. L. A. Mishkina and T. S. Safonova, *Khim. Geterotsikl. Soedin.* **1970**, 1101.

536. E. M. Peresleni, L. A. Mishkina, K. F. Turchin, T. Ya. Filipenko, T. S. Safonova, and Yu. N. Sheinker, *Khim. Geterotsikl. Soedin.* **1979**, 114.

537. B. V. Unkovskii, L. A. Ignatova, M. M. Donskaya, and M. G. Zaitseva, *Problems of Organic Synthesis*, Nauka, Moscow, 1965, p. 202 (In Russian).

538. B. V. Unkovskii, L. A. Ignatova, M. G. Zaitseva, and M. M. Donskaya, *Khim. Geterotsikl. Soedin.* **1965**, 586.

539. B. V. Unkovskii, L. A. Ignatova, and M. G. Zaitseva, *Khim. Geterotsikl. Soedin.* **1969**, 889.

540. G. I. Ovechkina, L. A. Ignatova, M. A. Ratomskaya, and B. V. Unkovskii, *Khim. Geterotsikl. Soedin.* **1971**, 1258.

541. G. I. Ovechkina, L. A. Ignatova, and B. V. Unkovskii, *Zh. Vses. Khim. Ova.* **1971**, *16*, 585.

542. W. Metlesics, G. Silverman, V. Toome, and L. H. Sternbach, *J. Org. Chem.* **1966**, *31*, 1007.

543. Y. Sato, T. Tanaka, and T. Nagasaki, *J. Pharm. Soc. Jpn.* **1970**, *90*, 629.

544. R. Y. Ning, I. Douvan, and L. H. Sternbach, *J. Org. Chem.* **1970**, *35*, 2243.

545. S. Petersen, H. Heitzer, and L. Born, *Justus Liebigs Ann. Chem.* **1974**, 2003.

546. A. K. Arens, I. K. Jurgevica, F. A. Grunsbergs, and I. P. Lencbergs, *Latv. PSR Zinat. Akad. Vestis, Kim. Ser.* **1970**, 323.

547. R. Shapiro and N. Chaterjie, *J. Org. Chem.* **1970**, *35*, 447.

548. Dz. V. Bite, S. P. Valtere, and A. K. Arens, *Latv. PSR Zinat. Akad. Vestis, Kim. Ser.* **1969**, 109.

549. V. J. Grinsteins, A. E. Sausins, and S. P. Valtere, *Latv. PSR Zinat. Akad. Vestis, Kim. Ser.* **1972**, 441.

550. R. V. Lamon, W. J. Humphlett, and W. P. Blum, *J. Heterocycl. Chem.* **1967**, *4*, 349.

551. C. M. Roussel, R. Gallo, M. Chanon, and J. Metzger, *Bull. Soc. Chim. Fr.* **1971**, 1902.

552. J. L. Garraway, *J. Chem. Soc.* **1964**, 4004.

553. B. V. Unkovskii, L. A. Ignatova, M. M. Donskaya, L. M. Andreev, and L. L. Khoroshilova, *Khim. Geterotsikl. Soedin.* **1968**, 991.

554. B. Oddo and Q. Mingoia, *Gazz. Chim. Ital.* **1927**, *57*, 465.

555. W. Asker, A. Mustafa, K. Hilmy, and M. A. Alam, *J. Org. Chem.* **1958**, *23*, 2002.

556. R. A. Abramovitch, E. M. Smith, M. Humber, B. Purtschert, P. C. Srinavasan, and G. M. Singer, *J. Chem. Soc., Perkin Trans. 1*, **1974**, 2589.

557. A. Mustafa and M. K. Hilmy, *J. Chem. Soc.* **1952**, 1339.

558. H. Watanabe, Ch.-L. Mao, I. T. Barnish, and Ch. R. Hauser, *J. Org. Chem.* **1969**, *34*, 919.

559. G. Heyes, G. Holt, and A. Lewis, *J. Chem. Soc., Perkin Trans. 1*, **1972**, 2351.

560. D. E. Balode and R. E. Valters, *Latv. PSR Zinat. Akad. Vestis, Kim. Ser.* **1980**, 227.

561. F. Ramirez and A. Paul, *J. Am. Chem. Soc.* **1955**, *77*, 3337.

562. A. Nakamura and S. Kamiya, *Chem. Pharm. Bull. Tokyo* **1968**, *16*, 1466.

563. A. Nakamura and S. Kamiya, *Chem. Pharm. Bull. Tokyo* **1969**, *17*, 425.

564. W. Metlesics, T. Anton, M. Chaykovsky, V. Toome, and L. H. Sternbach, *J. Org. Chem.* **1968**, *33*, 2874.

565. P. Aeberli, P. Eden, J. H. Gogerty, W. J. Houlihan, and G. Penberthy, *J. Med. Chem.* **1975**, *18*, 177.

566. W. J. Houlihan and V. A. Parrino, *J. Heterocycl. Chem.* **1981**, *18*, 1549.

567. P. M. Kochergin and M. N. Shchukina, *Zh. Obshch. Khim.* **1956**, *26*, 2905.
568. P. M. Kochergin, *Zh. Obshch. Khim.* **1960**, *30*, 1529.
569. I. A. Mazur and P. M. Kochergin, *Khim. Geterotsikl. Soedin.* **1970**, 508.
570. I. A. Mazur and P. M. Kochergin, *Khim. Geterotsikl. Soedin.* **1970**, 512.
571. R. S. Schadbolt, *J. Chem. Soc., Sect. C* **1971**, 1667.
572. L. M. Alekseeva, E. M. Peresleni, Yu. N. Sheinker, P. M. Kochergin, A. N. Krasovskii, and B. V. Kurmaz, *Khim. Geterotsikl. Soedin.* **1972**, 1125.
573. H. Singh and S. Singh, *Tetrahedron Lett.* **1970**, 585.
574. H. Singh and S. Singh, *Indian J. Chem.* **1971**, *9*, 918.
575. R. S. Schadbolt, *J. Chem. Soc., Sect. C* **1971**, 1669.
576. A. N. Krasovskii and P. M. Kochergin, *Khim. Geterotsikl. Soedin.* **1967**, 899.
577. A. E. Alper and A. Taurins, *Can. J. Chem.* **1967**, *45*, 2903.
578. H. Ogura, T. Itoh, and Y. Shimada, *Chem. Pharm. Bull. Tokyo* **1968**, *16*, 2167.
579. H. Ogura, T. Itoh, and K. Kikuchi, *J. Heterocycl. Chem.* **1969**, *6*, 797.
580. H. Alper, *J. Chem. Soc., Chem. Commun.* **1970**, 383.
581. H. Alper and A. E. Alper, *J. Org. Chem.* **1970**, *35*, 835.
582. H. Alper, E. C. H. Keung, and R. A. Partis, *J. Org. Chem.* **1971**, *36*, 1352.
583. H. Singh and S. Singh, *Indian J. Chem.* **1973**, *11*, 311.
584. E. G. Knysh, A. N. Krasovskii, and P. M. Kochergin, *Khim. Geterotsikl. Soedin.* **1971**, 1128.
585. E. G. Knysh, A. N. Krasovskii, and P. M. Kochergin, *Khim. Geterotsikl. Soedin.* **1972**, 25.
586. M. I. Yurchenko, B. V. Kurmaz, and P. M. Kochergin, *Khim. Geterotsikl. Soedin.* **1972**, 996.
587. M. I. Yurchenko, P. M. Kochergin, and A. N. Krasovskii, *Khim. Geterotsikl. Soedin.* **1974**, 693.
588. H. Alper and B. H. Lipshutz, *J. Org. Chem.* **1973**, *38*, 3742.
589. P. M. Kochergin, A. M. Tsyganova, and L. M. Viktorova, *Khim. Geterotsikl. Soedin.* **1967**, 93.
590. P. M. Kochergin, A. M. Tsyganova, L. M. Viktorova, and E. M. Peresleni, *Khimiya Geterotsiklicheskikh Soedinenii, Sb. 1, Azotsoderzhashchie Geterotsikly*, Zinatne, Riga, 1967, p. 126.
591. G. Toth, G. Hornyak, and K. Lempert, *Chem. Ber.*, **1977**, *110*, 1492.
592. O. S. Anisimova, Yu. N. Sheinker, E. M. Peresleni, P. M. Kochergin, and A. N. Krasovskii, *Khim. Geterotsikl. Soedin.* **1976**, 676.
593. H. Alper, A. E. Alper, and A. Taurins, *J. Chem. Education* **1970**, *47*, 222.
594. H. Singh, S. Singh, and K. B. Lal, *Chem. Ind.* **1972**, 255.
595. E. van Loock, G. l'Abbe, and G. Smets, *J. Org. Chem.* **1971**, *36*, 2520.
596. E. van Loock, G. l'Abbe, and G. Smets, *Tetrahedron* **1972**, *28*, 3061.
597. C. E. Olsen and Ch. Pedersen, *Acta Chem. Scand.* **1973**, *27*, 2271.
598. C. E. Olsen and Ch. Pedersen, *Acta Chem. Scand.* **1973**, *27*, 2279.
599. J. F. McGarrity, *J. Chem. Soc., Chem. Commun.* **1974**, 558.
600. S. Treppendahl and P. Jakobsen, *Acta Chem. Scand.* **1980**, *B34*, 303.
601. V. A. Kaminskii and M. N. Tilichenko, *Khim. Geterotsikl. Soedin.* **1971**, 1149.
602. L. N. Donchak, V. A. Kaminskii, and M. N. Tilichenko, *Khim. Geterotsikl. Soedin.* **1975**, 239.
603. A. Belly, F. Petrus, and J. Verducci, *Bull. Soc. Chim. Fr.* **1973**, 1395.
604. R. Jacquier, L. L. Olive, C. Petrus, and F. Petrus, *Tetrahedron Lett.* **1975**, 2337.
605. R. Fusco and P. D. Croce, *Tetrahedron Lett.* **1970**, 3061.
606. Ch. Hedbom and E. Helgstrand, *Acta Chem. Scand.* **1970**, *24*, 1744.
607. S. I. Yakimovich and V. N. Nikolaev, *Zh. Org. Khim.* **1979**, *15*, 1100.
608. V. G. Yusupov, S. I. Yakimovich, S. D. Nasirdinov, and N. A. Parpiev, *Zh. Org. Khim.* **1980**, *16*, 415.

609. S. I. Yakimovich, V. N. Nikolaev, and T. I. Temnikova, *Zh. Org. Khim.* **1980**, *16*, 2235.
610. S. I. Yakimovich, V. N. Nikolaev, and E. Yu. Kutsenko, *Zh. Org. Khim.* **1982**, *18*, 762.
*611. S. I. Yakimovich, V. N. Nikolaev, and E. Yu. Kutsenko, *Zh. Org. Khim.* **1983**, *19*, 2333.
612. S. I. Selivanov, R. A. Bogatkin, and B. A. Ershov, *Zh. Org. Khim.* **1981**, *17*, 886.
*613. S. I. Selivanov, R. A. Bogatkin, and B. A. Ershov, *Zh. Org. Khim.*, **1982**, *18*, 909.
614. A. Amer and H. Zimmer, *J. Heterocycl. Chem.* **1981**, *18*, 1625.
615. J. Parrick, R. Wilcox, and A. H. Kelly, *J. Chem. Soc., Perkin Trans 1* **1980**, 132.
616. A. Babadjamian and J. Metzger, *J. Heterocycl. Chem.* **1975**, *12*, 643.
617. A. R. Gainsford, R. D. Pizer, A. M. Sargeson, and P. O. Whimp, *J. Am. Chem. Soc.* **1981**, *103*, 792.
618. D. Cram, In: *Steric Effects in Organic Chemistry*, Ed. M. S. Newman, John Wiley and Sons, New York, 1956, Chapter 5.
619. T. C. Bruice and S. J. Benkovic, *Bioorganic Mechanisms*, W. A. Benjamin, Inc., New York, 1966.
620. L. V. Pavlova and F. Yu. Rachinskii, *Uspekhi Khim.* **1968**, *37*, 1369.
621. D. V. Banthorpe, In: *The Chemistry of Amino Group*, Ed. S. Patai, Interscience Publishers, London, 1968, p. 585.
622. W. P. Jencks, *Catalysis in Chemistry and Enzymology*, McGraw-Hill Book Company, New York, 1969, Chapters 1 and 10.
623. A. J. Kirby, *Compr. Chem. Kinet.* **1972**, *10*, 104.
624. B. Capon, A. K. Ghosh, and D. McL. A. Grieve, *Acc. Chem. Res.* **1981**, *14*, 306.
625. B. Helferich and H. Liesen, *Chem. Ber.* **1950**, *83*, 567.
626. J. Hine, D. Ricard, and R. Perz, *J. Org. Chem.* **1973**, *38*, 110.
627. G. A. Rogers and Th. C. Bruice, *J. Am. Chem. Soc.* **1973**, *95*, 4452.
628. G. A. Rogers, and Th. C. Bruice, *J. Am. Chem. Soc.* **1974**, *96*, 2481.
629. T. Wieland and H. Hornig, *Justus Liebigs Ann. Chem.* **1956**, *600*, 12.
630. R. B. Martin, S. Lowey, E. L. Elson, and J. T. Edsall, *J. Am. Chem. Soc.* **1959**, *81*, 5089.
631. J. S. Harding and L. N. Oven, *J. Chem. Soc.* **1954**, 1536.
632. E. E. van Tamelen, *J. Am. Chem. Soc.* **1951**, *73*, 5773.
633. G. Fodor and J. Kiss, *J. Chem. Soc.* **1952**, 1589.
634. L. H. Welsh, *J. Org. Chem.* **1967**, *32*, 119.
635. T. Wieland and E. Bokelmann, *Justus Liebigs Ann. Chem.* **1952**, *576*, 20.
636. R. B. Martin and A. Parcell, *J. Am. Chem. Soc.* **1961**, *83*, 4835.
637. V. I. Minkin, L. P. Olekhnovich, and Yu. A. Zhdanov, *Zh. Vses. Khim. Ova.* **1977**, *22*, 274.
638. V. I. Minkin, L. P. Olekhnovich, and Yu. A. Zhdanov, *Molecular Design of Tautomeric Systems*, Rostov on Don University Publishing House, Rostov on Don, 1977 (In Russian).
639. V. I. Minkin, L. P. Olekhnovich, and Yu. A. Zhdanov, *Acc. Chem. Res.* **1981**, *14*, 210.
640. L. P. Olekhnovich, N. G. Furmanova, V. I. Minkin, Yu. T. Struchkov, O. E. Kompan, Z. N. Budarina, I. A. Yudilevich, and O. V. Eryuzheva, *Zh. Org. Khim.* **1982**, *18*, 465.
641. N. G. Furmanova, L. P. Olekhnovich, V. I. Minkin, Yu. T. Struchkov, O. E. Kompan, Z. N. Budarina, V. P. Metlushenko, and O. V. Eryuzheva, *Zh. Org. Khim.* **1982**, *18*, 474.
642. L. P. Olekhnovich, I. E. Mikhailov, V. I. Minkin, N. G. Furmanova, O. E. Kompan, Yu. T. Struchkov, and A. V. Lukash, *Zh. Org. Khim.* **1982**, *18*, 484.
643. K. Stich and H. G. Leeman, *Helv. Chim. Acta* **1963**, *46*, 1151.
644. V. K. Antonov, A. M. Shkrob, and M. M. Shemyakin, *Zh. Obshch. Khim.* **1965**, *35*, 1380.
645. A. M. Shkrob, Yu. I. Krylova, V. K. Antonov, and M. M. Shemyakin, *Zh. Obshch. Khim.* **1965**, *35*, 1389.
646. D. M. Wrinch, *Nature* **1936**, *137*, 411.

* See also additions in proof on page 168.

647. D. M. Wrinch, *Nature* **1936**, *138*, 241.
648. D. Wrinch, *Nature* **1963**, *199*, 564.
649. V. K. Antonov, G. A. Ravdel', and M. M. Shemyakin, *Chimia* **1960**, *14*, 374.
650. G. A. Ravdel', N. A. Krit, L. A. Shchukina, and M. M. Shemyakin, *Dokl. Akad. Nauk SSSR* **1961**, *137*, 1377.
651. A. Hofmann, A. J. Frey, and H. Ott, *Experientia* **1961**, *17*, 206.
652. A. Hofmann, A. J. Frey, H. Ott, and I. Rutshmann, *Zh. Vses. Khim. Ova.* **1962**, *7*, 468.
653. A. Hofmann, H. Ott, R. Griot, P. A. Stadler, and A. J. Frey, *Helv. Chim. Acta* **1963**, *46*, 2306.
654. H. Ott, A. J. Frey, and A. Hofmann, *Tetrahedron* **1963**, *19*, 1675.
655. M. M. Shemyakin and V. K. Antonov, *Pure Appl. Chem.* **1964**, *9*, 75.
656. M. M. Shemyakin, V. K. Antonov, A. M. Shkrob, V. I. Shchelokov, and Z. E. Agadzhanyan, *Tetrahedron* **1965**, *21*, 3537.
657. V. K. Antonov, Ts. E. Agadzhanyan, T. R. Telesnina, and M. M. Shemyakin, *Zh. Obshch. Khim.* **1965**, *35*, 2231.
658. V. K. Antonov, V. I. Shchelokov, and M. M. Shemyakin, *Zh. Obshch. Khim.* **1965**, *35*, 2239.
659. R. G. Griot and A. J. Frey, *Tetrahedron* **1963**, *19*, 1661.
660. V. K. Antonov, A. M. Shkrob, and M. M. Shemyakin, *Tetrahedron Lett.* **1963**, 439.
661. V. K. Antonov, A. M. Shkrob, V. I. Shchelokov, and M. M. Shemyakin, *Tetrahedron Lett.* **1963**, 1353.
662. V. K. Antonov, A. M. Shkrob, and M. M. Shemyakin, *Zh. Obshch. Khim.* **1967**, *37*, 2225.
663. G. I. Glover, R. B. Smith, and H. Rapoport, *J. Am. Chem. Soc.* **1965**, *87*, 2003.
664. A. M. Shkrob, Yu. I. Krylova, V. K. Antonov, and M. M. Shemyakin, *Tetrahedron Lett.* **1967**, 2701.
665. Yu. I. Krylova, A. M. Shkrob, V. K. Antonov, and M. M. Shemyakin, *Zh. Obshch. Khim.* **1968**, *38*, 2046.
666. M. Rothe, T. Toth, and D. Jacob, *Angew. Chem.* **1971**, *83*, 113.
667. G. Lucente, A. Romeo, S. Cerrini, W. Fedeli, and F. Mazza, *J. Chem. Soc., Perkin Trans. 1* **1980**, 809.
668. G. Lucente and A. Romeo, *J. Chem. Soc., Chem. Commun.* **1971**, 1605; S. Cerrini, V. Fedeli, and F. Mazza, *Ibid.* **1971**, 1607.
669. G. Lucente, F. Pinnen, and G. Zanotti, *Tetrahedron Lett.* **1978**, 1009; *See also* G. Lucente, F. Pinnen, A. Romeo, and G. Zanotti, *J. Chem. Soc., Perkin Trans. 1* **1983**, 1127.
670. G. Lucente, F. Pinnen, G. Zanotti, S. Cerrini, F. Mazza, A. L. Segre, and F. Fedeli, *J. Chem. Soc., Perkin Trans. 2* **1982**, 1169.
671. L. A. Cohen and B. Witkop, *J. Am. Chem. Soc.* **1955**, *77*, 6595.
672. M. Rothe and R. Steinberger, *Angew. Chem.* **1968**, *80*, 909.
673. M. Rothe and R. Steinberger, *Tetrahedron Lett.* **1970**, 649.
674. M. Rothe and R. Steinberger, *Tetrahedron Lett.* **1970**, 2467.
675. G. Lucente, F. Pinnen, G. Zanotti, S. Cerrini, W. Fedeli, and F. Mazza, *J. Chem. Soc., Perkin Trans. 1* **1980**, 1499.
676. D. S. Kemp, J. M. Duclos, Z. Bernstein, and W. M. Welch, *J. Org. Chem.* **1971**, *36*, 157.
677. H. Vorbrüggen and K. Krolikiewicz, *Chem. Ber.* **1975**, *108*, 2137.
678. U. Kramer, A. Guggisberg, M. Hesse, and H. Schmid, *Angew. Chem. Int. Ed. Engl.* **1977**, *16*, 861.
679. U. Kramer, A. Guggisberg, M. Hesse, and H. Schmid, *Angew. Chem. Int. Ed. Engl.* **1978**, *17*, 200.
680. M. Hesse, *Chimia* **1978**, *32*, 58.
681. A. Guggisberg, B. Dabrowski, U. Kramer, Ch. Heidelberger, M. Hesse, and H. Schmid, *Helv. Chim. Acta* **1978**, *61*, 1039.

682. A. Guggisberg, U. Kramer, Ch. Heidelberger, R. Charubala, E. Stephanou, M. Hesse, and H. Schmid, *Helv. Chim. Acta* **1978**, *61*, 1050.
683. U. Kramer, A. Guggisberg, M. Hesse, and H. Schmid, *Helv. Chim. Acta* **1978**, *61*, 1342.
684. U. Kramer, H. Schmid, A. Guggisberg, and M. Hesse, *Helv. Chim. Acta* **1979**, *62*, 811.
685. H. K. Reimschuessel, *Angew. Chem.* **1975**, *87*, 43.
686. N. J. Leonard, *Rec. Chem. Progr.*, **1956**, *17*, 243.
687. N. J. Leonard, *Acc. Chem. Res.* **1979**, *12*, 423.
688. A. Ohno and S. Oae, In: *Organic Chemistry of Sulfur*, Ed. S. Oae, Plenum Press, New York—London, 1977, 119.
689. J. C. Jochims, *Chem. Ber.* **1975**, *108*, 2320.
690. J. C. Jochims and A. Abu-Taha, *Chem. Ber.* **1976**, *109*, 139.
691. A. Alemagna and T. Bacchetti, *Gazz. Chim. Ital.* **1972**, *102*, 1068.
692. A. Alemagna and T. Bacchetti, *Gazz. Chim. Ital.* **1972**, *102*, 1077.
693. A. B. Tomchin and G. A. Shirokii, *Zh. Org. Khim.* **1977**, *13*, 404.
694. A. B. Tomchin and G. A. Shirokii, *Zh. Org. Khim.* **1979**, *15*, 855.
695. M. M. Shemyakin and L. A. Shchukina, *Zh. Obshch. Khim.* **1948**, *18*, 1925.
696. D. P. Vitkovskii and M. M. Shemyakin, *Zh. Obshch. Khim.* **1951**, *21*, 547.
697. L. A. Shchukina and M. M. Shemyakin, *Zh. Obshch. Khim.* **1956**, *26*, 1708.
698. L. A. Shchukina, Yu. B. Shvetsov, and M. M. Shemyakin, *Zh. Obshch. Khim.* **1951**, *21*, 346.
699. L. F. Fieser and M. Fieser, *J. Am. Chem. Soc.* **1948**, *70*, 3215.
700. G. A. R. Kon, A. Stevenson, and J. F. Thorpe, *J. Chem. Soc.* **1922**, *121*, 650.
701. S. S. Deshapande and J. F. Thorpe, *J. Chem. Soc.* **1922**, *121*, 1430.
702. B. Singh and J. F. Thorpe, *J. Chem. Soc.* **1923**, *123*, 113.
703. L. Bains and J. F. Thorpe, *J. Chem. Soc.* **1923**, *123*, 1206.
704. E. W. Lanfear and J. F. Thorpe, *J. Chem. Soc.* **1923**, 1683.
705. E. W. Lanfear and J. F. Thorpe, *J. Chem. Soc.* **1923**, *123*, 2865.
706. I. B. Blagoeva, B. J. Kurtev, and I. G. Pojarlieff, *J. Chem. Soc., Perkin Trans. 2* **1979**, 1115.
707. G. Illuminati and L. Mandolini, *Acc. Chem. Res.* **1981**, *14*, 95.
708. A. J. Kirby, *Adv. Phys. Org. Chem.* **1980**, *17*, 183.
709. K. B. Wiberg and H. W. Holmquist, *J. Org. Chem.* **1959**, *24*, 578.
710. H. O. Larson and G. S. K. Sung, *Austr. J. Chem.* **1962**, *15*, 261.

Additions in proof: References to some more important papers published in 1983 and 1984 are added below. They supplement with new data the main list references marked with an asterisk.

*471. For ring-chain tautomerism of monooximes of 1,4-diketones see A. Maccioni, P. P. Piras, A. Plumitallo, and G. Podda, *Gazz. Chim. Ital.* **1983**, *113*, 91.
*473. K. N. Sawhney and Th. L. Lemke, *J. Org. Chem.* **1983**, *48*, 4326.
*486. For NMR study in solid state of both these isomers ((151A) and (151B), page 119) see H. Kessler, H. Oschkinat, G. Zimmermann, H. Möhrle, M. Biegholdt, W. Arz, and H. Förster, *Chem. Ber.* **1984**, *117*, 702.
*500. For ring-chain tautomerism of 4-methylamino-1-(phenyl or 3-pyridyl)-1-butanones see S. Brandänge, L. Lindblom, A. Pilotti, and B. Rodriguez, *Acta Chem. Scand.* **1983**, *B 37*, 617.
*611. S. I. Yakimovich, V. N. Nikolaev, and S. A. Blohtina, *Zh. Org. Khim.* **1984**, *20*, 1371.
*613. S. I. Selivanov, K. G. Golodova, Ya. A. Abbasov, and B. A. Ershov, *Zh. Org. Khim.* **1984**, *20*, 1494.

Intramolecular Reversible Addition Reactions to the C=N Group

Despite the lower reactivity of azomethine groups compared to carbonyl groups, intramolecular reversible nucleophilic additions of OH, NH, and SH groups often proceed. Moreover, equilibria of these systems are frequently shifted toward the ring isomers to a considerably greater extent than in cases of corresponding C=O analogues. Evidently the higher basicity of the nitrogen atom favors the formation of the protonated $C=\overset{+}{N}HR$ group, to which the nucleophilic part of the molecule is easily added [1].

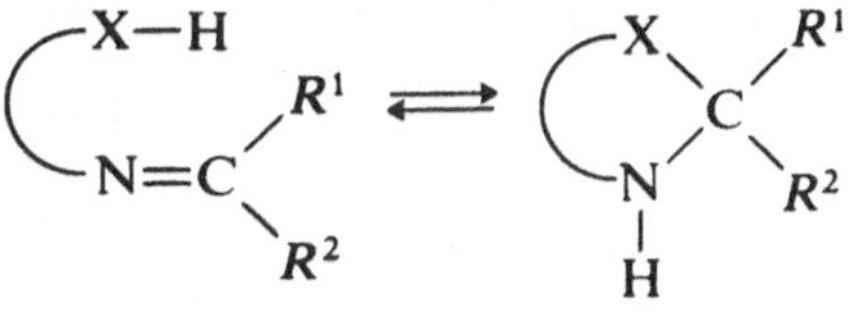

(1A)   **(1B)**

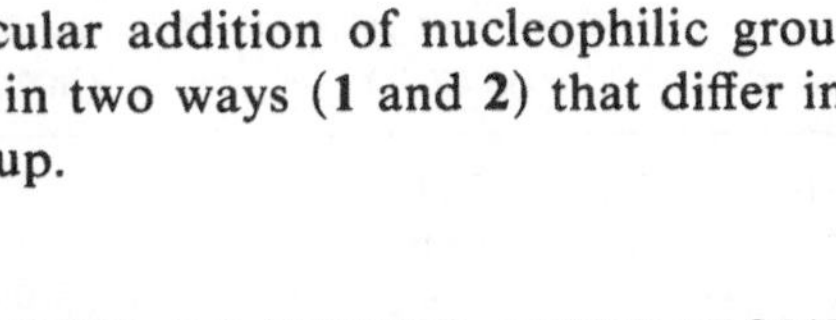

(2A)   X = O, NR, S   **(2B)**

The intramolecular addition of nucleophilic groups XH to a C=N group can proceed in two ways (**1** and **2**) that differ in the orientation of the azomethine group.

3.1. OH-DERIVATIVES OF IMINES, HYDRAZONES, OXIMES, AND NITRONES

3.1.1. *N*-(Hydroxyalkyl)- and *N*-(hydroxyaryl)-imines

In solutions of *N*-hydroxyethylimines of aldehydes and ketones the equilibrium involving Schiff's base and 1,3-oxazolidine **(3A)** ⇄ **(3B)** is observed [2–7].

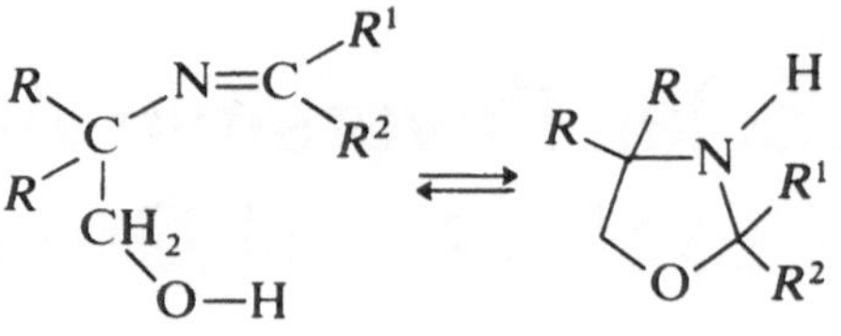

$$R \underset{R}{\overset{R}{\diagdown}} \underset{\underset{O-H}{\overset{|}{CH_2}}}{\overset{N=C\overset{R^1}{\diagdown}}{C}} \rightleftharpoons R \underset{R}{\overset{R}{\diagdown}} \underset{O}{\overset{N}{\diagdown}} \overset{H}{\underset{R^2}{\diagdown}} R^1$$

(3A)   (3B)

Using ^{1}H-NMR, a quantitative investigation of a series of 4,4-dimethyl-2-(3- or 4-X-phenyl)oxazolidines (3, R = Me, R^1 = H, R^2 = 3- or 4-XC_6H_4) was carried out. The equilibrium constants were deduced from the intensities of the signals of the proton adjacent to the aryl groups (R^1 = H) of both tautomeric forms [4, 5, 7, 8].

Increasing the electron-withdrawing effect of substituents X in the aryl group shifts the equilibrium toward the cyclic form (see Table 29). A linear correlation was detected for the ring-chain equilibrium constants and σ or, better, σ^+-coefficients of the substituents X in the aryl group [5]:

$$\lg K_T/(K_T)_0 = 0.86\sigma, \qquad r = 0.973;$$

$$\lg K_T/(K_T)_0 = 0.54\sigma^+, \qquad r = 0.999.$$

Table 29. The Influence of the Polar Properties of Substituent X on Ring-Chain Equilibrium Constants in Solutions of 4,4-Dimethyl-2-(3- or 4-X-phenyl)oxazolidines [5] and 2-(4-X-Phenyl)-3,4-dihydro-2H-1,3-benzoxazines [9] as Determined by ^{1}H-NMR

	$K_T = [B]/[A]$	
Substituent X in aryl group	(3A) $\rightleftarrows$ (3B) R = Me, R^1 = H R^2 = XC_6H_4 (in CCl_4)	(5A) $\rightleftarrows$ (5B) R^1 = H, R^2 = XC_6H_4 (in $CDCl_3$)
4-Me$_2$N	0.21	—
4-MeO	0.66	—
4-i-Pr	—	0.11
4-Me	1.18	—
H	1.71	0.19
3-MeO	1.96	—
4-Cl	2.16	—
4-Br	—	0.25
3-Cl	2.72	—
3-NO$_2$	4.0	0.96
4-NO$_2$	—	1.04

Alper with co-workers [8] determined a ring-chain equilibrium constant for a solution of 4,4-dimethyl-2-(4-tolyl)oxazolidine, the toluene ring of which was complexed with chromium tricarbonyl (3, R = Me, R^1 = H,

$R^2 = Me$, K_T = 4.3 in CCl$_4$). The high value of the

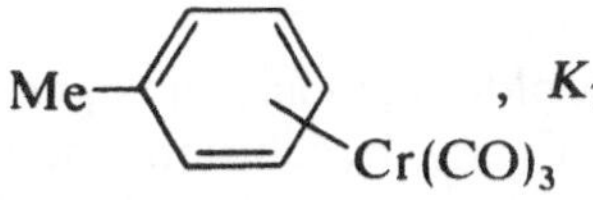

equilibrium constant can be explained by the strong electron-withdrawing effect of the aryl group involved in the complex.

Condensation reactions of 3-aminopropanol with carbonyl compounds gave either Schiff's bases or tetrahydro-1,3-oxazines or mixtures of both isomers [10, 11].

4-Aminobutanol in a reaction with formaldehyde formed hexahydro-1,3-oxazepine while condensation products with benzaldehyde or acetophenone have been reported to possess the open structure [10].

A study of condensation reactions of substituted 2-aminoethanols and 3-aminopropanols with carbonyl compounds revealed [3, 10] that an elongation and branching of alkyl substituents at the carbon atom of the azomethine group (R^1 and R^2 in the formula (3)) stabilize the open form. Branching at the chain between the interacting functional groups on the other hand stabilizes the cyclic isomer.

Condensation of 2-aminophenols with substituted glyoxals yielded 2-substituted 2,2'-bis-benzoxazolines (4B) which form tautomeric equilibria (4A) ⇌ (4B) in solution [12, 13].

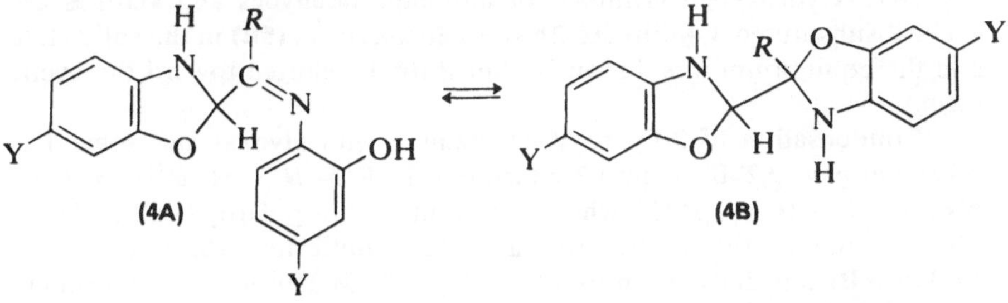

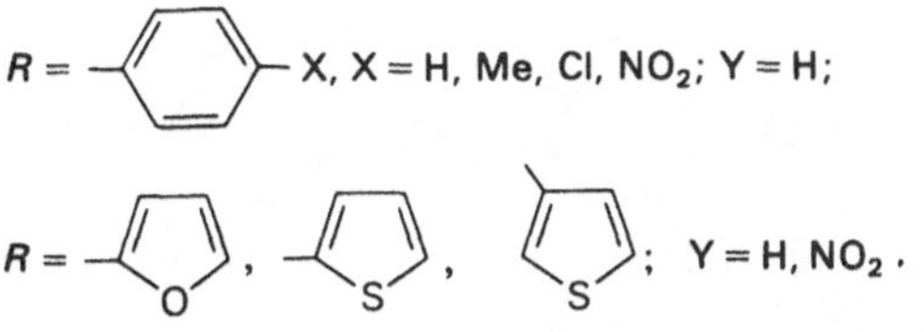

UV as well as ^{1}H-NMR investigations rendered possible a determination of the contents of the open isomer (4B) in solutions of arylderivatives (4, R = 4-XC$_6$H$_4$) [12]. As can be expected electron-withdrawing substituents X shift the equilibrium toward the ring form. However, an introduction of a nitro group into the aryl ring of the 2-aminophenol ($Y = NO_2$) stabilizes the open form. The additions of bases to solutions of (4) displaces the equilibrium toward the more acidic open form.

N-(2-Hydroxybenzyl)imines of aromatic aldehydes have an open structure in the solid state (5A, $R^1 = H$, $R^2 = 4\text{-}XC_6H_4$). In solutions the equilibrium (5A) ⇌ (5B) occurs [9, 14] (see Table 29). In this case also a linear free energy relationship with σ or, better, σ^+-coefficients of the substituents X of the phenyl group was found [5, 9]:

$$\lg K_T/(K_T)_0 = 0.76\sigma, \qquad r = 0.993;$$

$$\lg K_T/(K_T)_0 = 0.68\sigma^+, \qquad r = 0.999.$$

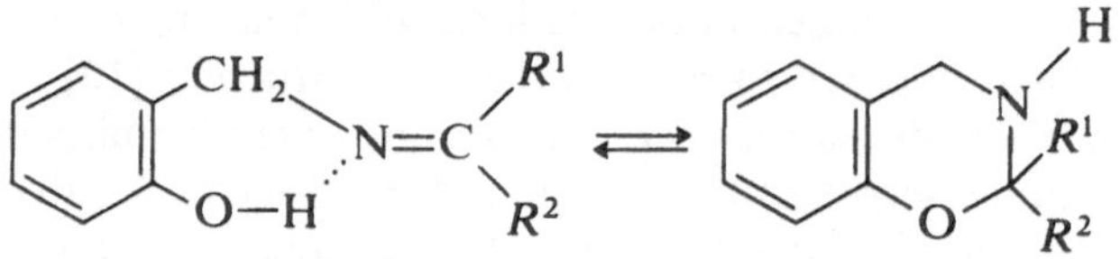

(5A)   (5B)

The open isomers (5A) are stabilized by the intramolecular hydrogen bonds between the phenolic hydroxyl and the azomethine nitrogen atom. This has been confirmed by a linear correlation between the chemical shifts of the hydroxyl protons and σ-coefficients of the substituents X ($R^2 = 4\text{-}XC_6H_4$).

N-(2-Hydroxybenzyl)imines of aliphatic aldehydes and ketones are cyclic 2-substituted 3,4-dihydro-2H-1,3-benzoxazines (5B) in the solid state and the equilibrium in solution is almost totally shifted toward the cyclic form.

Condensation of 2-hydroxybenzylamine with glyoxal and some 1,2-diketones gave 2,2'-bis-benz-1,3-oxazines (6B, $R^1 = R^2 = H$; $R^1 = H$, $R^2 = Me$; R^1, $R^2 = (CH_2)_4$) [15], which in solvents of low polarity (CCl_4, $CDCl_3$, PhCl) maintain the cyclic structure. An equilibrium (6A, $R^1 = R^2 = Me$) ⇌ (6B) was detected in pyridine ($K_T = 0.25$) as well as in chlorobenzene at 120°C.

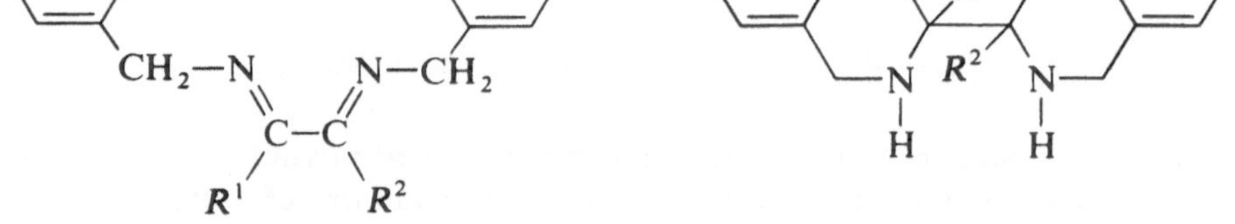

(6A)   (6B)

2,4-Diaryl-2,3-dihydro-1H-naphth[1,2-e]-1,3-oxazines obtained from 1-(α-aminobenzyl)-2-naphthol with aromatic aldehydes possess the ring structure in the solid state, but in solution they equilibrate [16]. Equilibration of protonated open and cyclic isomers were observed in trifluoroacetic acid. As has been shown before an introduction of electron-withdrawing substituents in the aryl group of the initial benzaldehydes displaces the equilibrium in solutions toward the cyclic form.

In different solvents the equilibrium constants (3A) ⇌ (3B) and (5A) ⇌ (5B) vary considerably. However no correlation either with the dipole moment, the dielectric constant, or Kosower's Z-constants [17] was observed [9]. Paukstelis and Hammaker [4] showed that the ability of the solvent to form intermolecular hydrogen bonds with the hydroxyl group of the open isomer (3A) decisively influences the equilibrium position. In fact a good correlation was observed between the di-tert-butylcarbinol OH band frequency shifts ($\Delta\nu_{O-H}$, assuming $\Delta\nu_{O-H} = 0$ in CCl_4) in IR spectra of di-tert-butylcarbinol solutions in different solvents and the enthalpies of ring-chain tautomeric interconversions of 4,4-dimethyl-2-phenyloxazolidine (3A, R = Me, R^1 = H, R^2 = Ph) ⇌ (3B) in the same solvents. In this case the frequency shift $\Delta\nu_{O-H}$ is caused by the formation of intermolecular hydrogen bonds t-Bu_2CHO—H···solvent and therefore utilized as a measure of solvent proton accepting ability. Increasing proton accepting ability of the solvent displaces the equilibrium toward the open form (3A) (see Table 30).

Several examples of the influence of N-protonation on ring-chain equilibria of N-hydroxyalkylimines were studied. Thus, using UV spectros-

Table 30. Influence of Solvent Proton Accepting Properties on Ring-Chain Equilibrium in Solutions of 4,4-Dimethyl-2-phenyloxazolidine (3A, R = Me, R^1 = H, R^2 = Ph) ⇌ (3B) [4]

Solvent	$\Delta\nu_{O-H}$ [a] (cm^{-1})	$K_T = [B]/[A]$ [b]	$-\Delta H$ (kcal/mol)	$-\Delta S$ (e.u.)
Carbon tetrachloride	0	1.89	1.154	2.7
Chloroform	8	1.72	1.269	3.0
Dimethylcarbonate	69	0.80	0.675	2.7
Acetonitrile	87	0.62	1.145	4.7
Acetone	110	0.42	0.097	2.0
Tetrahydrofuran	156	0.51	−0.425	0.0
Dimethylacetamide	193	0.12	−0.967	1.2
DMSO	224	0.10	−1.213	0.7

[a] Hydroxyl band shift in IR spectra of di-tert-butylcarbinol solution as the reference of solvent proton accepting ability.
[b] Determined by ^{1}H-NMR method at 38°C.

copy it was shown [18] that a protonation of oxazolidines (CH_2Cl_2 solution in the presence of a great excess of CF_3COOH) leads to ring opening and formation of immonium salts **(7A)**. Using equimolar amounts of CF_3COOH oxazolidine salts **(7B)** were isolated.

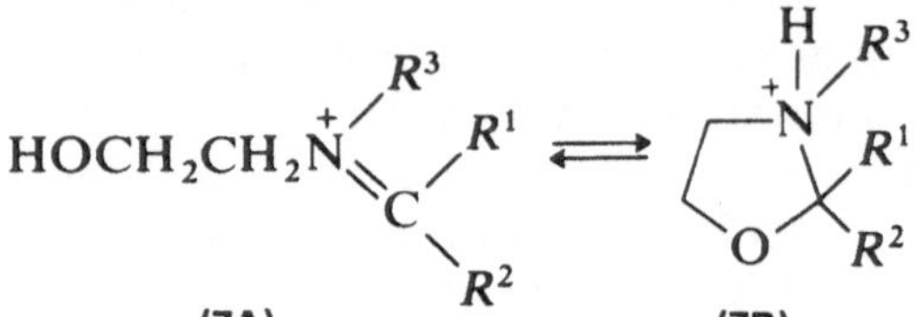

Following ^{1}H-NMR investigations [19] *N*-methylated 1,3-oxazaheterocycles (**8**, $n = 1, 2$) in CCl_4 solution at room temperature do not equilibrate with open forms. In CF_3COOH solution, however, an equilibrium **(9A)** $\rightleftarrows$ **(9B)** has been observed ($\sim$10% **(9A)** at room temperature). In the ^{1}H-NMR spectrum of the protonated *N*-methyloxazolidine, two singlets were observed in the $CH_2=$ region which led the authors to propose the presence of a second open tautomer (**9A'**, $n = 1$). However, the formation of the diprotonated form **(9A'')** seems to be more likely.

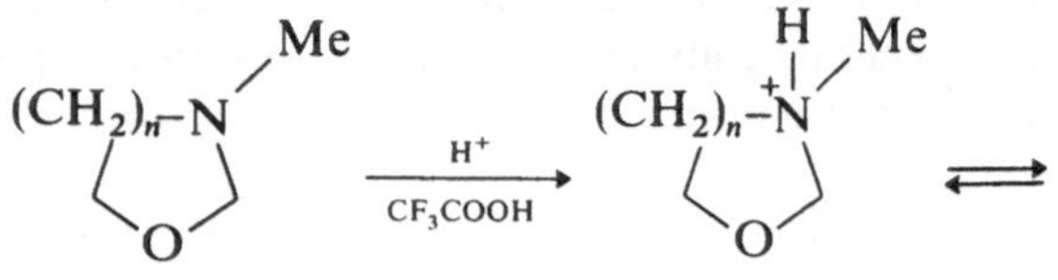

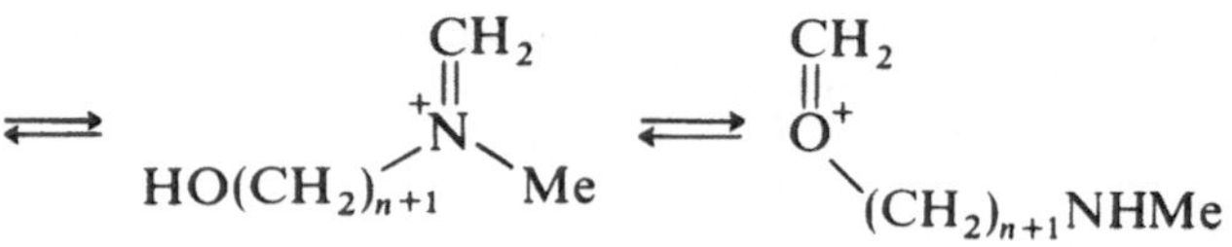

^{1}H-NMR investigations also showed that protonating the compounds (**5**, $R^1 = H$, $R^2 =$ alkyl, aryl; $R^1 = R^2 =$ alkyl) in CF_3COOH yielded the immonium ions of open structure. Protonated ring isomers form only from formaldehyde and trichloroacetaldehyde imines [20].

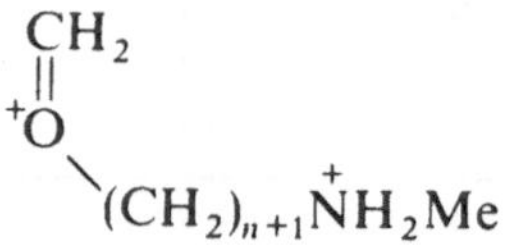

Yakimovich and co-workers [21, 22] gave an interesting example of ring-chain tautomerism in the series of the N-hydroxyalkylimines of β-ketoesters. Here, two equilibria, i.e., **(10A)** ⇌ **(10B)** and **(10A)** ⇌ **(10C)** have been observed. The situation is even more complicated by the possibility of forming configurational isomers, i.e., syn-**(10A)**, anti-**(10A)**, E-**(10C)**, Z-**(10C)**, as well as two stereoisomers for **(10B)**.

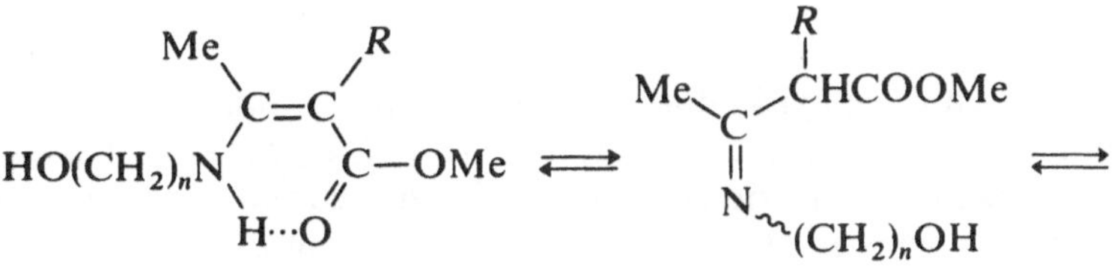

Z-**(10C)**   **(10A)**

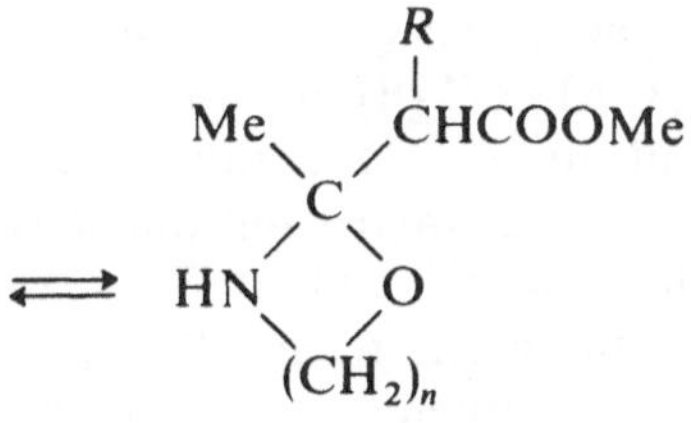

(10B)

It has been shown by ^{1}H-NMR spectroscopy that compounds (**10**, $R = H$, $n = 2$ or 3) exist in the enamine forms Z-(**10C**, $n = 2$) and Z- and E-(**10C**, $n = 3$). Imine **(10A)** and cyclic **(10B)** were observed only in derivatives containing alkyl substituents (R = Me, Et, i-Pr). Going from N-(2-hydroxyethyl)imines (**10**, $n = 2$) to N-(3-hydroxypropyl)imines (**10**, $n = 3$), the proportion of the cyclic form decreases.

3.1.2. N-Hydroxyalkylhydrazones

Yoffe and Potekhin [23, 24] simultaneously with Dorman [25] observed ring-chain tautomerism of N-(2-hydroxyalkyl)hydroazones of aldehydes and ketones. In this case the cyclic isomer is formed as a result of an intramolecular addition of a hydroxyl group to the hydrazone C=N bond [26].

A series of papers published by Potekhin et al. [27–34] are related to this tautomerism **(11A)** ⇌ **(11B)**, the results being compiled in a monograph [35]. The composition of the equilibrium mixtures was deduced from molecular refraction data and ^{1}H-NMR spectra.

(11A) ⇌ **(11B)**

N-(2-Hydroxyalkyl)hydrazones of aldehydes and ketones mono-substituted at the nitrogen atom ($11, R^3 = H$) [27, 28, 32] as well as products of a condensation of lactic acid hydrazide with carbonyl compounds [30] possess open structure **(11A)** and are not capable of cyclization.

Freshly distilled condensation products of N-alkyl-N-(2-hydroxyalkyl)hydrazines with aldehydes and ketones in most cases are cyclic perhydro-1,3,4-oxadiazines **(11B)**. On standing as well as in solution equilibrium mixtures **(11A)** ⇌ **(11B)** are formed.

The introduction of one or two methyl groups ($R^4 = R^5 = $ Me) in the hydroxyethyl group as well as increasing the steric bulk of the alkyl substituent at the nitrogen atom ($R^3 = $ Me $<$ Et $<$ Pr $<$ i-Pr $<$ t-Bu) displaces the equilibrium toward the cyclic form [28, 32]. It has been said [33] that introduction of alkyl substituents at the nitrogen atom weakens the intramolecular hydrogen bond between the hydroxyl group and nitrogen atom in **(11A)**, thus destabilizing the open isomer. The decrease of open isomer rotation entropy brought about by the substituent at the nitrogen atom may also be of significance.

Pinacoline N-aryl-N-(2-hydroxyethyl)hydrazones ($11, R^1 = $ Me, $R^2 = $ t-Bu, $R^3 = 4$-XC$_6$H$_4$, $R^4 = R^5 = $ H) show [31] equilibrium displacement toward the ring form corresponding to an increase in the electron-withdrawing effect of the substituents X in the aryl group (see Table 31).

A greater tendency of hydrazones of aliphatic aldehydes compared to that of ketones to form the cyclic **(11B)** has been explained [27] assuming that perhydro-1,3-4-oxadiazines with two alkyl substituents at $C_{(2)}$ possess an increased conformational energy. A greater electrophilicity of C=N bond carbon atom in aldehyde derivatives compared to the ketoanalogues [36, 37] may also be an important factor favoring intramolecular cyclization.

The condensation products of N-methyl-N-(2-hydroxy-2-methyl-1-propyl)hydrazine with isobutyric aldehyde have found special interest since for the first time both isomers, i.e., ($11A, R^1 = $ H, $R^2 = $ i-Pr, $R^3 = R^4 = R^5 = $ Me) and **(11B)** have been isolated [32]. A rapid distillation of the reaction mixture gave first the low boiling 2-isopropyl-4,6,6-trimethylperhydro-1,3,4-oxadiazine **(11B)** and then the high boiling open isomer **(11A)**. A slow distillation gave exclusively the ring isomer. The same investigators found

Table 31. Ring-Chain Equilibrium Constants of *N*-(2-Hydroxyalkyl)hydrazones of Aldehydes and Ketones (11A) ⇌ (11B), Determined by ¹H-NMR for Neat Liquids at 25°C [27-29, 31-33]

R^1	R^2	R^4	R^5	$K_T = [B]/[A]$ when $R^3 =$								
				H	Mo	Et	Pr	i-Pr	t-Bu	Ph	4-MeC₆H₄	4-NO₂C₆H₄
H	Me	H	H	0	1.0	13.3	7.3	8.1	32ᵃ	19ᵇ	19ᵇ	19ᵇ
H	Et	H	H	0		9.0						
H	i-Pr	H	H		0.49	4.6						
Me	Me	H	H	0	2.3	1.4	1.8	2.7	13.3ᵃ	5.7ᵇ	4.3ᵇ	9ᵇ
Me	t-Bu	H	H					0.11	24ᵃ	0.32ᵇ	0.15ᵇ	0.67ᵇ
Et	t-Bu	H	H			0.05						
H	Me	H	Me	0	7.3	50			24ᵃ			
Me	Me	H	Me	0	3.2	4						
H	i-Pr	H	Me		0.67							
H	t-Bu	H	Me		0.05	0.1			24ᵃ			
H	Me	Me	Me	0	5.7							
H	i-Pr	Me	Me		0.79							
Me	Me	Me	Me	0	0.05							

ᵃ Determined by a refractometric method.
ᵇ Determined for 15-20% solution in chloroform.

conditions for a conversion of the perhydro-1,3,4-oxadiazine into the open isomer by keeping the cyclic isomer at about 100° for several hours, leading to the enrichment in the open isomer in the equilibrium mixture. This is followed by a subsequent rapid distillation. Thus, it appears that increasing temperature displaces the equilibrium **(11A)** ⇌ **(11B)** toward the open isomer (see also [28]).

The thermodynamic parameters of the equilibrium **(11A)** ⇌ **(11B)** have been determined by a refraction method. The value $-\Delta H = 2.2 \pm 0.2$ kcal/mol turned out to be near to that found for the product of isobutyric aldehyde condensation with N-aminoephedrine (2.41 ± 0.11 kcal/mol) [25], but the value $-\Delta S = 7.6 \pm 0.5$ e.u. was significantly lower (9.15 ± 0.35 e.u.). However, it is rather difficult to compare these results because of essential differences in the structure of the initial hydrazinoalcohols.

Dorman [25] by condensation of N-methyl-N-(1-hydroxy-1-phenyl-2-propyl)hydrazine (N-amino-1-ephedrine) with aliphatic aldehydes obtained equilibrium mixtures of hydrazones **(12A)** and cyclic isomers 2-alkyl-4,5-dimethyl-6-phenyl-1,3,4-oxadiazines **(12B)**.

(12A)   (12B)

Equilibrium constants for **(12A)** ⇌ **(12B)** in tetrachloroethylene were measured by ^{1}H-NMR spectroscopy at different temperatures; 0.1% of CF_3COOH was added to the solutions to increase the rate of tautomerization. Increasing the dielectric constant of the solvent, raising the solution temperature, as well as extending the steric bulk of the alkyl substituent R shifts the equilibrium in favor of the open isomer (see Table 32).

Potekhin and Safronov [34] obtained a very interesting and unexpected result studying the N-benzyl-N-(2-hydroxyethyl- or 2-hydroxypropyl)hydrazones of benzaldehydes and acetophenones (**13**, $R^1 = $ H, Me; $R^2 = $ H, Me).

Unlike the hydrazones of aliphatic aldehydes and ketones (see Table 31) discussed so far, hydrazones of the benzaldehyde (**13**, $R^1 = $ H) are less prone to form cyclic isomers than derivatives of acetophenone (**13**, $R = $ Me). Thus, benzaldehyde N-benzyl-N-(2-hydroxyethyl- and 2-hydroxypropyl)hydrazones and their derivatives substituted in the aryl group

Table 32. Ring-Chain Equilibrium (12A) $\rightleftarrows$ (12B) Constants and Thermodynamic Parameters, Determined by ^{1}H-NMR in a Solution of Tetrachloroethylene and 0.1% CF_3COOH [25]

	$K_T = [B]/[A]$ at t, °C				$-\Delta H$	$-\Delta S$
R	22.5	35	50	82	(kcal/mol)	(e.u.)
H	0.67					
Me	1.61	1.39	1.13	0.70	2.93	8.94
Et	1.23	1.00	0.76	0.55	2.88	9.31
i-Pr	0.59	0.53	0.43	0.30	2.41	9.15
t-Bu	0.26	0.23	0.18	0.15	2.05	9.62

exist exclusively as open isomers (13A, $R^1 = H$, $R^2 = H$, Me) in the solid state and in tetrachloroethylene solution. The stabilization of (13A) was thought to be due to a conjugation which includes the benzene ring [34].

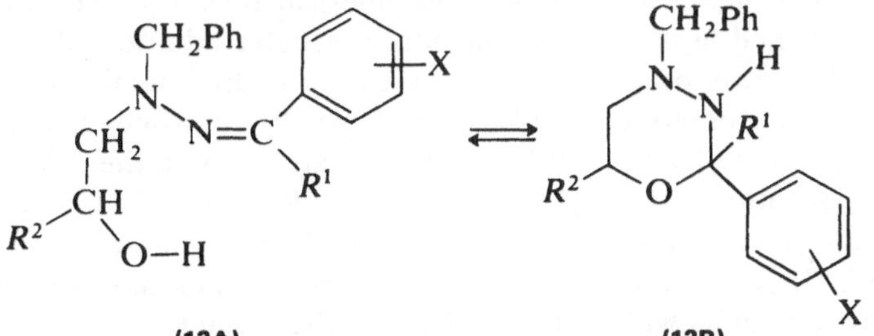

(13A)   (13B)

Table 33. Ring-Chain Equilibrium (13A, $R^1 = $ Me, $R^2 = $ H) $\rightleftarrows$ (13B) Constants and Thermodynamic Parameters Determined by ^{1}H-NMR using Tetrachloroethylene [34]

X	$K_T[B]/[A]$ at 30°C	$-\Delta H$ (kcal/mol)	$-\Delta S$ (e.u.)
3-NO$_2$	2.92	3.04	7.89
3-Br	2.09	4.47	13.4
4-Br	1.29	3.56	11.2
4-Cl	1.14	3.68	12.0
H	0.752	4.49	10.8
4-Me	0.472	3.39	12.8
4-MeO	0.287	3.15	12.9
3-NHAc[a]	0.192	—	—
4-NHAc[a]	0.10	—	—

[a] In solution of acetone-d$_6$.

An equilibrium **(13A)** ⇌ **(13B)** is observed in the case of derivatives of acetophenone (R^1 = Me). A twisting of the benzene ring out of the plane of the azomethine group, probably due to the methyl group, disturb the conjugation and favors the cyclic structure **(13B)**.

Equilibrium constants for hydrazones (**13**, R^1 = Me, R^2 = H) in tetra-chloroethylene have been determined at different temperatures (30, 50, 70, and 90°C) by ^{1}H-NMR spectroscopy; thermodynamic parameters were deduced. As is easily seen from Table 33, electron-withdrawing substituents in the benzene ring shift the equilibrium toward the cyclic form, a good correlation being observed with σ or σ^+ coefficients of the substituents X:

$$\lg K_T/(K_T)_0 = 1.03\sigma, \qquad r = 0.980, \qquad s = 0.08;$$

$$\lg K_T/(K_T)_0 = 0.74\sigma^+, \qquad r = 0.986, \qquad s = 0.06.$$

3.1.3. Hydroxyalkyl and Hydroxyiminoalkyl Nitrones

Several examples of ring-chain tautomerism are known where the cyclic isomer is formed as a result of an intramolecular addition of a hydroxyl group to the C=N bond of nitrones, yielding hydroxylamine derivatives.

No cyclic isomers of aryl-N-(2-hydroxyalkyl)nitrones (**14A**, R = Ar) could be observed by IR and ^{1}H-NMR spectroscopy in the solutions [38]. However, solutions of alkylderivatives (**14**, R = Me, C_6H_{11}) showed an equilibrium **(14A)** ⇌ **(14B)** which is strongly displaced in favor of the open form [39]. Chemical reactions, as for example acylation, gave derivatives of the open form as well as those of the cyclic isomer.

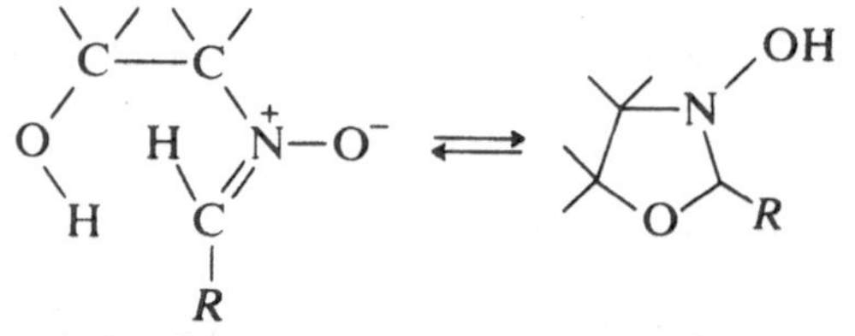

(14A)   (14B)

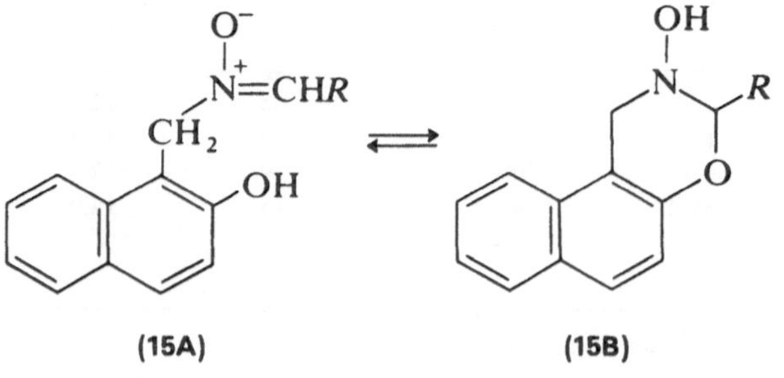

(15A)   (15B)

An analogous ring-chain equilibrium was detected [40] in solutions of N-(2-hydroxy-1-naphthyl)methylnitrones **(15A)** $\rightleftarrows$ **(15B)** in DMSO-d_6 and methanol-d_4. ^{1}H-NMR investigations showed that the equilibrium depends on the substituent R. An introduction of an aryl substituent (**15**, R = Ar) instead of hydrogen or a methyl group (**15**, R = H, Me) stabilizes the open structure. No cyclic isomers of compounds (**15**, R = 4-XC$_6$H$_4$, X = NO$_2$, NMe$_2$) have been detected in solutions (see also [41]).

Volodarskii and co-workers using ^{1}H-NMR spectroscopy have been able to show that products of condensation of syn-α-hydroxylaminooximes with aliphatic aldehydes [42] and acetone [43] in alcoholic solution form an equilibrium mixture of N-(syn-2-hydroxyiminoalkyl)nitrones and 5-hydroxydihydro-4H-1,2,5-oxadiazines **(16A)** $\rightleftarrows$ **(16B)**. Derivatives of formaldehyde, however, are exclusively cyclic (**16B** R^3 = R^4 = H). The equilibrium position also depends on the structure of the substituents R^1 and R^2. If the oxadiazine ring is condensed with the five-membered indane ring (**16**, R^1, R^2 = $-o$-C$_6$H$_4$CH$_2$-, R^3 = H, R^4 = Me) the open form predominates, but in the case of a six-membered tetraline ring (**16**, R^1, R^2 = $-o$-C$_6$H$_4$CH$_2$CH$_2$-, R^3 = H, R^4 = Me) the equilibrium is shifted toward the cyclic form (see Table 34).

(16A)   **(16B)**

Increasing the temperature shifts the equilibrium toward the open form. It has been shown by UV spectroscopy that aqueous alkaline solutions contain the anion of the open isomer.

Table 34. Isomeric Composition of N-(2-hydroxyiminoalkyl)nitrones **(16A)** $\rightleftarrows$ **(16B)** in Ethanol [42, 43]

R^1	R^2	R^3	R^4	Composition
Ph	Me	H	H	B
	$-o$-C$_6$H$_4$CH$_2$CH$_2$-	H	H	B
Ph	Me	H	Me	A $\rightleftarrows$ B (27% A at 28°C)
	$-o$-C$_6$H$_4$CH$_2$-	H	Me	A $\rightleftarrows$ B, A > B
	$-o$-C$_6$H$_4$CH$_2$CH$_2$-	H	Me	A $\rightleftarrows$ B, A < B
Ph	H	Me	Me	A $\rightleftarrows$ B, 37% A
Ph	Me	Me	Me	A $\rightleftarrows$ B, 20% A
	$-o$-C$_6$H$_4$CH$_2$CH$_2$-	Me	Me	A $\rightleftarrows$ B, 24% A

The intramolecular addition of a hydroxyl group to a nitrone C=N bond has been observed also in the case of nitrone C-hydroxyalkylderivatives.

Thus, N-methyl-N-(tetrahydro-2-pyranyl)hydroxylamine **(17B)** at room temperature is stable, but during vacuum distillation (150°C, 0.3 mm Hg) it is partially converted into the α-(4-hydroxybutyl)-N-methylnitrone **(17A)** [44, 45].

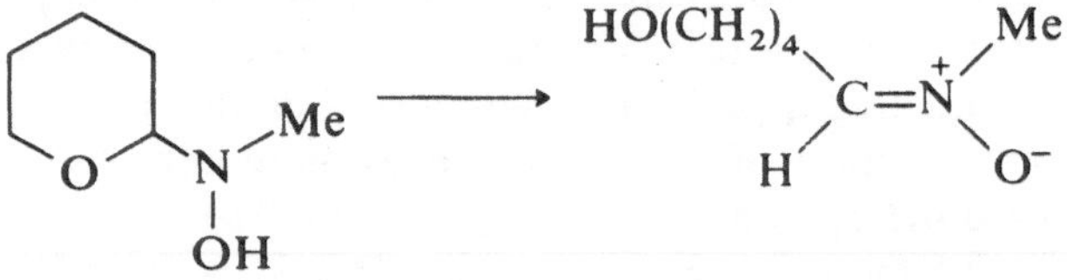

(17B)   **(17A)**

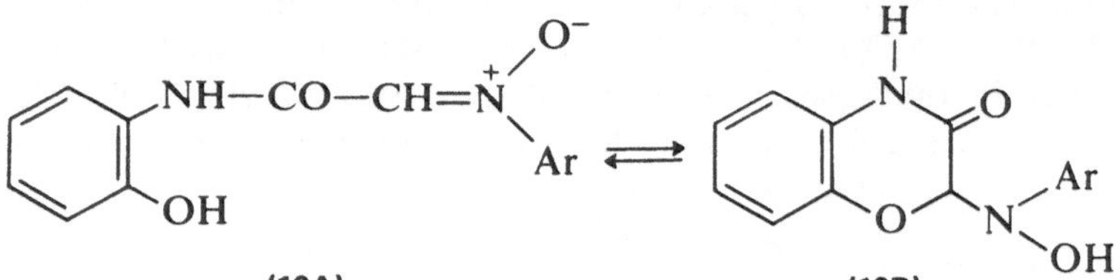

(18A)   **(18B)**

The intramolecular addition of a phenolic hydroxyl to a nitrone C=N bond in **(18A)** leads to the formation of 2-(N-arylhydroxylamino)-3-benzo[e]morpholinone **(18B)** [46]. The isomerization **(18A) → (18B)** proceeds in different solvents at 20°C. The reverse isomerization is carried out by heating to 165° or by an action of trace amounts of an acid.

3.1.4. Imines, Hydrazones, and Oximes of Hydroxyaldehydes and Hydroxyketones

The ring-chain tautomerism of the N-hydroxyalkyl- or N-hydroxyarylalkylimines and hydrazones of aldehydes and ketones (type **(1A)** ⇌ **(1B)**) has been studied rather thoroughly but little attention has been paid so far to the series of imines and hydrazones of hydroxyketones and hydroxyaldehydes (type **(2A)** ⇌ **(2B)**).

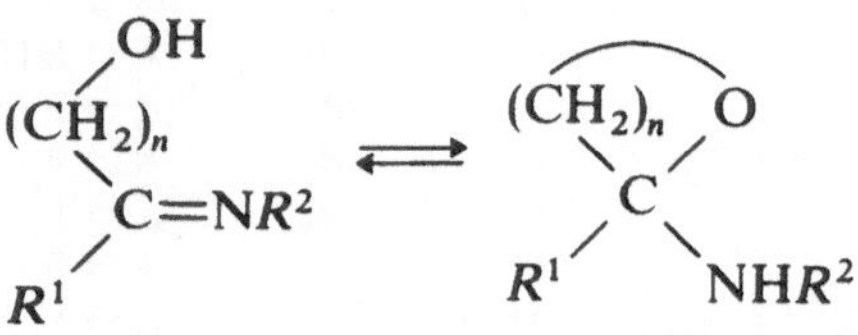

(19A)   **(19B)**

Table 35. Ring-Chain Equilibrium **(19A)** ⇄ **(19B)** Constants,
Determined by ^{1}H-NMR at 30°C [50]

| | | | $K_T = [B]/[A]$ | |
R^1	R^2	n	Neat liquid	Solution in CCl_4
H	Me	3	0.91	6.25
H	t-Bu	3	0.28	2.0
H	Me	4	12.5	>100
H	Et	4	11.1	>100
H	i-Pr	4	6.3	>100
H	t-Bu	4	5.6	>100
Me	Me	3	0.50	4.0
Me	Et	3	0.48	3.55
Me	i-Pr	3	0.06	0.39
Me	Me	4	1.27	7.2
Me	i-Pr	4	0.08	0.67

Potekhin, Zhdanov, and coauthors recently reported on a series of investigations [47–51] thus eliminating this lack of information. They investigated a large number of N-alkyl and N-arylimines of 4-hydroxybutanal (**19**, R^1 = H, n = 3), 5-hydroxypentanal (**19**, R^1 = H, n = 4), 5-hydroxy-2-pentanone (**19**, R^1 = Me, n = 3), and 6-hydroxy-2-hexanone (**19**, R^1 = Me, n = 4) which were obtained as equilibrium mixtures **(19A)** ⇄ **(19B)** (see Table 35).

Analogous to hydroxyaldehydes and hydroxyketones the six-membered cyclic isomers of the corresponding imines are more stable. Sterically demanding alkyl substituents at the nitrogen atom shift the equilibria toward the open form. The ring form of imines of hydroxyaldehydes is more stable than that of imines of hydroxyketones.

It has been pointed out [50] that steric peculiarities of the five- and six-membered ring exercise a strong influence on the stability of the cyclic form. A displacement of the equilibria toward the open form, passing from the imines of hydroxyaldehydes (**19**, R^1 = H) to those of hydroxyketones (**19**, R^1 = Me), was explained by an increased destabilizing repulsion between the 1,3-diaxial groups in the tetrahydropyrane ring or by a cis-1,2-interaction in the tetrahydrofuran ring. Differences in the electrophilicity of the C=N bond in aldehyde and ketone derivatives have already been discussed.

In the solid state as well as in solution N-arylimines of 5-hydroxypentanal exist exclusively as cyclic 2-arylaminotetrahydropyranes (**19B**, R^1 = H, R^2 = 4-XC$_6$H$_4$, n = 3) [48]. Even heating solutions to 90°C does not lead to the formation of open isomers **(19A)**. The unexpected stability of the cyclic form of N-arylimines as compared with that of N-alkylimines was attributed to a conjugation of the lone pair of the nitrogen atom and the benzene ring in the cyclic form (19B, $R^2 = $ Ar). Conjugation in the open form is disturbed because of the noncoplanarity of the system C=N—Ar.

One may also suggest that during the transformation (19B) → (19A) a heterolytic rupture of the C—O bond occurs simultaneously forming a $C=\overset{+}{N}HR^2$ group in the intermediate (20). The lone pair of the nitrogen atom that is involved in this transformation may be hindered by conjugation with the benzene ring ($R^2 = $ Ar) and thus disfavors the ring opening.

$$(19B) \rightleftarrows \overset{R^1}{\underset{(20)}{\bar{O}(CH_2)_n\overset{|}{C}=\overset{+}{N}HR^2}} \rightleftarrows (19A)$$

Solvents have a significant influence on the position of the equilibria of these N-alkylimines. In nonpolar solvents the equilibrium is strongly displaced in favor of the cyclic form. Thus, the equilibrium constant of 5-hydroxy-2-pentanone N-methylimine (19, $R^1 = R^2 = $ Me, $n = 3$) changes from 5.6 to 0.66 on passing from benzene to nitromethane. Increasing the temperature shifts the equilibrium toward the open form [50].

Whitting and Edvard [37] were the first to investigate the structure of 2,4-dinitrophenylhydrazones of 5-hydroxy-2-pentanone and 6-hydroxy-2-hexanone (19, $R^1 = $ Me, $R^2 = $ 2,4-$(NO_2)_2C_6H_3NH$, $n = 3,4$). Using spectroscopic methods they were able to show that these compounds are exclusively open structured in the solid state and also in ethanol, acetone, or DMSO.

Systematic studies of a large number of alkyl, dialkyl, and arylhydrazones of 4-hydroxybutanal, 5-hydroxypentanal [48, 51], 5-hydroxy-2-pentanone, and 6-hydroxy-2-hexanone [47] showed that they possess exclusively the open structure independently of the substitution pattern of the hydroxycarbonyl compound, the substituent at the nitrogen atom, the properties of the solvent, and the temperature. Using IR, UV, and ^{1}H-NMR spectroscopy no indication of the presence of cyclic isomers was obtained in spite of an extensive variation of the solvent (acetonitrile, benzene, dimethylformamide, dioxane, pyridine, deuterium oxide) and the temperature (30–90°C). The authors suggest [47] that the stability of the open isomers may possibly be caused, first, by a destabilization of the cyclic form due to the nonbonded interactions and, second, by the stabilization of the open form by means of hydrogen bonds involving the hydroxyl group.

Similar to 5-hydroxypentanal hydrazones, arylhydrazones of D-glucose also possess an open structure in ethanol or in DMSO [51]. Nitroarylhydrazones of D-glucose, however, in DMSO show equilibria with the cyclic pyranozic form which according to the authors [51] is caused by the nitro group of the benzene ring stabilizing the cyclic structure by the increase of conjugation in the system —NH—Ar.

Similar to the hydrazones oximes of 5-hydroxypentanal [48], 5-hydroxy-2-pentanone, and 6-hydroxy-2-hexanone [47] neat and in solutions exist exclusively as the open isomers (**19A**, R^1 = H, R^2 = OH, n = 3; R^1 = Me, R^2 = OH, n = 3, 4).

3.1.5. Imines of Ketocarboxylic Acids

There are well-known examples of intramolecular reversible addition of the carboxylic group to the C=N bond. Indications of a ring-chain tautomerism of N benzylideneanthranilic acid **(21A)** ⇌ **(21B)** have been given [52]; however, no spectroscopic evidence is available.

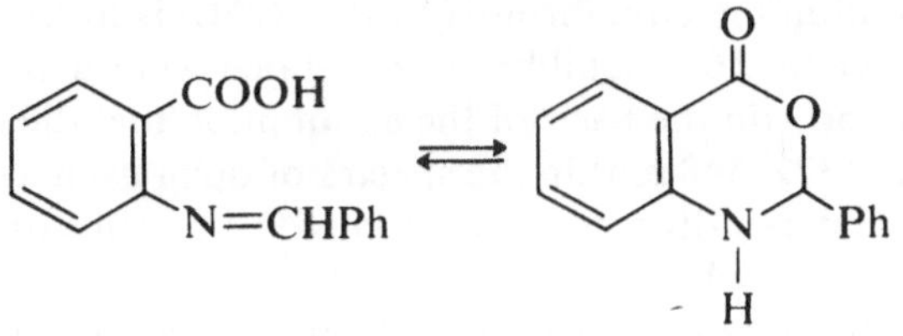

(21A)   (21B)

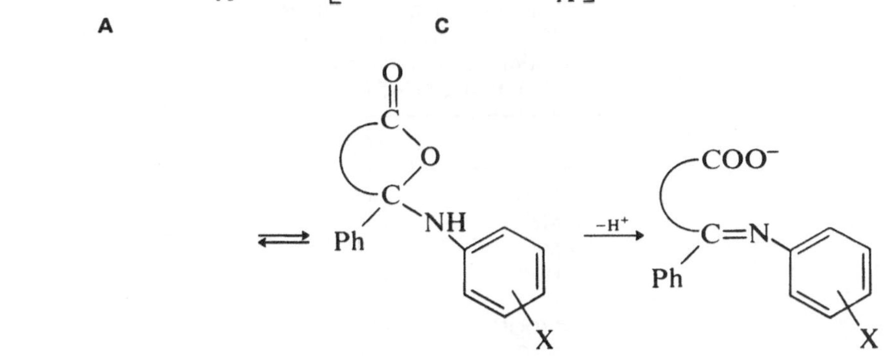

A   C

B   D

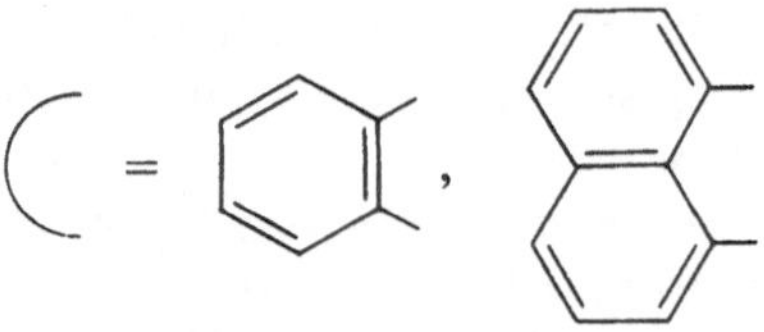

(22)   **(23)**

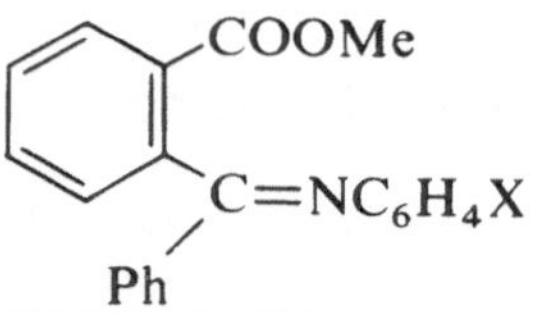

(24)

3-Arylamino-3-phenylphthalides [53–56] are cyclic **(22B)** in the solid state and in solution the equilibrium **(22A)** $\rightleftarrows$ **(22B)** is displaced toward the ring form (see Table 36). Equilibrium constants were determined [56, 57] by UV spectroscopy on the basis of the assumption that the intensity of the long-wave band (322–348 nm) in the spectra of open isomers **(22A)** is equal to that of the corresponding model compounds, i.e., anils of 2-methoxycarbonylbenzophenone **(24)**.

Electron-withdrawing substituents X in the benzene ring increase the electrophilicity of the azomethine carbon atom, thus stabilizing the cyclic form. The equilibrium constants approximately correlate with σ-coefficients of the substituents X:

$$\lg K_T/(K_T)_0 = 0.58\sigma, \qquad r = 0.94$$

Table 36. Ring-Chain
Equilibrium Constants of 3-
Arylamino-3-phenylphthalides
(22A) $\rightleftarrows$ **(22B)** in Dioxan
Solution as Determined by
UV at 20°C [57]

X	$K_T = [B]/[A]$
4-MeO	2.69
4-Me	3.83
3-Me	5.21
H	4.78
4-I	4.67
4-EtOOC	7.94
4-MeCO	9.61

Table 37. Solvent Influence on Ring-
Chain Equilibrium of 3-Phenylamino-3-
phenylphthalide (**22A**, X = H) ⇌ (**22B**)
[56]

Solvent	$K_T = [B]/[A]$
Dichloroethane	>100
Tetrahydrofuran	5.26
Dioxan	4.78
Ethanol	1.64
DMSO	1.15

A better correlation was observed with σ^+ coefficients [57]:

$$\lg K_T/(K_T)_0 = 0.35\sigma^+, \qquad r = 0.97$$

This observation (found in various reversible intramolecular additions to the azomethine bond [5, 9]) was explained assuming that the equilibrium (**22A**) ⇌ (**22B**) goes through bipolar intermediate (**22C**) in which the electron-donating substituent X exerts a direct interaction on the nitrogen atom bearing the positive charge.

Increasing proton acceptor properties of the solvent shifts the equilibrium (**22A**) ⇌ (**22B**) toward the open form (see Table 37). A similar regularity was already discussed for 4,4-dimethyl-2-phenyloxazolidine [4] (see Table 30). The stabilization of the open isomer in proton acceptor solvents is brought about by a formation of intermolecualr hydrogen bonds, COOH···solvent.

3-Arylamino and 3-alkylamino-3-phenylperinaphthalides (**23B**) are cyclic in the solid state as well as in solutions (IR and UV evidence) [58, 59]. The reason for the stability of the cyclic form (**23B**) is similar to that already discussed for 8-acyl-1-naphthoic and 2-acylbenzoic acids (see p. 39, Fig. 5).

Bases convert aminolactones (**22B**) and (**23B**) to anions of open structure (**22D**) and (**23D**).

3.1.6. Fused Systems

Grandberg, Dashkevich, and Ivanova [60, 61] investigating analogues of the alkaloid physovenin demonstrated by means of UV and ^{1}H-NMR spectroscopic data that substituted 2,3,3a,8a-tetrahydrofuro[2,3-b]indols in acidic media open the furanidine ring (**25B**) → (**26**).

Electron-donating substituents, especially at nitrogen atom (R^5), increasing the basicity of nitrogen atom and decreasing the electrophilicity of the atom $C_{(8a)}$ in the molecule (**25B**), facilitate this ring opening, thus

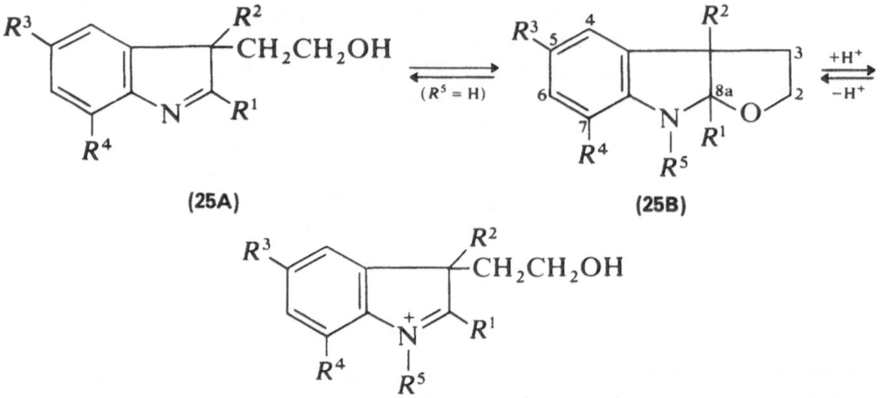

(25A)   **(25B)**

(26)

$R^1 = R^2 = $ Me and other alkyl

affording a less acidic medium. The introduction of alkoxygroups at $C_{(5)}$ of the aryl ring (**25**, $R^1 = R^2 = $ Me, $R^3 = $ MeO, PhCH$_2$O, $R^4 = R^5 = $ H) permits observation of the equilibrium **(25A)** ⇄ **(25B)** in neutral, hydroxyl-containing solvents. Apparently, the electron-donating alkoxy group decreases the electrophilicity of the atom $C_{(8a)}$ and polarizes the $C_{(8a)}$—O bond sufficiently for a furanidine ring opening without a preliminary protonation of the nitrogen atom. Electron-withdrawing substituents in the aryl ring (R^3 or $R^4 = $ Br) or at the nitrogen atom ($R^5 = $ Ph) stabilize the cyclic structure, and the equilibrium **(25B)** ⇄ **(26)** is obtained at lower pH values.

3.2. N-H-DERIVATIVES OF IMINES, HYDRAZONES, OXIMES, AND NITRONES

3.2.1. N-Aminoalkylimines

In the reactions of ethylenediamines with aldehydes and ketones mixtures of Schiff's bases **(27A)** and isomeric imidazolidines **(27B)** are formed. However, it was not possible to confirm the existence of a tautomeric equilibrium **(27A)** ⇄ **(27B)** in solution [62]. A ^{1}H-NMR study proved that the amount of the cyclic isomer in the reaction mixture decreases with increasing steric demand of the alkyl substituent R^1 (when $R = $ H) or if $R^1 = $ Ph.

Condensation of N-methylethylenediamine with aromatic aldehydes gave a mixture of the open and cyclic isomers (**27A**, $R = R^3 = $ H, $R^1 = $ 4-XC$_6$H$_4$, $R^2 = $ Me) and **(27B)** [63]. The amount of the cyclic isomer

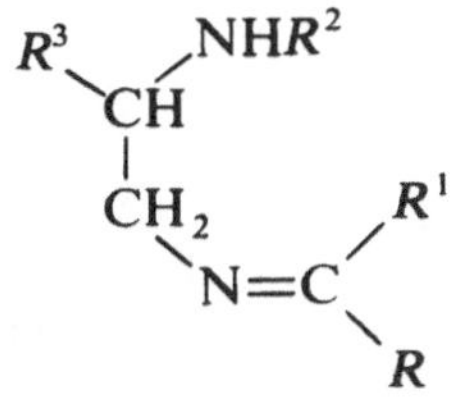

(27A) 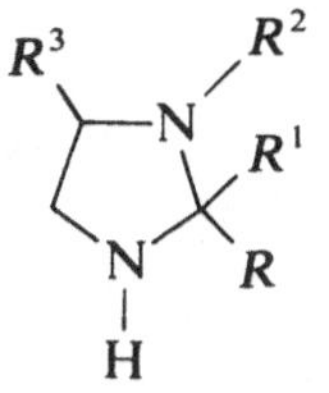   (27B)

increases with increasing electron-withdrawing properties of the substituent X: when X = NMe_2 the mixture contains 50% of the cyclic isomer, X = NHAc: 62%, X = H: 80%, X = NO_2: 100%. However, ^{1}H-NMR spectra in different solvents (CCl_4, DMSO) gave no evidence for an equilibrium (27A) ⇌ (27B) displacement. Obviously, one can conclude that isomers (27A) and (27B) in the usual solvents at room temperature do not undergo interconversion.

Products of condensation of 1,3-diaminopropane and its alkylsubstituted derivatives with formaldehyde have been shown to be cyclic hexahydropyrimidines (28B, $R^2 = R^3 = $ H) by IR and ^{1}H-NMR technique [64]. The ring structure is destabilized in the condensation product of 1,3-diaminopropane with acetone. 1-Benzylideneamino-3-*tert*-butylaminopropane exists in the open form (28A, $R^1 = t$-Bu, $R^2 = $ Ph, $R^3 = $ H).

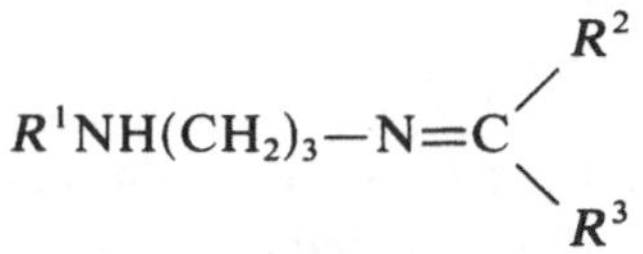

(28A) 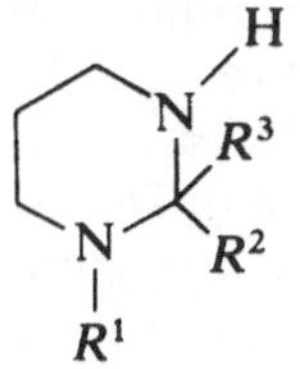   (28B)

Products of condensation of 2,4-diamino-2-methylpentane and 4-amino-2-methylamino-2-methylpentane with aldehydes have been shown to be hexahydropyrimidines. An analogous condensation with ketones led to the open or cyclic isomers or their mixtures depending on the structure of the ketone [65].

Reactions of 4-amino-5-aminomethyl-2-methyl (or phenyl) pyrimidine with benzaldehydes yield open (29A) or cyclic (29B) products or mixtures (X = F) [66]. However, no correlation has been found between the structure of the products (i.e., substituents X) and the isomer preferentially formed.

^{1}H-NMR investigations showed [19, 67] that 1,3-dimethylimidazolidine (30, n = 1) and 1,3-dimethylperhydropyrimidine (30, n = 2) in CF_3COOH form equilibrium mixtures of protonated isomers (31A) ⇌ (31B) (when n = 1, K_T = 4.5).

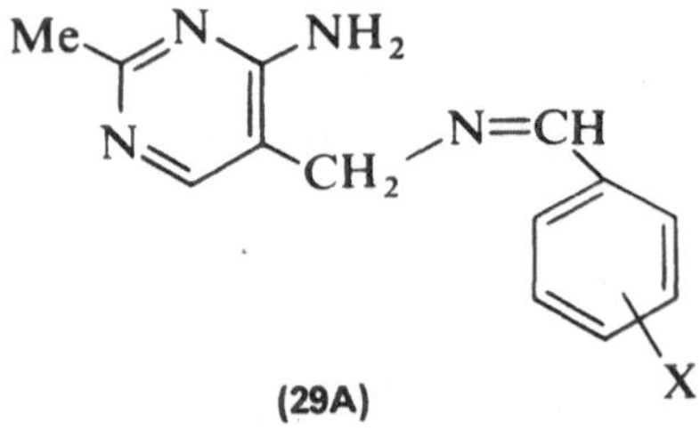

(29A)

X = 4-Cl, 4-Br, 4-F

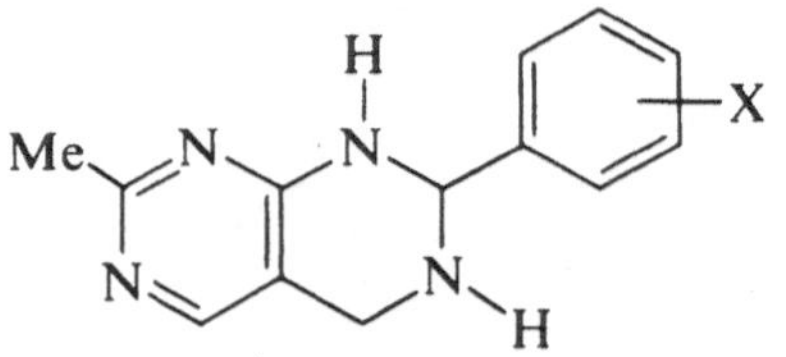

(29B)

X = 4-F, 4-MeO, 4-NO₂,
2,4-Cl₂, 3,4-Cl₂

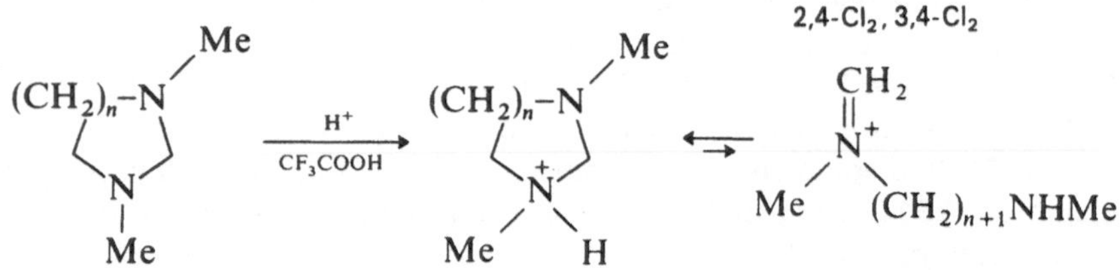

(30)   **(31B)**   **(31A)**

The rate of tautomerization is highest for the five-membered system which disagrees with Baldwin's rules [68–70]. The coalescence temperature of the equilibrium (**31A**, $n = 1$) ⇌ (**31B**) is found to be 90°C which corresponds to a free activation energy of 16.9 kcal/mol.

A study of the ¹H-NMR spectra of 5-nitro-1,2,3,4-tetrahydropyrimidines in CF₃COOH established [71] that the 2-unsubstituted derivatives (**32**, R^2 = H) are stable in acid but the 2-alkyl and 2-arylderivatives (**32**, R^2 = alkyl, aryl) transform into diprotonated open isomers E-**(33)** and Z-**(33)**. The authors suggest [71] that in CF₃COOH an equilibrium exists between a N-monoprotonated ring form, a diprotonated pyrimidine of little stability, and two more stable diprotonated open isomers E-**(33)** and Z-**(33)**. Thus, the equilibrium is strongly shifted toward **(33)**.

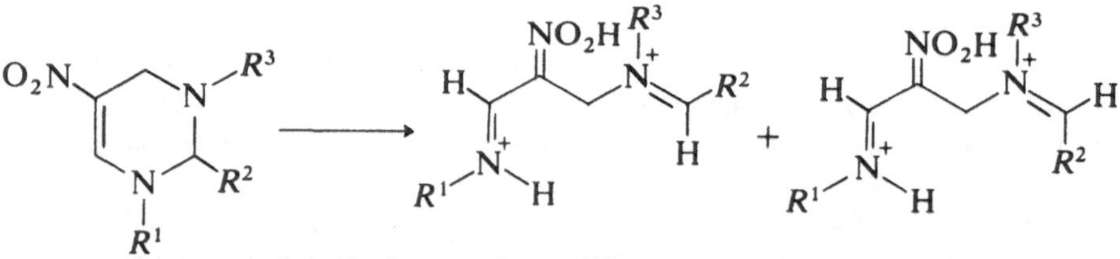

(32)   E-**(33)**   Z-**(33)**

3.2.2. N-Aminoalkyl and N-Aminoacyl Hydrazones and Related Compounds

Ring-chain tautomerism caused by an intramolecular addition of an amino group to the hydrazone C=N bond of the general type **(34A)** ⇌ **(34B)**

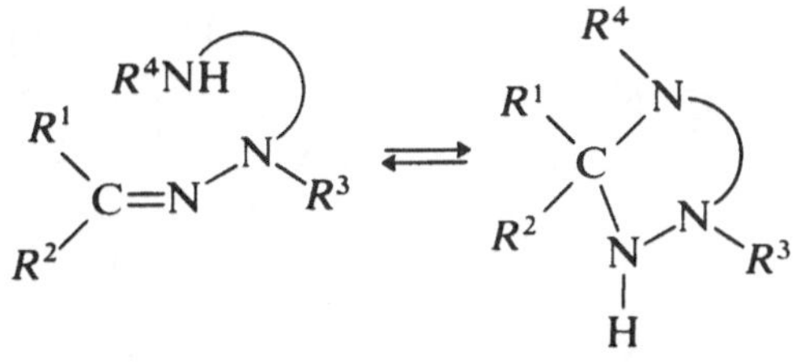

(34A)   **(34B)**

was investigated for a rather large number of compounds of different structure, and as a rule the formation of five- and six-membered rings is observed.

2-Phenylsemicarbazones of acetone and acetophenone are obtained [72], both in the open form (**35A**, R^1 = Me, R^2 = Me, Ph, R^3 = Ph) and as the cyclic 1,2,4-triazolidin-3-one (**35B**). The isomerization (**35A**) → (**35B**) is sensitive to acidic catalysis as are the isomerizations of 4-alkyl-2-phenyl-semicarbazones of aliphatic and alicyclic ketones [73]. Introduction of a t-butyl group at the terminal nitrogen atom (R^4 = t-Bu) prevents cyclization.

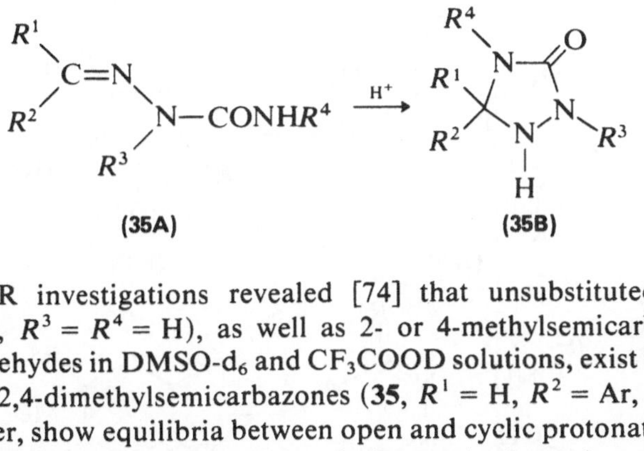

(35A)   **(35B)**

^{1}H-NMR investigations revealed [74] that unsubstituted semicar-bazones (**35**, $R^3 = R^4 =$ H), as well as 2- or 4-methylsemicarbazones of aromatic aldehydes in DMSO-d_6 and CF$_3$COOD solutions, exist in the open form. Some 2,4-dimethylsemicarbazones (**35**, $R^1 =$ H, $R^2 =$ Ar, $R^3 = R^4 =$ Me), however, show equilibria between open and cyclic protonated isomers in CF$_3$COOD. Electron-withdrawing substituents in the aryl group stabilize the cyclic form.

An equilibration being effected only after protonation of a nitrogen atom was detected recently [75] for 1-(2-propylidene)benzamidrazone. An analogous equilibrium (**36A**) ⇌ (**36B**) was obtained [76] in solutions of 1-(2-propylidene)-2-methylacetamidrazone hydroiodide which in the solid state possesses the cyclic structure (**36B**). The contents of (**36A**) decreases in the sequence DMFA-d_7 > DMSO-d_6 > CDCl$_3$. The conjugate base possessing the cyclic structure does not show a ring-chain equilibrium in solution.

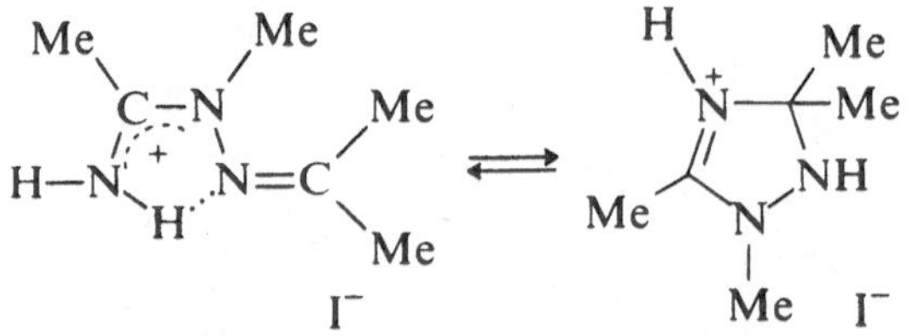

(36A)   (36B)

Walter and Rohloff [77] using ^{1}H-NMR spectroscopy studied in detail the ring-chain tautomerism of aldehyde and ketone 2-methylthiosemicarbazone S,S,S-trioxides **(37A)** $\rightleftarrows$ **(37B)**. A methyl or benzyl group at the nitrogen atom (R^3 = Me, PhCH$_2$) displaces the equilibrium toward the cyclic form compared with N-unsubstituted derivatives (**37**, R^3 = H). Sterically-demanding 2,6-dimethyl or 2,6-diisopropylphenyl substituents, however, destabilize the cyclic form. Increasing solvent polarity (CDCl$_3$, CD$_3$CN, DMSO) favors the ring form. Basic or acidic solvents as well as raising the temperature shift the equilibrium toward the open form. Thermodynamic and activation parameters of the interconversions were determined [77].

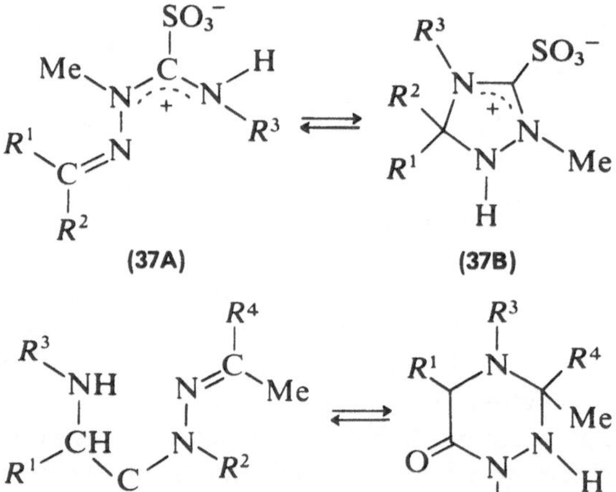

(37A)   (37B)

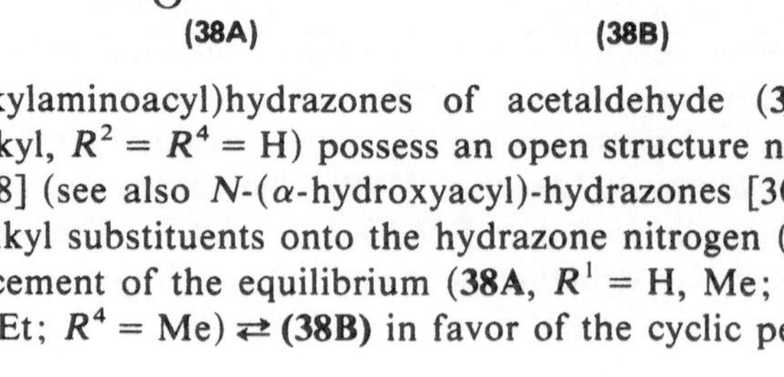

(38A)   (38B)

N-(α-Alkylaminoacyl)hydrazones of acetaldehyde (**38A**, R^1 = H, alkyl, R^3 = alkyl, $R^2 = R^4$ = H) possess an open structure neat as well as in solution [78] (see also N-(α-hydroxyacyl)-hydrazones [30]). However, introducing alkyl substituents onto the hydrazone nitrogen (R^2 = Me, Pr) causes displacement of the equilibrium (**38A**, R^1 = H, Me; R^2 = Me, Pr; R^3 = H, Me, Et; R^4 = Me) $\rightleftarrows$ **(38B)** in favor of the cyclic perhydro-1,2,4-

triazine-6-one [79]. Equilibrium constants, determined by the ^{1}H-NMR method, show that the increasing steric bulk of substituents R^2 displaces the equilibrium toward the ring form; an analogous substitution at R^3 (Me < Et < i-Pr) acts in the opposite direction. It seems noteworthy that an introduction of a methyl group at the α-position of aminoacid (R^1 = Me) also destabilizes the cyclic form somewhat. Increasing temperature and solvent polarity displaces the equilibrium toward the open form.

Ring-chain equilibria (39A) $\rightleftarrows$ (39B) of N-(2-hydroxyiminoalkyl)hydrazones have been described [80]. Here, the cyclic isomer is formed as a result of an intramolecular addition of a hydroxylamine nitrogen atom to the hydrazone C=N bond, the hydroxylamino group changing into a nitrone function.

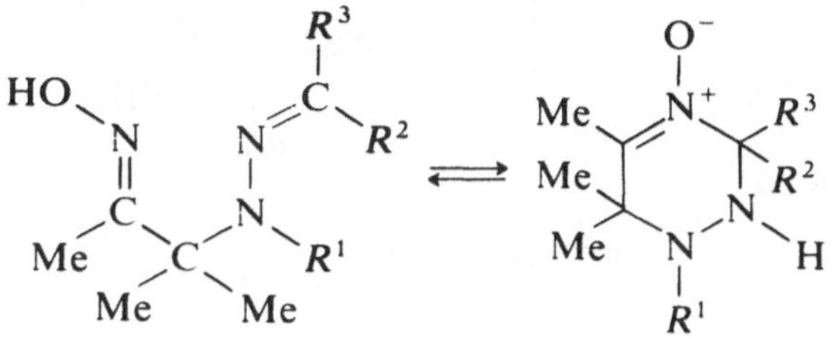

(39A)   (39B)

3.2.3. 2-Benzimidoyl-benzamides and -benzenesulfonamides

Ring-chain tautomerism caused by the intramolecular addition of CONHR and SO$_2$NHR groups to the C=N group of imines and anils was observed in a series of 2-benzimidoyl-benzamides and -benzenesulfonamides (40) and (41).

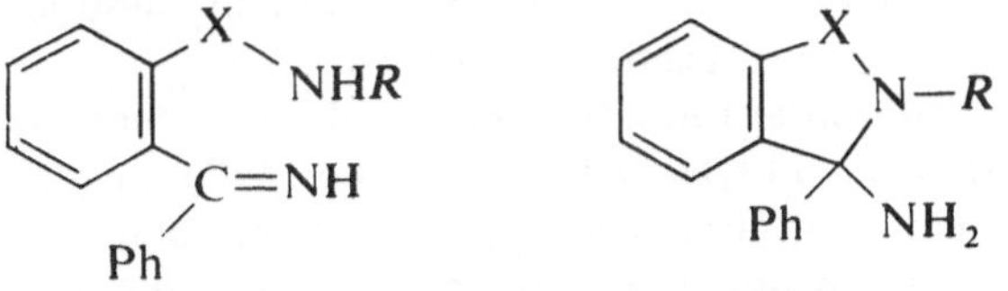

(A)   (B)

(40, X = CO)
(41, X = SO$_2$)

Both imine and anil isomers of 2-benzoylbenzamides were obtained. 3-Amino-2-methyl-3-phenylisoindolinone (40B, R = Me) is formed in a condensation reaction of the dilithium salt of N-methylbenzamide with benzonitrile [81]. In an analogous reaction the open N-phenyl-2-benzimidoylbenzamide (40A, R = Ph) was obtained. The isomerization

(40A) → (40B) proceeds easily in $CDCl_3$ at room temperature or on heating in ethanol.

Two reactions of 2-R^1-3-aryl-3-chloroisoindolinones **(42)** with primary amines (R^2NH_2) have been observed [82, 83]. Depending on the structure of the substituents R^1 and R^2 either the chlorine atom of **(42)** is substitued forming 2-R^1-3-amino-3-arylisoindolinones **(43)** or an addition occurs at the carbonyl group of **(42)**, leading to the N-alkyl-2-aroylbenzamide anils **(44A,** R^1 = Ph**)** and the cyclization products, i.e., 2-R^2-3-phenylamino-3-phenylisoindolinones **(44B,** R^1 = Ph**)**, subsequently [82, 83].

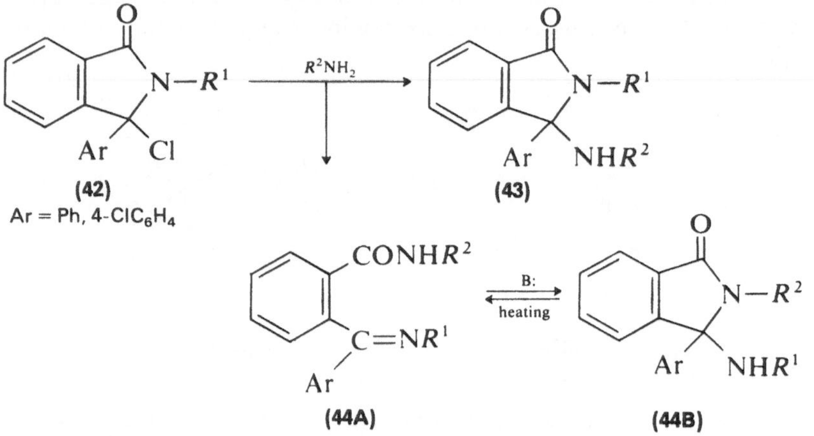

(42)
Ar = Ph, 4-ClC_6H_4

(43)

(44A)

(44B)

Increasingly sterically demanding alkyl substituents R^2 hinder the cyclization **(44A,** R^1 = Ph**)** → **(44B)** of N-alkyl-2-benzoylbenzamide anils: in ethanol at room temperature the N-benzylamide isomerizes slower than the N-propylamide, but N-isopropyl- and N-t-butylamides are stable under these conditions. A cyclization was carried out in these cases in boiling ethanol in the presence of triethylamine.

In dioxan solution at room temperature both isomers, i.e., **(44A,** Ar = Ph, R^1 = Ph, R^2 = Pr, PhCH$_2$, i-Pr, t-Bu**)** and **(44B)** are stable. No equilibrium was formed even after several days at room temperature.

The isomerization **(44A)** → **(44B)** is favored by increasing polarity of the C=N bond, i.e., by an introduction of electron-withdrawing substituents R^1 at the nitrogen atom. Therefore reactions of 2-(4-nitrophenyl) and 2-(2-pyridyl)-3-chloro-3-phenylisoindolinones **(42)** with isopropylamine gave 2-isopropyl-3-(4-nitrophenylamino- or 2-pyridylamino)-3-phenylisoindolinones **(44B,** Ar = Ph, R^1 = 4-$NO_2C_6H_4$, 2-pyridyl, R^2 = i-Pr**)** instead of the expected open isomers **(44A)**.

2-($tert$-Butyl)-3-(4-chlorophenyl)-3-phenylaminoisoindolinone **(44B,** Ar = 4-ClC_4H_4, R^1 = Ph, R^2 = t-Bu**)** have been obtained from 2-($tert$-

butyl)-3-chloro-3-(4-chlorophenyl)isoindolinone **(42)** and aniline [83, 84].
A thermal isomerization **(44B)** → **(44A)** was carried out for the first time by
a short heating at 220°C. The reverse isomerization **(44A)** → **(44B)** proceeds
in boiling ethanol in the presence of triethylamine. Both isomers are stable
in dioxan at room temperature.

A thermal isomerization **(44B)** → **(44A)** for 2-(*n*- or *sec*-alkyl)-3-pheny-
lamino-3-phenylisoindolinones could not be achieved [82]. Obviously, the
above-mentioned example of isomerization is favored by the sterically
demanding *tert*-butyl substituent in position 2 of isoindolinone thus destabilizing the cyclic structure **(44B)** [83].

Condensation of dilithium salts of *N*-substituted benzenesulfonamides
with benzonitrile yielded the cyclic *N*-methyl-2-benzoylbenzensulfonamide
imine **(41B, R = Me)**, while the *N*-phenylderivative was obtained as a
mixture of both isomers **(41A, R = Ph)** and **(41B)** with a predominance of
the open isomer [81].

Similarly *N*-benzyl **(41B, R = PhCH$_2$)** and *N*-phenylderivatives **(41B,
R = Ph)** have been obtained and the last mentioned compound was isomerized in both directions: the conversion **(41B)** → **(41A)** could be effected at
140°C or by recrystallization from CCl$_4$. The reverse reaction was observed
by heating in ethanol in the presence of triethylamine. Attempts to carry
out a thermal ring opening of the *N*-benzyl-derivative **(41B, R = PhCH$_2$)**
were not successful. Evidently, the presence of a phenyl substituent at the
sulfonamide nitrogen atom destabilizes the cyclic structure to a greater
extent.

The introduction of sterically demanding *sec*- and *tert*-alkyl substituents
at the sulfonamide nitrogen atom stabilizes the open structure. Therefore
cyclic isomers of sulfonamides **(41A, R = i-Pr, t-Bu)** have not been observed
[85]. In solutions of **(41, R = PhCH$_2$, Ph)** the ring-chain equilibrium is
shifted toward the cyclic form.

Surprisingly, acylation of the open isomers **(41A, R = i-Pr, t-Bu, Ph)**
as well as the cyclic isomer **(41B, R = Ph)** yielded only cyclic *N*-acylderiva-
tives **(45B)**.

(45A)   **(45B)**

Evidently, the introduction of an electron-withdrawing acyl group at
the imine nitrogen atom of the sulfonamide molecule **(41A)** increases the

electrophilicity of the C=N carbon atom favoring an intramolecular addition of the SO$_2$NHR group to this bond even with bulky substituents at the nitrogen atom.

In solution the equilibrium (45A) ⇌ (45B) is observed. Increasing solvent polarity shifts the equilibrium toward the cyclic form (45B).

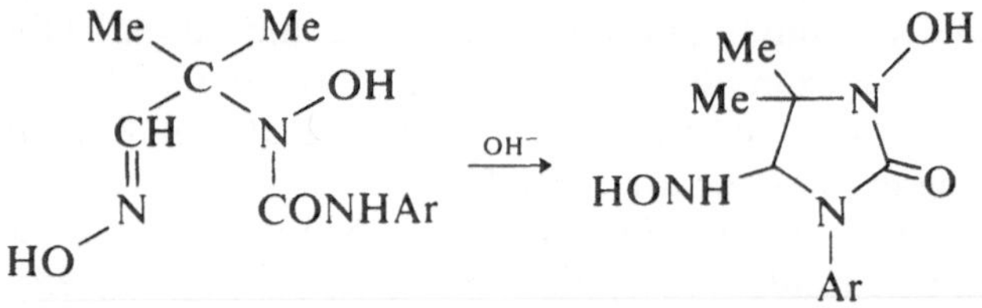

(46A)   (46B)

An interesting example of an intramolecular addition of the NH of a carbamide group to the aldoxime C=N bond leading to 2-imidazolidine (46B) has been reported [86, 87] for N-arylcarbamoyl-N-(1-oximino-2-methyl-2-propyl)hydroxylamine (46A). The cyclization requires alkaline catalysis.

3.2.4. N-(2- and 3-Hydroxyiminoalkyl)nitrones

Koptyug, Volodarskii, and Putsykin, studying the condensation of anti-α-hydroxylaminooximes (47) with aldehydes and ketones, have observed an interesting type of ring-chain tautomerism [88, 89]. N-(2-Hydroxyiminoalkyl)nitrones (48A) are the primary products of these reactions. Their cyclization into 1-hydroxy-Δ^3-imidazoline-3-oxides (48B) may be considered as an intramolecular addition of the NOH nitrogen to the nitrone C=N bond transforming the hydroxylamino group into a nitrone function and the initial nitrone into a hydroxylamino group on the other side.

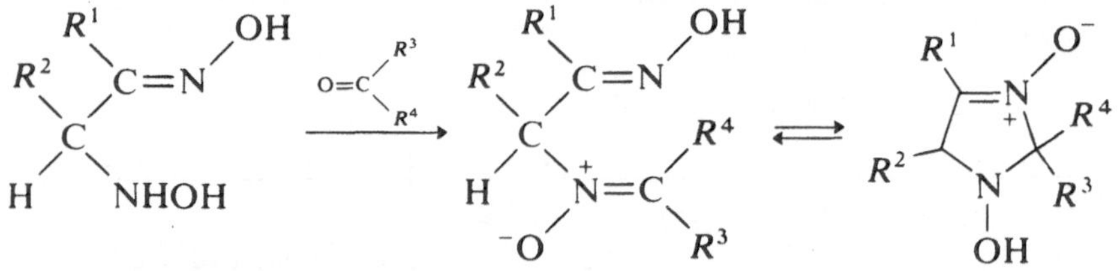

(47)   (48A)   (48B)

Although ring formation occurred as a result of an addition of an oxime group to a nitrone C=N bond, the molecule still contains both functional groups.

The condensation of anti-α-hydroxylaminooximes with benzaldehyde gave only N-(2-hydroxyiminoalkyl)-α-phenylnitrones (**48A**, R^1 = Me, Ph; R^2 = H, Me; $R^1 = R^2 = (CH_2)_4$; $R^3 = H$; R^4 = Ph) which are stabilized by a conjugation of the nitrone group with the benzene ring [45, 90–92].

Heating aromatic anti-α-hydroxylaminooximes (**47**, R^1 = Ph, R^2 = H, Me) with acetaldehyde [92, 93], acetone [94], or orthoformic ester [95] led to 2-substituted derivatives of 1-hydroxy-4-phenyl-Δ^3-imidazoline-3-oxides (**48B**, R^1 = Ph, R^2 = Me, R^3 = H, R^4 = Me, OEt; $R^3 = R^4$ = Me).

Products of condensation of aliphatic (**47**, $R^1 = R^2$ = Me) and alicyclic (**47**, R^1, $R^2 = (CH_2)_n$, n = 4, 5) anti-α-hydroxylaminooximes with acetaldehyde in solution (chloroform, methanol, DMSO) form equilibrium mixtures (**48A**) $\rightleftarrows$ (**48B**). Two stereoisomeric ring forms (**48B**) differing in configuration (cis and trans) of the substituents R^2 and R^4 (R^3 = H) and one open form (**48A**, R^3 = H) have been established by ^{1}H-NMR spectroscopy [88].

Passing from CDCl$_3$ to more polar solvents shifts the equilibrium toward the open form. A linear correlation between the logarithm of the equilibrium constant and the dipole moment of the solvent has been observed [88] for the equilibrium (**48A**, $R^1 = R^2 = R^4$ = Me, R^3 = H) $\rightleftarrows$ (**48B**).

Increasing the temperature shifts the equilibrium in DMSO toward the open form. A linear dependence of lg K_T on $1/T$ was detected [88]. Thermodynamic parameters ($-\Delta H$ = 1.69 kcal/mol and $-\Delta S$ = 4.1 e.u.) are near to those obtained for the tautomeric equilibria of oxazolidines (**3A**) $\rightleftarrows$ (**3B**) (see Table 30) [4].

In the series of N-(2-hydroxyiminocycloalkyl)α-methyl-nitrones (**48**, R^1, $R^2 = (CH_2)_n$, n = 3, 4, 5; R^3 = H, R^4 = Me) a decreasing size of the cycloalkane ring stabilizes the open form (**48A**): in solution the equilibrium of the cycloheptyl-derivative (n = 5) is totally displaced toward the cyclic form (**48B**), the corresponding cyclohexane derivative (n = 4) shows K_T = 9 and when n = 3 no cyclic isomer has been found in deuteriomethanolic solution.

Surprisingly, the condensation of N-(1-hydroxyimino-2-cyclo-pentyl)hydroxylamine with formaldehyde forms an equilibrium mixture (K_T = 3) in solution, thus differing fundamentally from the condensation product with acetaldehyde (K_T = 0). The replacing of the methyl group in the nitrone function (**48A**, R^4 = Me) with a hydrogen atom (**48A**, R^4 = H) increases the electrophilicity of the carbon atom of the C=$\overset{+}{N}$—$\overset{-}{O}$ bond facilitating the cyclization even in case of a less favorable structure of the connecting link.

Six-membered rings are formed from N-(3-hydroxyiminoalkyl)-nitrones. Recently it has been shown [96, 97] that condensation products of β-hydroxylaminooximes with formaldehyde are cyclic 1-hydroxy-1,2,5,6-tetrahydropyrimidine-3-oxides (**49B**, $R^5 = R^6 = $ H); with acetone the open structure N-(3-hydroxyiminoalkyl)-α, α-dimethylnitrone (**49A**, $R^5 = R^6 = $ Me) was obtained, but the condensation product with acetaldehyde was shown to give an equilibrium mixture (**49A**, $R^5 = $ H, $R^6 = $ Me) $\rightleftarrows$ (**49B**) in solution.

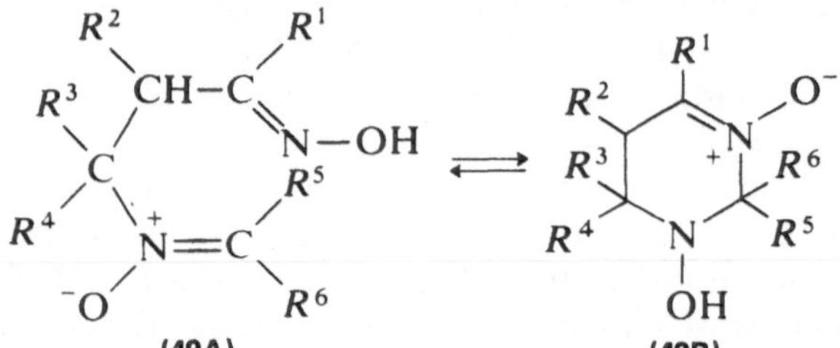

(49A)   (49B)

Introduction of methyl groups ($R^3 = R^4 = $ Me) into the chain between interacting groups shifts the equilibrium toward the cyclic form. A cyclohexane ring (**49**, R^1, $R^2 = (CH_2)_4$, $R^3 = R^4 = R^5 = $ H, $R^6 = $ Me) [97] resulted in an additional displacement of the equilibrium toward the ring form (compared with (**49**, $R^1 = R^2 = $ Me), containing two methyl groups instead of the tetramethylene bridge [96]).

Summarizing, one can state that introduction of methyl substituents at the α-carbon of a nitrone group decreases its ability for intramolecular additions as follows:

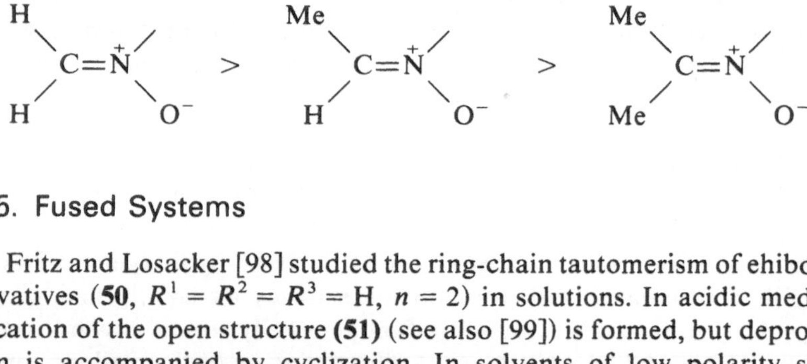

3.2.5. Fused Systems

Fritz and Losacker [98] studied the ring-chain tautomerism of ehiboline derivatives (**50**, $R^1 = R^2 = R^3 = $ H, $n = 2$) in solutions. In acidic medium a dication of the open structure (**51**) (see also [99]) is formed, but deprotonation is accompanied by cyclization. In solvents of low polarity some ehiboline derivatives (**50**, $R^1 = $ H, $R^2 = $ H, Me; $R^3 = $ MeO) exist in the ring form (**50B**), but in methanol solution an equilibrium was observed. D ring opening is facilitated by 1) unsubstituted nitrogen atoms ($R^1 = R^2 = $ H); 2) a methoxy group in the benzene ring ($R^3 = $ MeO) which lowers the electrophilicity of the C=N carbon atom in (**50A**), and 3) increasing the size of ring D. In solution, ehiboline itself (**50B**, $R^1 = R^2 = R^3 = $ H, $n = 2$)

is cyclic; for $n = 3$ an equilibrium **(50A)** $\rightleftarrows$ **(50B)** was observed, but for $n = 4$ no cyclic isomer could be detected.

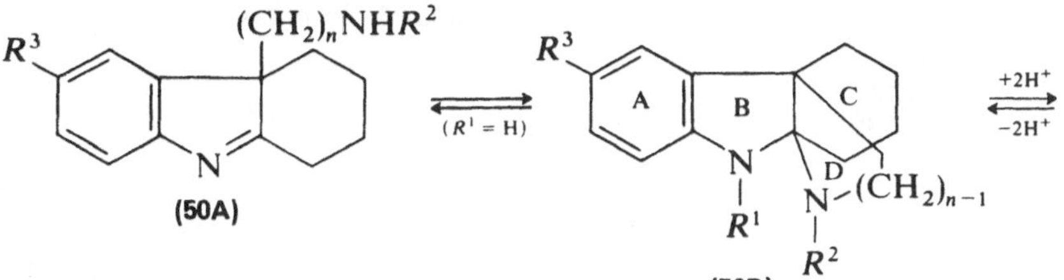

Similar influences of substituents, solvents, and medium acidity on the stability of pyrrolidine or piperidine derivatives was observed [100–102] also for substituted 2,3,3a,8a-tetrahydropyrrolo[2,3-*b*]- and 1,2,3,4,4a,9a-hexahydropyrido-[2,3-*b*]-indoles, which are analogues of the alkaloid physostigmine (eserine).

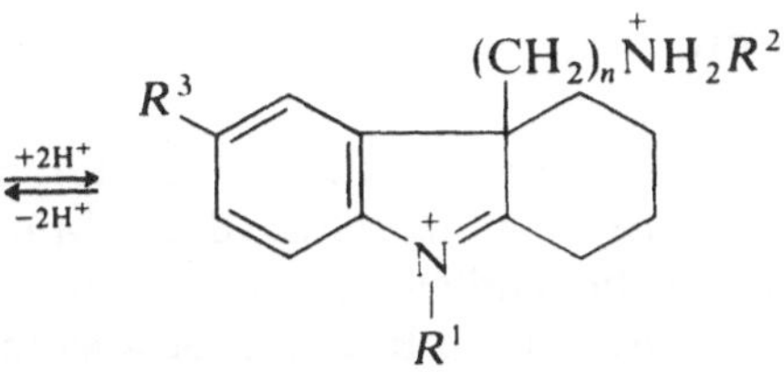

Comparing the conditions for ring opening of these compounds with those for 2,3,3a,8a-tetrahydrofuro[2,3-*b*]indole derivatives **(25B)** already discussed led the authors to the conclusion that opening of a pyrrolidine ring is more difficult (proceeds at lower pH values) than opening of the furanidine ring [60, 102]. Compared to a C—O bond the less polarized C—N bond gives rise to this difference. A general conclusion can be made: Cyclic isomers being formed by intramolecular addition of NH groups to a C=N bond possess a greater stability than those from an addition of the less nucleophilic OH⁻ group.

3.3. S-H-DERIVATIVES OF IMINES AND HYDRAZONES

Condensation products of 2-mercaptoethylamine [103–105], 3-mercapto-2-butylamine [103] and 2-mercaptoaniline [106–108] with carbonyl compounds possess exclusively the cyclic structure **(52)**. The formation of open derivatives in the course of alkylation **(52)** → **(54)** or oxidation **(52)** → **(55)** in strong alkaline media is an indirect evidence for anions of open structure **(53)** [104, 106].

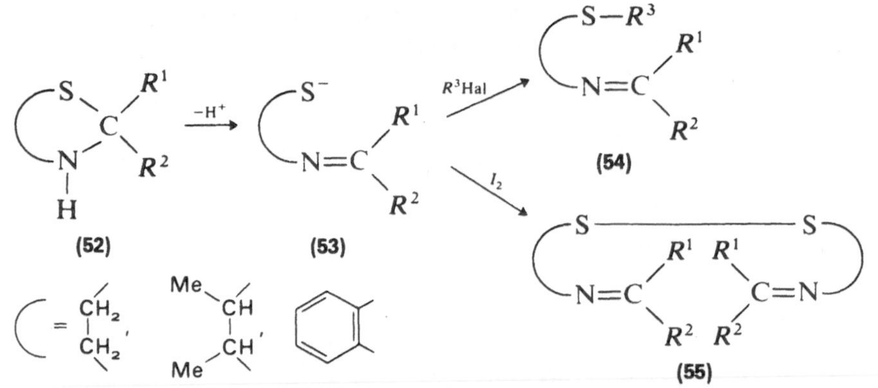

(52) (53) (54) (55)

3-Mercaptopropylamine in reactions with formaldehyde and cyclohexanone forms the corresponding tetrahydro-1,3-thiazines, but with benzaldehyde a mixture of ring and chain isomers with a predominance of the Schiff's base was obtained [109]. Reaction of 4-mercaptobutylamine with cyclohexanone yielded the corresponding hexahydro-1,3-thiazepine [110].

The ring-chain equilibrium of mercaptoalkylimines was used [111] for an explanation of the mechanism of the ring contraction of 5,6-dihydro-1,4-thiazine under the action of elemental sulfur.

Potekhin and Shevchenko [112–117] investigated in detail the ring-chain tautomerism of aldehyde and ketone N-(2-mercaptoalkyl)hydrazones (56A) $\rightleftarrows$ (56B) yielding perhydro-1,3,4-thiadiazines as a result of an intramolecular addition of a mercapto group to the hydrazone C=N bond. ^{1}H-NMR spectroscopy allowed a determination of the equilibrium constants at different temperatures in the range of 30–90°C and thermodynamic parameters ΔH and ΔS were calculated (see Table 38).

Table 38. Ring-Chain Equilibrium (56A, R^1 = Me) $\rightleftarrows$ (56B) Constants and Thermodynamic Parameters Determined by ^{1}H-NMR using Tetrachloroethylene

R^2	R^3	R^4	R^5	R^6	$K_T = [B]/[A]$ at 50°C	$-\Delta H$ (kcal/mol)	$-\Delta S$ (e.u.)	Refs.
Me	H	H	H	H	0.11[a]	3.20	14.3	114
Me	H	H	Me	H	0.29[a]	3.78	13.6	114
Me	H	Me	H	H	0.46	5.07	17.2	117
Et	H	Me	H	H	0.16	4.13	16.3	117
t-Bu	Me	H	H	H	1.09	7.53	22.9	116
t-Bu	Me	H	Me	H	2.63	6.81	19.4	116
t-Bu	Me	H	Me	Me	1.09	5.69	17.7	116
t-Bu	Pr	H	H	H	2.22	6.17	17.4	116

[a] At 52°C.

(56A)   (56B)

First, it should be noted that, compared to the similarly substituted
N-(2-hydroxyalkyl)hydrazones (11), N-(2-mercaptoalkyl)hydrazones (56)
show a significantly greater preference for cyclization to give 1,3,4-
thiadiazines. Thus, oxygen analogues (11), bearing no alkyl substituents at
the nitrogen atom ($R^3 = H$), are not at all capable of forming cyclic isomers
[27, 28, 32]. It has been pointed out [114, 116] that the energy of the
intramolecular SH$\cdots$N bond is low compared to that of the OH$\cdots$N bond.
In addition, the ring strain decreases on passing from the oxygen containing
heterocycle (11B) to its sulfur analogue (56B) (see also [118]). Evidently,
increasing nucleophilicity in the sequence $O < N < S$ [119] plays a significant role also.

The introduction of alkyl substituents at the nitrogen atom (56, $R^3 =$
Me, Pr) (similar to that already observed [27-29, 32, 33, 35] in the series
of N-alkyl-N-(2-hydroxyethyl)hydrazones (11B, $R^3 =$ alkyl)) strongly
stabilizes the cyclic form. Thus, the majority of N-alkylderivatives (56,
$R^3 =$ Me, Et, Pr, i-Pr) exist as stable cyclic isomers (56B), and a tautomeric
equilibrium can be observed only if bulky substituents R^2 are present
($R^2 = t$-Bu, see Table 38). The cyclic form is stabilized as well by subsequent
introduction of methyl groups in positions 5 or 6 of the 1,3,4-thiadiazine
ring ($R^4 =$ Me or $R^5 =$ Me), methyl substituents in position 5 ($R^4 =$ Me)
exerting a stronger influence. However, the introduction of the second
methyl group in position 6 of the thiadiazine ring ($R^5 = R^6 =$ Me) destabilizes the cyclic isomer, which can be explained [116] by destabilizing nonbonded interactions between syn-diaxial groups (2-Me and 6-Me) in the chair
conformation of the 1,3,4-thiadiazine.

An increase in the number and steric bulk of substituents in position
2 displaces the equilibrium toward the open form as already observed for
N-(2-hydroxyalkyl)hydrazones (11).

Ring-chain equilibria (56A) $\rightleftarrows$ (56B) were detected [115] in the series
of N-(2-mercaptoalkyl)hydrazones of aromatic aldehydes and acetophenones (56, $R^1 =$ H, Me; $R^2 = C_6H_4X$). Surprisingly, on passing from the
benzaldehyde hydrazone (56, $R^2 =$ Ph, $R^3 =$ Pr, $R^5 =$ Me, other $R =$ H)
($K_T = 0.64$, in tetrachloroethylene at 90°C) to the acetophenone hydrazone
(56, $R^1 =$ Me, other substituents are the same) ($K_T = 11.1$) the equilibrium

is displaced toward the cyclic form. A similar effect was observed [34] also for oxygen analogues **(11)**. Compared to alkylderivatives **(56,** R^2 = alkyl), aryl analogues (R^2 = aryl) equilibrate **(56A)** ⇄ **(56B)** considerably more slowly, requiring more than 5 hours at 90°C in tetrachloroethylene containing 0.1% CF_3COOH.

In the series **(56,** R^1 = H, R^2 = 4-XC_6H_4, R^3 = Pr, R^5 = Me) the dependence of the equilibrium constants K_T on the polar character of substituents X is insignificant, and not even an approximate correlation has been observed between $\lg K_T$ and σ- or σ^+-coefficients [115].

Mayer and Lauerer [120] using ^{1}H-NMR spectroscopy were able to detect the equilibrium **(57A)** ⇄ **(57B)** in solutions of condensation products of dithiocarbazine acids with aldehydes and ketones. The anions, being formed in alkaline media, possess the open structure **(58)**.

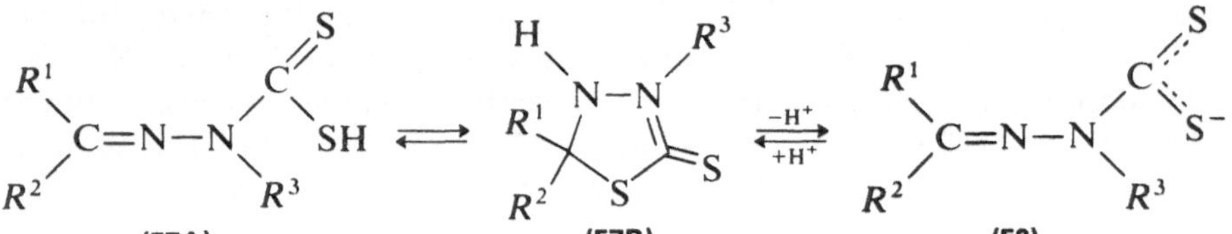

The authors [120] assumed an equilibrium **(59A)** ⇄ **(59B)** in acidic solutions of N_1-thioacylhydrazones and **(61A)** ⇄ **(61B)** in the case of N_2-substituted thiosemicarbazones of aldehydes and ketones.

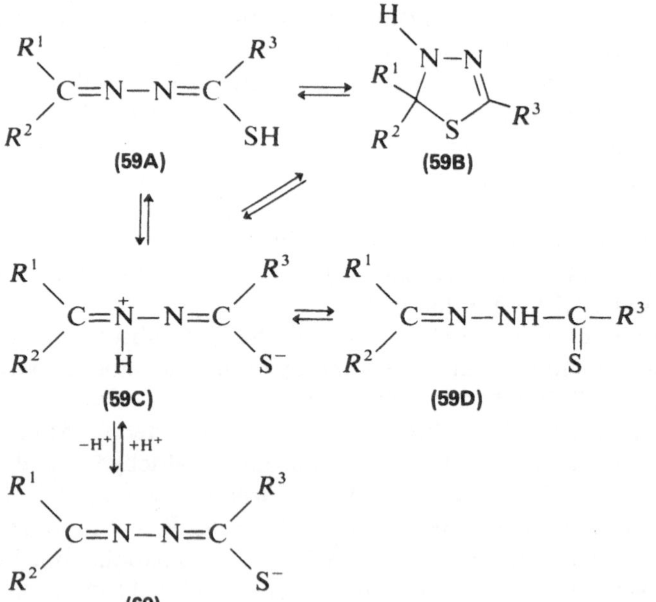

(61A)   **(61B)**

Later, it was demonstrated [121] that the condensation products of acetone and 2-butanone with thiobenzhydrazide are cyclic (**59B**, R^1 = Me, R^2 = Me, Et, R^3 = Ph), but for benzaldehyde phenylthioacetylhydrazone the open structure was supported by spectroscopic methods, the bipolar structure (**59C**, R^1 = H, R^2 = Ph, R^3 = PhCH$_2$) being preferred by the authors [122]. ^{1}H-NMR investigations showed that an equilibrium (**59C**) ⇌ (**59B**) occurs rapidly in CDCl$_3$, benzene, or acetone-d$_6$ solutions. Passing from acetone (K_T = 0.33) to benzene(K_T = 5), the equilibrium is strongly displaced in favor of the cyclic isomer (**59B**).

Acetone thiosemicarbazones have been investigated in more detail by ^{1}H-NMR spectroscopy [123]. Thiosemicarbazone (**61**, R^1 = R^2 = Me, R^3 = R^4 = H) and its 2-methyl (R^3 = Me) and 4-methyl (R^4 = Me) derivatives are open structured (**61A**) in DMSO-d$_6$ solution, but cyclic (**61B**) in CF$_3$COOD. In these solvents acetone 2,4-dimethylthiosemicarbazone exists exclusively in the ring form (**64B**, R^1 = R^2 = R^3 = R^4 = Me).

As a general rule cyclic isomers of imine mercaptoderivatives form exclusively open structured anions (**53**), (**58**), (**60**).

(62B)   **(62A)**   **(62C)**

An interesting example of a tautomeric equilibrium between two cyclic forms (**62B**) ⇌ (**62C**) was presented by Zelenin and coworkers [124]. The cyclic (**62B**) obtained in a solution of 2,4-pentanedione thiobenzhydrazone is the result of an intramolecular addition of a NH group to a C=O bond. The second cyclic isomer (**62C**) can be considered as being formed as the result of an intramolecular addition of a SH group to a C=N bond. Therefore interconversions (**62B**) ⇌ (**62C**) most likely proceed through an intermediate hydrazone (**62A**) which nevertheless could not be detected by ^{1}H and ^{13}C-NMR methods. In the sequence CDCl$_3$, CCl$_4$, CD$_3$OD, and DMFA-d$_7$ the equilibrium shifts toward (**62B**).

3.4. MISCELLANEOUS

Recently, using IR, ^{1}H, ^{13}C, and ^{29}Si-NMR spectroscopy a new type of ring-chain tautomerism of 2-trimethylsiloxyphenyl isocyanate **(63B)** $\rightleftarrows$ **(63A)** $\rightleftarrows$ **(63C)** was detected [125]. Here the Me$_3$SiO group is added intramolecularly either to the C=O **(63C)** or the C=N **(63B)** bond of the isocyanate. Rather high activation energies have been reported for the transitions **(63A)** → **(63B)** and **(63A)** → **(63C)**.

(63C)   **(63A)**   **(63B)**

An analogous reversible rearrangement was also established [126] for the trimethylsilyl ester of 2-isocyanatobenzoic acid **(64A)** $\rightleftarrows$ **(64B)**, but the second cyclic isomer **(64C)** could not be detected. The cyclic **(64B)** was isolated; it was stable at room temperature and even on boiling in CCl$_4$ for 3 hours. The equilibrium **(64A)** $\rightleftarrows$ **(64B)** only occurs on melting of **(64B)**. Increasing temperature shifts the equilibrium toward the open isomer **(64A)**. Equilibrium constant $K_T = [(64B)]^2/[(64A)]^2 = 0.174$ at 128°C has been reported assuming that the isomerization is bimolecular. Thermodynamic parameters have been found: $\Delta H = -6.66$ kcal/mol and $\Delta S = -20.4$ e.u. A comparatively high activation energy $\Delta H^{\neq} = 22.8$ kcal/mol was detected for the transition **(64B)** $\rightleftarrows$ **(64A)**.

(64A)   **(64B)**   **(64C)**

Recently Medyantseva, Minkin, and co-workers [127–131] have observed and by using UV, IR, and ^{1}H-NMR spectroscopy thoroughly studied an interesting type of tautomerism **(65A)** $\rightleftarrows$ **(65A′)** $\rightleftarrows$ **(65B)**, which can be related to the ring-chain tautomerism or simultaneously to valence isomerism.

Formally the ring tautomers **(65B**, X = NR, O) are formed as a result of an intramolecular addition of a phenolic OH group to the C=N or C=O bond. However, the authors supposed that the transformation

(65A) ⇌ **(65A′)** precedes ring formation, and then a valence isomerization **(65A′)** → **(65B)** results in ring closure.

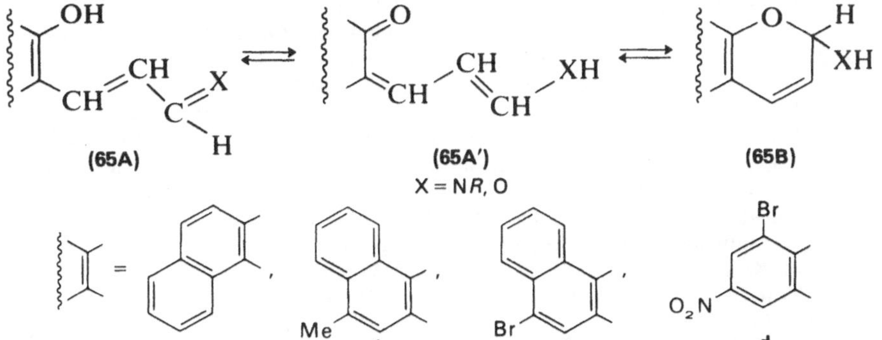

(65A)   **(65A′)**   **(65B)**

X = NR, O

a   b   c   d

E-3-(2-Hydroxy-1-naphthyl)propenal (**65a**, X = O) and its N-alkyl and N-aryl imines (**65a**, X = NR) possess cyclic structure (**65B**) in the solid state and in non-polar solvents, but on passing to the polar solvents the equilibrium is displaced in favor of the quinonoid tautomer (**65A′**). Some N-alkylimines (**65a**, X = NR, R = i-Pr, Bu, t-Bu) exist as open isomers (**65A′**) in the solid state. Cyclic isomer (**65B**) and two supposed E-Z-isomers of quinonoid structure (**65A′**) have been isolated for N-methylimine (**65a**, X = NMe) [128].

E-3-(1-Hydroxy-4-methyl-2-naphthyl)propenal (**65b**, X = O) [129] and its 4-bromoanalogue (**65c**, X = O) [130] exist preferentially in cyclic form (**65B**) in the solid state and in non-polar solvents. Using ^{1}H-NMR spectroscopy the equilibria (**65A′a, b, c**) ⇌ (**65Ba, b, c**) have been observed in DMSO solution [129, 130]. In the sequence of solvents acetonitrile, acetone, DMSO the equilibrium is shifted toward the open isomer (**65A′**).

Generally passing from the benzo[e]- (**65a**) to the benzo[c]-derivatives (**65b, c**) of cinnamic aldehyde the equilibrium is shifted toward the open quinonoid isomer (**65A′**).

3-Bromo-2-hydroxy-5-nitrocinnamic aldehyde (**65d**) possesses structure (**65A**, X = O) in the solid state and in DMSO solution, but its N-arylimines (**65d**, X = NAr) are cyclic (**65B**) in the solid state as well as in solvents of different polarity [131].

REFERENCES

1. W. P. Jencks, In: *Progress in Physical Organic Chemistry*, Ed. S. G. Cohen, A. Streitwieser, Jr., and R. W. Taft, Interscience Publishers, New York, 1964, Vol. 2, p. 63.
2. E. D. Bergmann, *Chem. Rev.* **1953**, *53*, 309.
3. E. D. Bergmann, E. Gil-Av, and S. Pinchas, *J. Am. Chem. Soc.* **1953**, *75*, 358.

*4. J. V. Paukstelis and R. M. Hammaker, *Tetrahedron Lett.* **1968**, 3557.

5. J. V. Paukstelis and L. L. Lambing, *Tetrahedron Lett.* 1970, 299.

6. M. F. Rennekampf, J. V. Paukstelis, and R. G. Cooks, *Tetrahedron* **1971**, *27*, 4407.

7. K. Pihlaja and K. Aaljoki, *Finn. Chem. Lett.* **1982**, 1.

8. H. Alper, L. S. Dinkes, and P. J. Lennon, *J. Organomet. Chem.* **1973**, *57*, C12–C14.

9. A. F. McDonagh and H. E. Smith, *J. Org. Chem.* **1968**, *33*, 1.

10. E. D. Bergman and A. Kaluszyner, *Rec. Trav. Chim.* **1959**, *78*, 315.

11. R. Kotani, T. Kuroda, T. Isozaki, and S. Sumoto, *Tetrahedron* **1969**, *25*, 4743.

12. E. Belgodere, R. Bossio, V. Parrini, and R. Pepino, *J. Heterocycl. Chem.* **1977**, *14*, 957.

13. E. Belgodere, R. Bossio, V. Parrini, and R. Pepino, *J. Heterocycl. Chem.* **1980**, *17*, 1629.

14. A. F. McDonagh and H. E. Smith *J. Chem. Soc., Chem. Commun.* **1966**, 374.

15. H. Kanatomi and I. Murase, *Bull. Chem. Soc. Jpn.* **1970**, *43*, 226.

16. H. E. Smith and N. E. Cooper, *J. Org. Chem.* **1970**, *35*, 2212.

17. E. M. Kosower, *J. Am. Chem. Soc.* **1958**, *80*, 3253.

18. H. Griengl, A. Bleikolm, W. Grubbaurer, and H. Söllradl, *Justus Liebigs Ann. Chem.* **1979**, 392.

19. J. B. Lambert and M. V. Majchrzak, *J. Am. Chem. Soc.* **1980**, *102*, 3588.

20. A. F. McDonagh and H. E. Smith, *J. Org. Chem.* **1968**, *33*, 8.

21. S. I. Yakimovich and V. N. Nikolaev, *Zh. Org. Khim.* **1981**, *17*, 1104.

22. S. I. Yakimovich, V. N. Nikolaev, and N. V. Koshmina, *Zh. Org. Khim.* **1982**, *18*, 1173.

23. B. V. Ioffe and A. A. Potekhin, *Tetrahedron Lett.* **1967**, 3505.

24. A. A. Potekhin and B. V. Ioffe, *Dokl. Akad. Nauk SSSR* **1968**, *179*, 1120.

25. L. C. Dorman, *J. Org. Chem.* **1967**, *32*, 255.

26. Yu. P. Kitaev and B. I. Buzykin, *Hydrazones*, Nauka, Moscow, 1974, p. 95 (In Russian).

27. A. A. Potekhin, *Zh. Org. Khim.* **1971**, 7, 16.

28. A. A. Potekhin and B. D. Zaitsev, *Khim. Geterotsikl. Soedin* 1971, 301.

29. A. A. Potekhin and M. N. Vikulina, *Khim. Geterotsikl. Soedin.* **1971**, 1167.

30. A. A. Potekhin and V. M. Karel'skii, *Zh. Org. Khim.* **1971**, *7*, 2100.

31. B. L. Mil'man and A. A. Potekhin, *Khim. Geterotsikl. Soedin.* **1973**, 902.

32. A. A. Potekhin and T. F. Barkova, *Zh. Org. Khim.* **1973**, *9*, 1180.

33. A. A. Potekhin and E. A. Bogan'kova, *Khim. Geterotsikl. Soedin.* **1973**, 1461.

34. A. A. Potekhin and A. O. Safronov, *Zh. Org. Khim.* **1981**, *17*, 379.

35. B. V. Ioffe, M. A. Kuznetsov, and A. A. Potekhin, *Chemistry of Hydrazine Organic Derivatives*, Khimiya, Leningrad, 1979, p. 187 (In Russian).

36. J. T. Edvard, P. E. Morand, and I. Puskas, *Can. J. Chem.* **1961**, *39*, 2069.

37. J. E. Whitting and J. T. Edward, *Can. J. Chem.* **1971**, *49*, 3799.

38. W. Kliegel and H. Becker, *Chem. Ber.* **1977**, *110*, 2067.

*39. W. Kliegel and H. Becker, *Chem. Ber.* **1977**, *110*, 2090; See also W. Kliegel, B. Enders, and H. Becker, *Justus Liebigs Ann. Chem.* **1982**, 1712; W. Kliegel, B. Enders, and H. Becker, *Chem. Ber.* **1983**, *116*, 27.

40. H. Möhrle, M. Lappenberg, and D. Wendish, *Monatsh. Chem.* **1977**, *108*, 273.

41. H. Möhrle and K. Tröster, *Arch. Pharm.* **1981**, *314*, 836.

42. L. B. Volodarskii, Yu. G. Putsykin, and V. I. Mamatyuk, *Zh. Org. Khim.* **1969**, *5*, 355.

43. L. B. Volodarskii, A. Ya. Tikhonov, and L. A. Fust, *Izv. Sibirsk. Otd. Akad. Nauk SSSR, Ser. Khim.* **1971** (3), *7*, 91.

44. H. Ulrich and A. A. Sayigh, *Angew. Chem.* **1962**, *74*, 468.

45. J. Hamer and A. Macaluso, *Chem. Rev.* **1964**, *64*, 473.

46. Yu. V. Svetkin, M. A. Akmanova, and G. P. Plotnikova, *Zh. Org. Khim.* **1974**, *10*, 561.

47. S. L. Zhdanov, K. A. Ogloblin, and A. A. Potekhin, *Zh. Org. Khim.* **1975**, *11*, 1825.

* See also additions in proof on page 209.

48. A. A. Potekhin, S. L. Zhdanov, V. A. Gindin, and K. A. Ogloblin, *Zh. Org. Khim.* **1976**, *12*, 2090.

49. S. L. Zhdanov and A. A. Potekhin, *Khim. Geterotsikl. Soedin.* **1977**, 417.

50. A. A. Potekhin and S. L. Zhdanov, *Khim. Geterotsikl. Soedin.* **1979**, 1317.

51. A. A. Potekhin and S. L. Zhdanov, *Zh. Org. Khim.* **1979**, *15*, 1384.

52. H. R. Snyder, R. H. Levin, and P. F. Wiley, *J. Chem. Soc.* **1938**, *60*, 2025.

53. S. Wawzonek, H. Laitinen, and S. J. Kwiatkowski, *J. Am. Chem. Soc.* **1944**, *66*, 830.

54. R. E. Valters and S. P. Valtere, *Latv. PSR Zinat. Akad. Vestis, Kim. Ser.* **1969**, 753.

55. W. Flitsch, *Chem. Ber.* **1970**, *103*, 3205.

56. R. E. Valters and V. P. Ciekure, *Khim. Geterotsikl. Soedin.* **1972**, 502.

57. R. E. Valters and V. P. Ciekure, *Khim. Geterotsikl. Soedin.* **1975**, 1476.

58. R. E. Valters and V. R. Zinkovska, *Khim. Geterotsikl. Soedin.* **1973**, 1127.

59. V. R. Zinkovska and R. E. Valters, *Latv. PSR Zinat. Akad. Vestis, Kim. Ser.* **1974**, 207.

60. I. I. Grandberg and S. N. Dashkevich, *Khim. Geterotsikl. Soedin.* **1971**, 1194.

61. S. N. Dashkevich and I. I. Grandberg, *Dokl. Timiryazevsk. Sel'skokhozyaistv. Akad.* **1970**, *160*, 243.

62. C. Chapuis, A. Gavreau, A. Klaebe, A. Lattes, and J. J. Perie, *Bull. Soc. Chim. Fr.* **1973**, 977.

63. S. Witek, A. Bielawska, and J. Bielawski, *Heterocycles* **1980**, *14*, 1313.

64. R. F. Evans, *Austr. J. Chem.* **1967**, *20*, 1643.

65. W. Pöpel, G. Faust, H. Fürst, G. Dietz, and E. Carstens, *J. prakt. Chem.* **1968**, *38*, 339.

66. R. E. Harmon, J. L. Parson, and S. K. Gupta, *J. Org. Chem.* **1969**, *34*, 2760.

67. J. B. Lambert and M. W. Majchrzak, *J. Am. Chem. Soc.* **1979**, *101*, 1048.

68. J. E. Baldwin, *J. Chem. Soc., Chem. Commun.* **1976**, 734.

69. J. E. Baldwin, J. Cutting, W. Dupont, L. Kruse, L. Silberman, and R. C. Thomas, *J. Chem. Soc., Chem. Commun.* **1976**, 736.

70. J. E. Baldwin, R. C. Thomas, L. I. Kruse, and L. Silberman, *J. Org. Chem.* **1977**, *42*, 3846.

71. H. Piotrovska, W. Sas, and T. Urbanski, *Tetrahedron* **1977**, *33*, 1979.

72. H. Schildknecht and G. Hatzmann, *Justus Liebigs Ann. Chem.* **1969**, *724*, 226.

73. K. Pilgram, R. D. Skiles, and G. E. Pollard, *J. Heterocycl. Chem.* **1976**, *13*, 1257.

74. M. Uda and S. Kubota, *J. Heterocycl. Chem.* **1978**, *15*, 807.

75. V. A. Khrustalev, K. N. Zelenin, V. P. Sergutina, and V. V. Pinson, *Khim. Geterotsikl. Soedin.* **1980**, 1138.

*76. K. N. Zelenin, V. A. Khrustalev, V. P. Sergutina, and V. V. Pinson, *Zh. Org. Khim.* **1981**, *17*, 1825; see also V. A. Khrustalev, V. P. Sergutina, K. N. Zelenin, and V. V. Pinson, *Khim. Geterotsikl. Soedin.* **1982**, 1264.

77. W. Walter and Ch. Rohloff, *Justus Liebigs Ann. Chem.* **1977**, 485.

*78. P. S. Lobanov, A. N. Poltorak, and A. A. Potekhin, *Zh. Org. Khim.* **1978**, *14*, 1086.

79. P. S. Lobanov, A. N. Poltorak, and A. A. Potekhin, *Zh. Org. Khim.* **1980**, *16*, 2297.

80. V. A. Dokichev and A. A. Potekhin, *Zh. Org. Khim.* **1977**, *13*, 2617.

81. H. Watanabe, Ch.-L. Mao, I. T. Barnish, and Ch. R. Hauser, *J. Org. Chem.* **1969**, *34*, 919.

*82. R. E. Valters, *Khim. Geterotsikl. Soedin.* **1973**, 762.

83. R. E. Valters and G. A. Karlivans, *Khim. Geterotsikl. Soedin.* **1976**, 1207.

84. R. E. Valters and G. A. Karlivans, *Zh. Org. Khim.* **1976**, *12*, 238.

85. R. E. Valters, D. E. Balode, R. B. Kampare, and S. P. Valtere, *Khim. Geterotsikl. Soedin.* **1981**, 1209.

86. T. G. Kharlamova, Yu. A. Baskakov, and Yu. G. Putsykin, *Khim. Geterotsikl. Soedin.* **1975**, 715.

---

* See also additions in proof on page 209.

87. T. G. Kharlamova, Yu. G. Putsykin, and Yu. A. Baskakov, *Khim. Geterotsikl. Soedin.* **1976**, 1255.

88. Yu. G. Putsykin and L. B. Volodarskii, *Izv. Sibirsk. Otd. Akad. Nauk SSSR, Ser. Khim.* **1969** (4), 9, 86.

89. L. B. Volodarskii, *Khim. Geterotsikl. Soedin.* **1973**, 1299.

90. V. A. Koptyug, L. B. Volodarskii, and I. K. Baeva, *Zh. Obshch. Khim.* **1964**, *34*, 151.

91. L. B. Volodarskii and Yu. G. Putsykin, *Zh. Org. Khim.* **1967**, *3*, 1686.

92. L. B. Volodarskii and L. Ya. Tikhonov, *Zh. Org. Khim.* **1970**, *6*, 307.

93. L. B. Volodarskii, A. N. Lysak, and V. A. Koptyug, *Khim. Geterotsikl. Soedin.* **1966**, 766.

94. L. B. Volodarskii, A. N. Lysak, and V. A. Koptyug, *Khim. Geterotsikl. Soedin.* **1968**, 334.

95. L. B. Volodarskii and E. I. Vityaeva, *Zh. Org. Khim.* **1972**, *8*, 1887.

96. A. Ya. Tikhonov and L. B. Volodarskii, *Khim. Geterotsikl. Soedin.* **1977**, 252.

97. A. Ya. Tikhonov, L. B. Volodarskii, and O. M. Sokhatskaya, *Khim. Geterotsikl. Soedin.* **1979**, 1265.

98. H. Fritz and P. Losacker, *Justus Liebigs Ann. Chem.* **1967**, *709*, 135.

*99. M. Taniguchi and T. Hino, *Tetrahedron* **1981**, *37*, 1487.

100. I. I. Grandberg and T. A. Ivanova, *Dokl. Timiryazevsk. Sel'skokhozyaistv. Akad.* **1970**, *160*, 232.

101. A. H. Jackson and A. E. Smith, *J. Chem. Soc.* **1964**, 5510.

102. I. I. Grandberg, T. A. Ivanova, and N. G. Yarishev, *Khim. Geterotsikl. Soedin.* **1970**, 1276; I. I. Grandberg and T. A. Ivanova, *Khim. Geterotsikl. Soedin.* **1970**, 1489.

103. E. D. Bergmann and A. Kaluszyner, *Rec. trav. chim.* **1959**, *78*, 315.

104. G. W. Stacy and P. L. Strong, *J. Org. Chem.* **1967**, *32*, 1487.

105. G. W. Stacy and P. L. Strong, *J. Heterocycl. Chem.* **1968**, *5*, 101.

106. F. J. Goetz, *J. Heterocycl. Chem.* **1967**, *4*, 80.

107. F. J. Goetz, *J. Heterocycl. Chem.* **1968**, *5*, 501.

108. F. J. Goetz, *J. Heterocycl. Chem.* **1968**, *5*, 509.

109. E. D. Bergmann and A. Kaluszyner, *Rec. trav. chim.* **1959**, *78*, 327.

110. E. D. Bergmann and A. Kaluszyner, *Rec. trav. chim.* **1959**, *78*, 331.

111. F. Asinger, A. Saus, and D. Neuray, *Justus Liebigs Ann. Chem.* **1952**, *759*, 121.

112. S. M. Shevchenko, T. Ya. Vakhitov, and A. A. Potekhin, *Khim. Geterotsikl. Soedin.* **1978**, 1427.

113. A. A. Potekhin, S. M. Shevchenko, T. Ya. Vakhitov, V. A. Gindin, *Khim. Geterotsikl. Soedin.* **1978**, 1568.

114. S. M. Shevchenko and A. A. Potekhin, *Khim. Geterotsikl. Soedin.* **1979**, 1637.

115. A. A. Potekhin, V. V. Sokolov, and S. M. Shevchenko, *Khim. Geterotsikl. Soedin.* **1981**, 1040.

116. A. A. Potekhin, S. M. Shevchenko, T. Ya. Vakhitov, and V. A. Gindin, *Khim. Geterotsikl. Soedin.* **1981**, 1217.

117. A. A. Potekhin and S. M. Shevchenko, *Khim. Geterotsikl. Soedin.* **1981**, 1355.

118. M. I. Page, *Chem. Soc. Rev.* **1973**, *2*, 295.

119. R. E. Valters, *Uspekhi Khimii* **1982**, *51*, 1374.

120. K. H. Mayer and D. Lauerer, *Justus Liebigs Ann. Chem.* **1970**, *731*, 142.

121. K. N. Zelenin, V. A. Khrustalev, V. V. Pinson, and V. V. Alekseev, *Zh. Org. Khim.* **1980**, *16*, 2237; *See also* K. N. Zelenin, V. V. Alekseev, and V. A. Khrustalev, *Khim. Geterotsikl. Soedin.* **1983**, 769.

*122. V. V. Alekseev, V. A. Khrustalev, and K. N. Zelenin, *Khim. Geterotsikl. Soedin.* **1981**, 1569.

123. M. Uda, S. Kubota, *J. Heterocycl. Chem.* **1979**, *16*, 1273.

---

* See also additions in proof on page 209.

*124. V. A. Khrustalev, K. N. Zelenin, and V. V. Alekseev, *Zh. Org. Khim.* **1981**, *17*, 2451; see also S. I. Yakimovich, K. N. Zelenin, V. N. Nikolaev, N. V. Koshmina, V. V. Alekseev, and V. A. Khrustalev, *Zh. Org. Khim.* **1983**, *19*, 1875.

*125. M. G. Kuznetsova, N. V. Mironova, A. V. Kisin, V. P. Kozyukov, V. S. Nikitin, and N. V. Alekseev, *Zh. Obshch. Khim.* **1981**, *51*, 1096.

126. N. L. Chikina, Yu. V. Kolodyazhnii, V. P. Kozyukov, N. V. Mironova, and O. A. Osipov, *Zh. Obshch. Khim.* **1981**, *51*, 1803.

127. I. M. Andreeva, O. M. Babeshko, E. A. Medyantseva, and V. I. Minkin, *Zh. Org. Khim.* **1979**, *15*, 1899.

128. I. M. Andreeva, E. A. Bondarenko, N. V. Volbushko, M. I. Knyazhanskii, E. A. Medyantseva, A. V. Metelitsa, V. I. Minkin, and B. Ya. Simkin, *Khim. Geterotsikl. Soedin.* **1980**, 1035.

129. I. M. Andreeva, E. M. Bondarenko, E. A. Medyantseva, R. G. Pudeyan, and V. I. Minkin, *Khim. Geterotsikl. Soedin.* **1982**, 610.

130. I. M. Andreeva, E. M. Bondarenko, E. A. Medyantseva, and V. I. Minkin, *Khim. Geterotsikl. Soedin.* **1983**, 181.

131. O. M. Babeshko, E. A. Medyantseva, O. T. Lyashik, and V. I. Minkin, *Khim. Geterotsikl. Soedin.* **1982**, 1477; See also V. A. Bren', L. M. Sitkina, A. D. Dubonosov, V. I. Minkin, M. I. Knyazhanskii, *Dokl. Akad. Nauk SSSR*, **1983**, *272*, 1382; A. D. Dubonosov, L. M. Sitkina, V. A. Bren', A. Ya. Bushkov, V. I. Minkin, *Khim. Geterotsikl. Soedin.* **1984**, 1171.

Additions in proof: References to some more important papers published in 1983 and 1984 are added below. They supplement with new data the main list references marked with an asterisk.

*4. For the investigation of the ring-chain equilibria of type ((3A) ⇌ (3B), page 170) by ^{15}N-NMR spectroscopy see B. Ch. Chen, W. v. Philipsborn, and K. Nagarajan, *Helv. Chim. Acta* **1983**, *66*, 1537.

*39. W. Kliegel and L. Preu, *Justus Liebigs Ann. Chem.* **1983**, 1937; W. Kliegel and J. Graumann, *Justus Liebigs Ann. Chem.* **1984**, 1545.

*76. V. V. Pinson, V. A. Khrustalev, K. N. Zelenin, and Z. M. Matveeva, *Khim. Geterotsikl. Soedin.* **1984**, 1415.

*78. For ring-chain tautomerism of N-[2-(N-alkylamino)ethyl]-N-methylhydrazones of formaldehyde, acetaldehyde, and acetone see P. S. Lobanov, O. V. Solod, and A. A. Potekhin, *Zh. Org. Khim.* **1983**, *19*, 2310.

*82. For mass spectrometric investigation of ring-chain tautomerism of type ((44A) ⇌ (44B), page 194), see O. S. Anisimova, Yu. N. Sheinker, and R. E. Valters, *Khim. Geterotsikl. Soedin.* **1984**, 1080.

*99. M. Taniguchi, A. Gonsho, N. Nakagawa, and T. Hino, *Chem. Pharm. Bull. Tokyo* **1983**, *31*, 1856.

*122. For the detailed investigation of the equilibrium ((59A) ⇌ (59B), page 202) see K. N. Zelenin, V. V. Alekseev, and V. A. Khrustalev, *Zh. Org. Khim.* **1984**, *20*, 169.

*124. For further investigation of the equilibrium ((62B) ⇌ (62C), page 203) see K. N. Zelenin, V. A. Alekseev, V. A. Khrustalev, S. I. Yakimovich, V. N. Nikolaev, and N. V. Koshmina, *Zh. Org. Khim.* **1984**, *20*, 180.

*125. N. L. Chikina, Yu. V. Kolodyazhnii, V. P. Kozyukov, N. V. Mironova, and O. A. Osipov, *Zh. Obshch. Khim.* **1984**, *54*, 139.

Intramolecular Reversible Addition Reactions to the Other Groups

4.1. ADDITION TO THE C≡N GROUP

Numerous synthetic methods for the preparation of heterocyclic imines and amines have been developed [1, 2] in which the last stage involves an intramolecular nucleophilic addition of a hydroxy, amino, or mercapto group to a C≡N bond. Intermediates in these syntheses, i.e., hydroxy, amino, or mercaptonitriles (**1A**, X = O, NR, S) usually were not isolated.

Only a few aspects of intramolecular addition reactions to the cyano group will be discussed here since there are comprehensive reviews [1, 2]. The main attention will be paid to cyclization conditions, to structural influences on the process, and to the reversibility of the intramolecular addition reactions.

The hydroxy, amino, or mercaptonitriles (**1A**) which have been isolated could be cyclized to (**1B**) easily and irreversibly by heating in polar solvents [3-5], in the presence of acids [6-11], bases [12-25], or upon melting [3, 4, 8, 22, 26].

2-Cyanobenzamide [27] and 2-thioureidobenzonitrile [3] show double melting points because of thermal isomerization. After melting the substances undergo isomerization to (**1B**) which crystallize and melt again at higher temperatures.

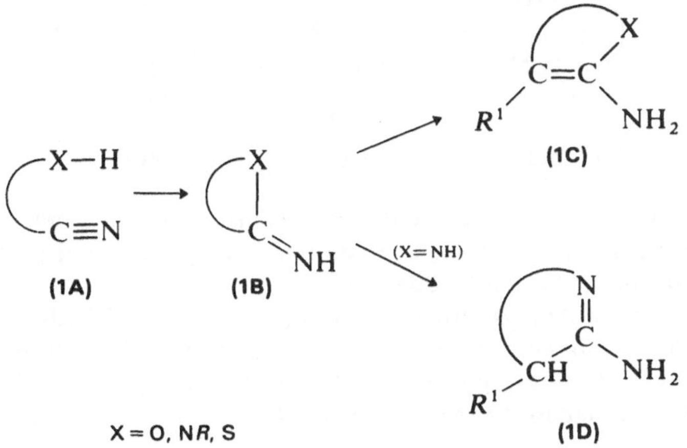

X = O, NR, S

The heterylimines (1B) often isomerize subsequently to heterylamines (1C) or (1D), the double bond shifting into the ring.

Matsui [28] and Renson [29] with co-workers detected $\nu_{C=N}$ and $\nu_{C\equiv N}$ bands in the IR spectra of ω-cyanoalcohols (2A) and 2-hydroxymethylbenzonitrile (3A) which led them to assume a ring-chain tautomeric equilibrium (A) $\rightleftarrows$ (B). Additional corroboration for this conclusion can be obtained by spectroscopic investigations of hydroxynitrile (2) or (3) in different solvents. Numerous data [1, 2, 6] concerning cyclization reactions of hydroxynitriles confirm the irreversibility of the cyclization. Perhaps Matsui and Renson [28, 29] dealt not with an equilibrium but a mixture of isomers (compare [30]).

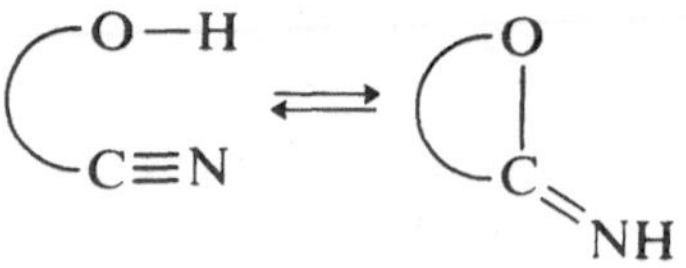

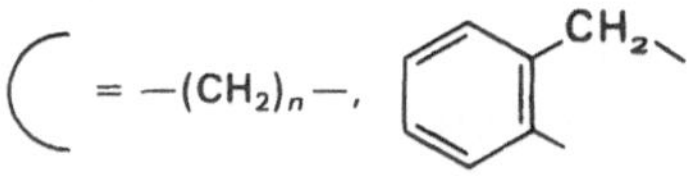

Mononitriles of phthalic [27], homophthalic [31], and other γ- or δ-dicarboxylic acids [2] easily isomerize into imides (4C). The isomerization proceeds via iminolactones (isoimides) (4B) of little stability [20, 31].

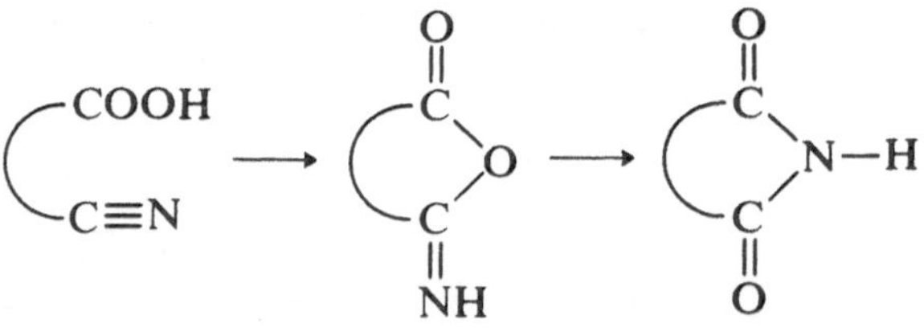

Investigation of IR spectra of 2-cyanobenzoic acid in different solvents [22] showed that in solutions at room temperature this acid is stable and the equilibrium (4A) $\rightleftarrows$ (4B) does not occur.

γ- and δ-Aminonitriles [7, 8, 14, 21, 32–35], β-hydrazinonitriles [12, 18, 36], cyanoketone hydrazones [4], and cyanocarboxylic acid hydrazides [5, 37–40] isomerize into the corresponding heterocyclic imines. 3-Cyanopropionamides (5A) [13, 15, 19, 41] and 2-cyanobenzamides (6A)

[22, 26, 27] isomerize either in the presence of bases (sodium alcoholate, triethylamine) or on heating into iminolactams **(5B)** and **(6B)**.

(A)   **(B)**

(5)   **(6)**

Braun and Tcherniac [27] were the first to observe the thermal isomerization of 2-cyanobenzamide (**6A**, R = H) → **(6B)** obtained in a reaction of phthalic acid diamide with acetanhydride. They detected that 2-cyanobenzamide after melting at 172–173°C crystallizes and melts anew at 200°C which corresponds to the melting point of phthalimide monoimine (**6B**, R = H). At 130–160°C an analogous isomerization takes place [42] for 2-cyanobenzanilide (**7A**, R = H) → **(7B)**.

An equilibrium (**7A**, X = H) $\rightleftarrows$ **(7B)** was observed [42] in polar aprotic solvents such as DMSO (K_T = 38.0 at 25°C), N-methylpyrrolidone (K_T = 17.4), dimethylacetamide (K_T = 6.6), and HMPT (K_T = 2.8). Cyclization of 2-cyanobenzanilide occurs under base catalysis: a linear relationship was observed [42] between the rate constant of cyclization and the concentration of the tributylamine in dimethylacetamide solution.

The kinetics of the isomerization of 2-cyanobenzamide in aqueous potassium hydroxide is first order regarding the hydroxide concentration [43] (and not second order as described earlier [44]).

An investigation [45] of the influence of solvents and substituents X on the isomerization (**7A**) → **(7B)** rate (see Table 39) showed that the rate constants in aqueous sulpholane in the presence of an acetate buffer (pH = 7.45) correlate sufficiently (r = 0.970) with the pK_a values of the corresponding benzanilides, i.e., the cyclization rate increases on passing from electron-donating to electron-withdrawing substituents. It follows that in a given medium the constant (k_1/k_2) for the equilibrium (**7A**) $\rightleftarrows$ **(7C)** exerts a strong influence on the total cyclization rate.

However, since this rate is defined as $k = k_1 \times k_3/k_2$, electron-withdrawing substituents X that increase 2-cyanobenzanilide (**7A**) N—H acidity may simultaneously decrease its nucleophilicity and thus slow down the transformation $(7C) \rightarrow (7D)$. If k_1/k_2 and k_3 in the cyclization $(7A) \rightarrow (7B)$ are comparable in magnitude, then a linear correlation between lg k and pK_a (i.e. the σ constants of substituents X) of corresponding benzanilides may not exist. The cyclization $(7A) \rightarrow (7B)$ in aqueous DMSO solution in the presence of biphthalate buffer (pH = 8.68) is most rapid for the N-unsubstituted 2-cyanobenzanilide $(7A, X = H)$ while increasing electron-donating as well as electron-withdrawing properties of the substituents X significantly lowers the cyclization rate (see Table 39). The dependence of lg k on the substituent constants σ is expressed by two straight crossing lines. As a consequence the rate-determining step of cyclizations of derivatives bearing electron-donating substituents is $(7A) \rightarrow (7C)$ while a rate control is exerted by the step $(7C) \rightarrow (7D)$ in the case of electron-withdrawing substituents.

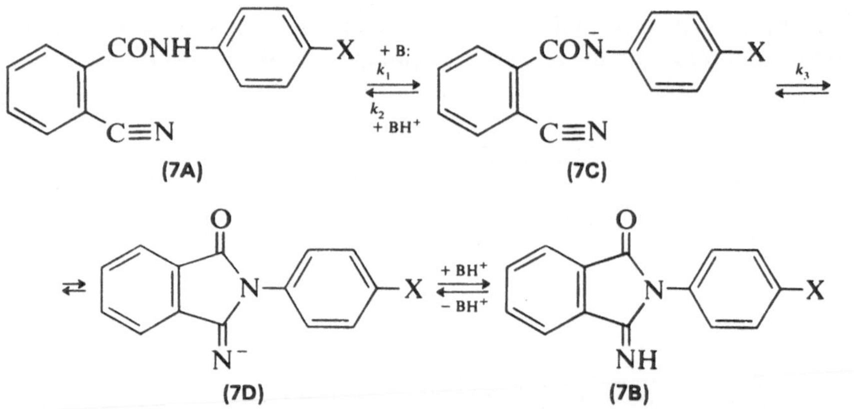

Table 39. Cyclization Rate Constants or 2-Cyanobenzanilides
$(7A) \rightarrow (7B)$ [45]

	$k \times 10^4 \, s^{-1}$	
X	Sulpholane-water (9:1), acetate buffer pH 7.45, 32.4°C	DMSO-water (9:1), biphthalate buffer pH 8.68, 19.7°C
COOMe	—	0.23
Br	8.8	0.66
Cl	5.9	1.09
F	5.1	1.18
H	2.1	4.36
Me	1.7	3.97
MeO	1.5	3.30
Me$_2$N	—	0.21

In the IR spectrum of 2-cyanobenzamide Flett [46] observed amide-I
(1753 cm^{-1}) and amide-II $(1667 \text{ cm}^{-1}, \text{ KBr pellet})$ bands being unusually
high for benzamides. The same is reported in Bellamy's monograph [47].
Later it was established [26] that the IR spectrum of a 2-cyanobenzamide
[27] showed amide-I and amide-II bands at significantly lower frequencies
$(1658 \text{ and } 1626 \text{ cm}^{-1}, \text{ KBr pellet})$. Apparently, Flett's compound was
phthalimide monoimine, the spectrum of which in dioxan shows two strong
bands $\nu_{C=O}$ 1737 cm^{-1} (A = 4.6 pract. un.) and $\nu_{C=N}$ 1671 cm^{-1} (3.0 pract.
un.) (see Fig. 11) [22, 26].

Increasing steric encumbrance of substituents R at the nitrogen atom
of N-monosubstituted 2-cyanobenzamides (6A) hampers the isomerization
(6A) → (6B) which proceeds under base catalysis or thermally [22]. In case
of a $tert$-alkyl substituent (t-Bu, 1-adamantyl) the isomerization (6A) → (6B)
does not occur at all. No tautomeric equilibrium was observed in solutions
of 2-cyanobenzamides (6A) and their cyclic isomers (6B) in dioxan,
acetonitrile, or DMSO (IR spectroscopy).

2-Cyanobenzthioamide was isolated [48] as cyclic (8).

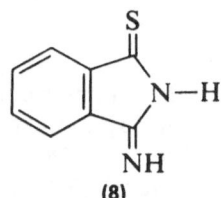

(8)

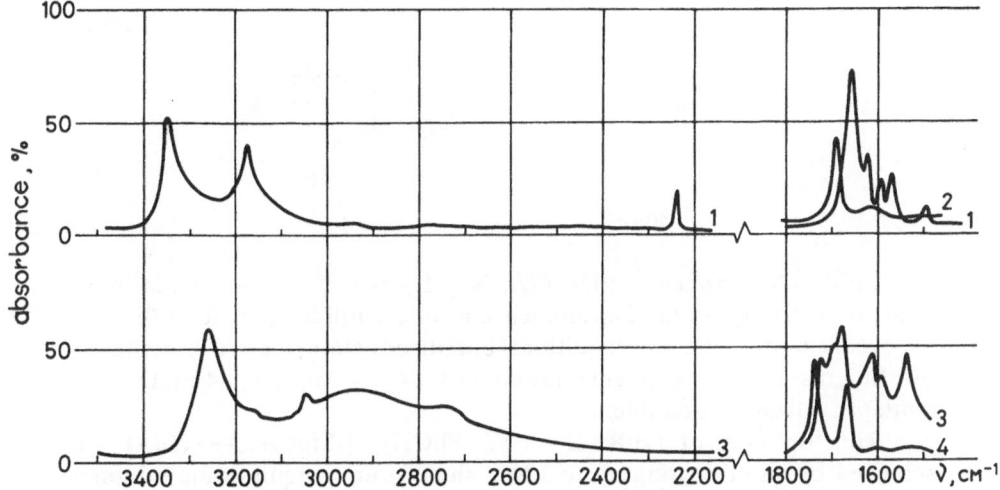

Figure 11. IR spectra of open and ring isomers of 2-cyanobenzamide: 1—2-cyanobenzamide
in the solid state, 2—its solution in dioxan, 3—3-iminoisoindolinone in the solid state, 4—its
solution in dioxan [26].

2-Ureido and 2-thioureidobenzonitriles [3, 17, 49] and α-cyanohydrine urethanes [16] cyclize similarly as 2-cyanobenzamides.

The direction of thermal isomerization of the 2-cyanobenzamides **(6A)** → **(6B)** differs from that of ketocarboxylic acid amides **(B)** → **(A)** (see section 2.1.3). This direction of thermal isomerization also has been observed for other hydroxy and aminoderivatives of nitriles. However, a case has been reported [35] where a reverse thermal isomerization **(9B)** → **(9A)** proceeds. Evidently the aromatic stabilization of the sym-triazine ring of the open isomer **(9A)** plays a decisive role.

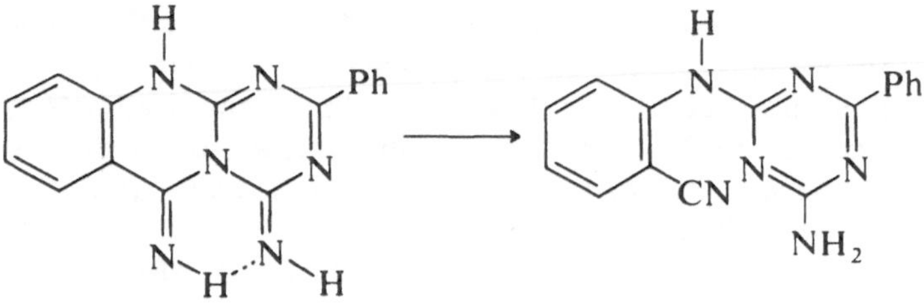

(9B)   **(9A)**

Both *N*-monosubstituted 2-cyanobenzenesulfonamide isomers **(10A)** and **(10B)** have been obtained [24, 25]. In the presence of bases 2-cyanobenzenesulfonamides **(10A,** R = Me, *i*-Pr, PhCH$_2$, 4-MeC$_6$H$_4$, 4-MeOC$_6$H$_4$) isomerize into 3-iminobenzisothiazoline-1,1-dioxides **(10B)**. *N*-(*t*-Butyl)- and *N*-(1-adamantyl)-sulfonamides do not cyclize because of steric hindrance.

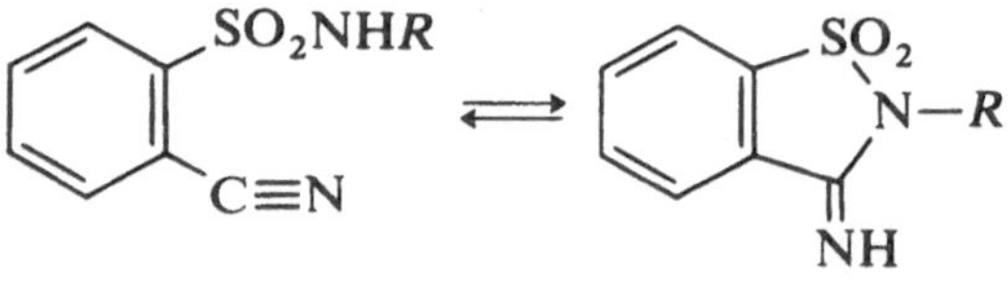

(10A)   **(10B)**

Unlike 2-cyanobenzanilide (**7A,** X = H) which forms isoindolinone under basic catalysis [25], 2-cyanobenzenesulfonanilide (**10A,** R = Ph) does not cyclize under the same conditions, but introduction of electron-donating substituents in to the phenyl group (**10,** R = 4-MeC$_6$H$_4$, 4-MeOC$_6$H$_4$) renders cyclization possible.

Ring opening of (**10B,** R = *i*-Pr, PhCH$_2$, 4-MeC$_6$H$_4$) → **(10A)** was achieved by short heating up to 220°C thus forming equilibrium mixtures **(10B)** ⇌ **(10A)**. Sterically demanding or electron-withdrawing substituents at the nitrogen atom shift the equilibrium toward the sulfonamide **(10A)**.

Table 40. Ring-Chain Equilibrium Constants **(10A)** ⇌ **(10B)**
Determined by IR Spectroscopy with Dioxan and 10%
Triethylamine at 20°C [25]

R	$K_T = [(10B)]/[(10A)]$	Time to reach equilibrium with (10A)
Me	>100	4 years
PhCH$_2$	2.1	36 h
Ph	0.15	8 min
4-MeC$_6$H$_4$	0.29	25 min
4-MeOC$_6$H$_4$	0.65	30 min

In dioxan at room temperature both isomers **(10A)** and **(10B)** are stable.
An equilibrium **(10A)** ⇌ **(10B)** is observed only after addition of triethylamine which transfers a proton. N-Isopropylderivatives (**10A**, $R = i$-Pr)
and **(10B)** and N-*tert*-alkylsulfonamides (**10A**, $R = t$-Bu, 1-adamantyl) do
not equilibrate. Equilibrium constants of the remaining derivatives were
determined by IR spectroscopy. As may be seen from Table 40 introduction
of electron-donating substituents (Me, MeO) into the aryl group of N-
arylsulfonamides shifts the equilibrium toward the cyclic form. Passing from
N-alkyl to N-arylsulfonamides the rate of equilibration strongly increases.

γ-Mercaptonitriles **(12A)** are transient species during the formation
from S-benzylderivatives **(11)** which cyclize immediately yielding imino-

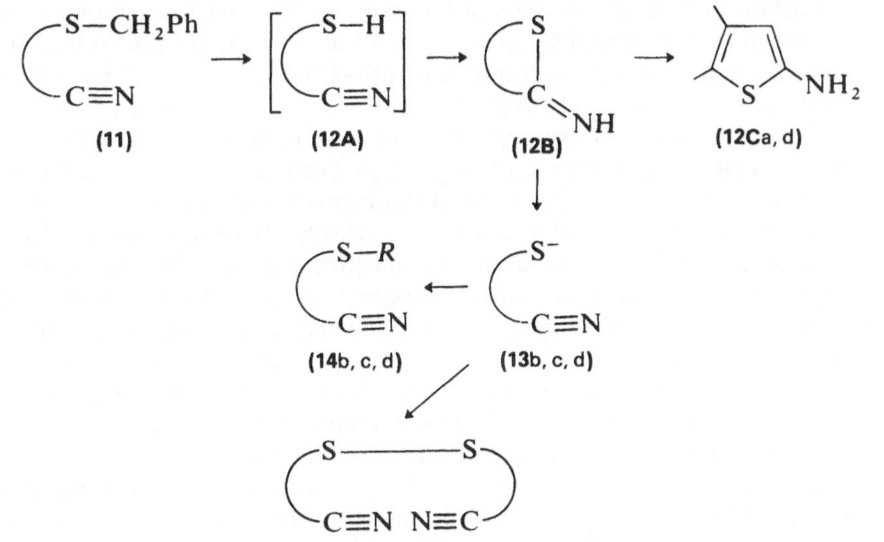

a [53]   b [29]   c [50]

d [51]   e [52]

dihydrothiophene derivatives **(12B)** [29, 50–53]. These isomerize to amino-thiophenes **(12C)** if an α-hydrogen atom is present relative to the C=N bond [51, 53].

Since iminodihydrothiophenes (**12B**, b, c, d) form both cyclic as well as open derivatives, Stacy [50, 52] and Renson [29] and coauthors assumed a ring-chain tautomerism **(12A)** ⇌ **(12B)**. For instance, alkylation of **(12B)** in the presence of base gave open S-alkylderivatives (**14b**, c, d) while oxidation yielded disulphides (**15b**, c, d). However, this only proves the existence of the open anion **(13)** in a basic medium. IR and ^{1}H-NMR spectroscopy failed to provide evidence for the presence of the open isomer **(12A)** either in the solid state or in solution, i.e., within the limits of these spectroscopic methods the tautomeric equilibrium **(12A)** ⇌ **(12B)** is not detectable.

Actually, there is no reliable evidence for the existence of a ring-chain equilibrium in neutral solutions of hydroxy, amino, and mercaptonitriles. No indication for an equilibrium has been obtained as well from an investigation of intramolecular electrophilic addition reactions to C≡N bond that proceed under acidic catalysis [1, 2].

The dependence of the cyclization of cyanocarboxylic acid chlorides **(16A)** → **(16B)** on the structure (a-m) has been thoroughly studied by Simchen and co-workers [54, 55] and the formation of five, six, or seven-membered aza-, diaza-, and thiaza-heterocycles has been reported. Cyclization proceeds at low temperatures in the presence of anhydrous hydro-halogenic acids. The cyclization is suggested to be preceded by addition of hydrochloric acid to the cyanogroup **(16A)** → **(17)**. Depending on the structure of the connecting chain two ways of cyclization are possible: **(17)** → **(18)** → **(16C)** and **(17)** → **(16B)**. Cyclization is favored energetically by the formation of either a conjugated enamide (**16Cb**, d, e, j) or aromatic systems **(16D)** as a result of prototropic transformations.

In most cases cyclization proceeds irreversibly. An equilibrium **(16Am)** ⇌ **(16Bm)** was observed during the formation of 3-chloro-2-benz[f]azepine-1-one **(16Bm)**.

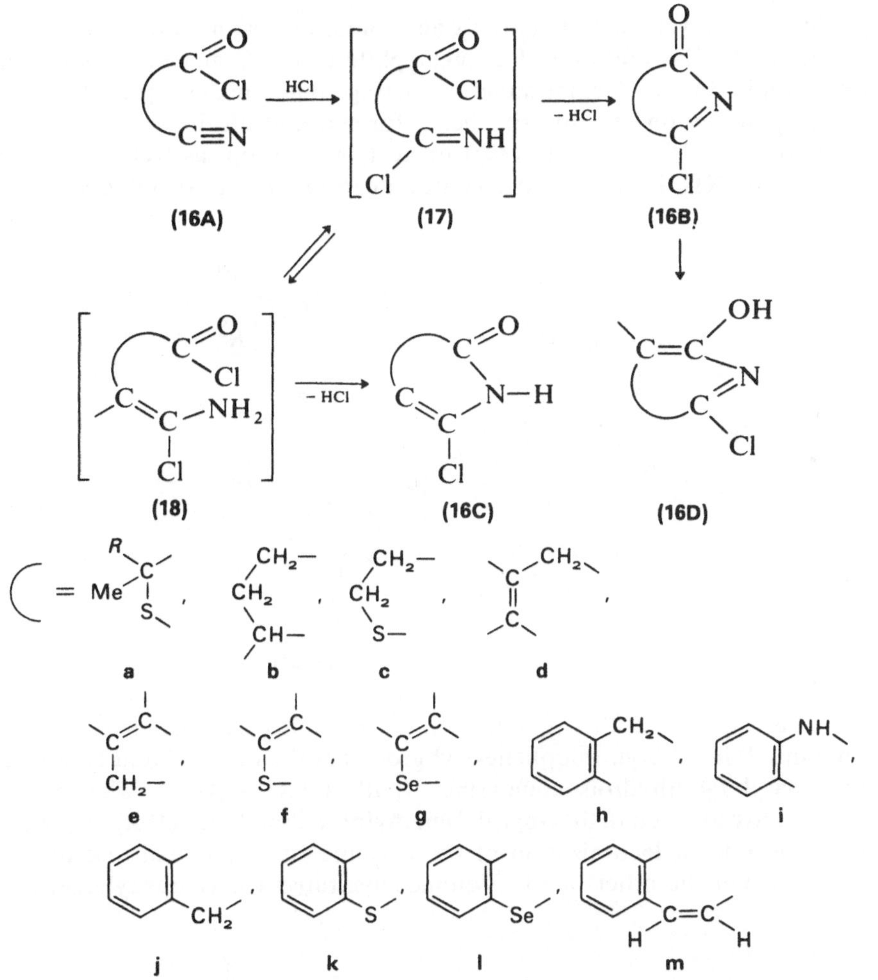

(16A) (17) (16B)

(18) (16C) (16D)

a   b   c   d

e   f   g   h   i

j   k   l   m

Surprisingly, a spontaneous isomerization in the reverse direction was found: 2-chloro-5,6-dihydro-1,3-thiazine-4-one **(16Bc)**, being incapable of a stabilization due to the formation of an aromatic or enamide system, transforms into the initial 3-thiocyanatopropionyl chloride **(16Ac)** after several days at room temperature [56].

4.2. ADDITION TO C=C AND C≡C GROUPS

Many examples of intramolecular additions of hydroxyl, amino, or mercaptogroups to ethylene or acetylene bonds are known [57, 58]. A particular group of reactions (strictly not isomeric transformations) proceeds by intramolecular addition of a nucleophile to C=C or C≡C bonds, an electrophilic intermolecular attack occurring simultaneously [59–61].

Depending on the ring size being formed and on the substituents at the ethylene (R^1, R^2, R^3 in (19)) or acetylene group, as well as on the nucleophile XH, intramolecular addition can proceed in two directions:

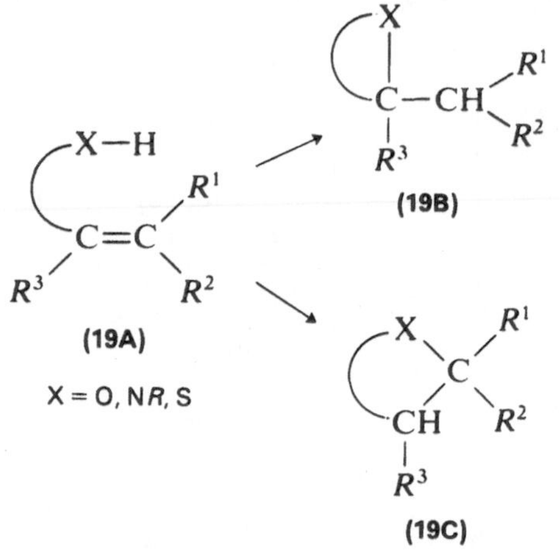

(19A)

X = O, NR, S

(19B)

(19C)

Thus, E and Z-2-(2-arylethenyl)benzoic acids having phenyl, 1- or 2-naphthyl, indol-3-yl, thiophene-3-yl groups at the double bond lactonize into 3-aryl-3,4-dihydroisocoumarines (path (19A) → (19C)) whereas 4-pyridylderivative led to 3-(4-pyridyl)methylphthalide (path (19A) → (19B)) [62]. The rate of lactonization of the Z-isomer is faster than that of the E-isomer. On the other hand, N-monosubstituted E-2-(2-phenylethenyl)-

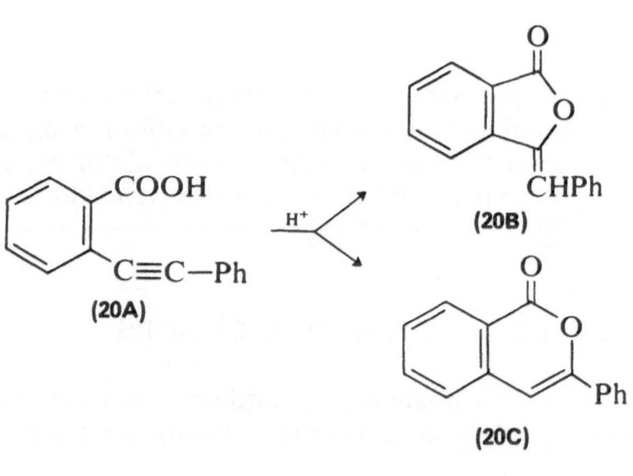

(20A)

(20B)

(20C)

benzamides cyclize into 2-alkyl-3-benzylisoindolinones (path
(19A) → (19B)) in the presence of a basic catalyst [63].

The acid-catalyzed isomerization of tolane-2-carboxylic acid **(20A)** gave
a mixture of 3-benzylidenephthalide **(20B)** and 3-phenylisocoumarine **(20C)**
[64].

In most cases the cyclization proceeds irreversibly under acidic or basic
(rather rarely neutral) conditions or on heating. The negligible propensity
to form equilibria obviously is due to the low polarity of C=C and C≡C
bonds.

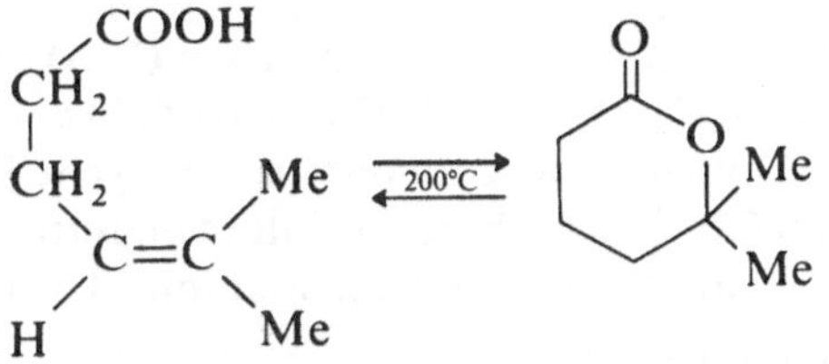

Linstead and Rydon [65] showed that 5-methyl-4-hexene-1-carboxylic
acid **(21A)** and 6,6-dimethyltetrahydro-2-pyranone **(21B)**, both being stable
at room temperature, gave on heating at about 200°C for 12–18 hours an
equilibrium mixture containing 40% of the acid **(21A)** and 60% of the
lactone **(21B)**.

Under milder conditions the equilibrium is observed only if substituents
increase the polarity of the double bond.

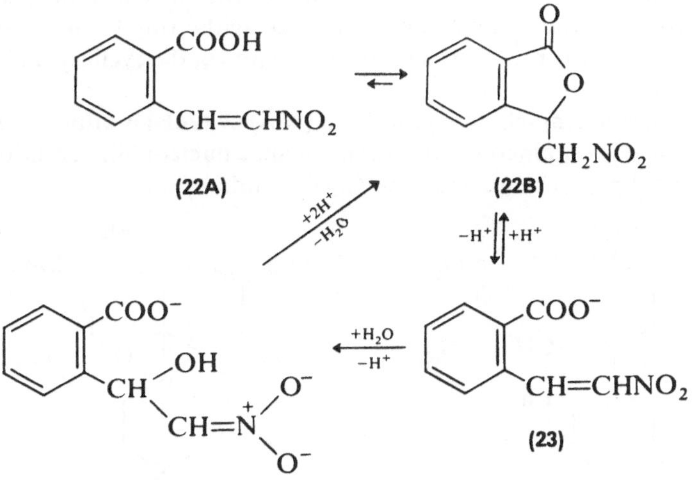

The condensation product of 2-formylbenzoic acid with nitromethane has been shown to be 3-nitromethylphthalide **(22B)** on the basis of IR and ^{1}H-NMR spectroscopic data [66] and not 2-(2-nitroethenyl)benzoic acid **(22A)** as reported earlier [67].

An investigation of the action of bases on 3-nitromethylphthalide revealed [68, 69] that ring opening occurs in aqueous alkaline solution to give the anion of 2-(2-nitroethenyl)benzoic acid **(23)** which, as indicated by its UV spectra, is unstable and rapidly transforms into the dianion **(24)**. Acidification of the solution leads to the regeneration of the initial 3-nitromethylphthalide **(23)** → **(22B)** or **(24)** → **(22B)**. A UV spectroscopic study established that at pH ~6 the equilibrium **(22A)** ⇌ **(22B)** is strongly displaced in favor of **(22B)** ($K_T \sim 10$). The acid **(22A)** could not be isolated.

3-Nitromethyl-2-phenylisoindolinone is also cyclic [66].

The equilibrium **(25A)** ⇌ **(25B)** is a result of an intramolecular addition of the carboxylic group to the 1,4-benzoquinone C=C bond [70]. The

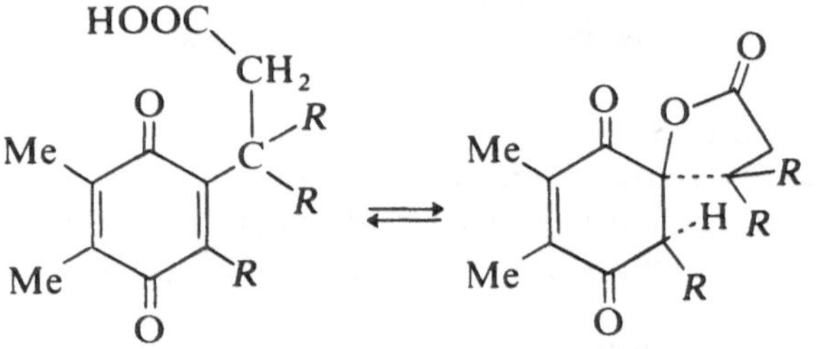

(25A)   **(25B)**

equilibrium largely depends on the acidity of the solvent: a decreasing pH value shifts the equilibrium toward the less acidic ring form. Methyl substituents (R = Me) are required for the formation of the cyclic form ("trialkyl lock" [70]).

One case of ring-chain equilibrium **(26A)** ⇌ **(26B)** is known [71] where the ring form is obtained by an intramolecular nucleophilic addition of the nitronic acid group to an enamine C=C bond.

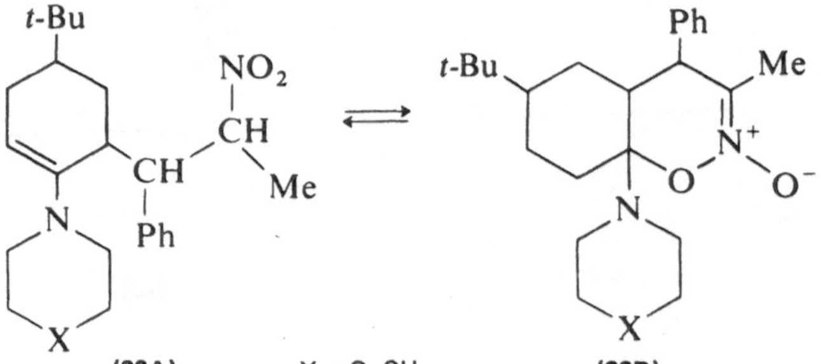

(26A)   X = O, CH$_2$ **(26B)**

Italian chemists [72] using IR, UV, ^{1}H, and ^{13}C-NMR spectroscopy have shown that 5-(o-aminophenylcarbamoyl)pyridazine-4-carboxylic acid, although possessing a bipolar structure (27A, X = NH) in the solid state, in DMSO almost completely changes into the spirocompound (27B). The equilibrium (27A) ⇌ (27B) position strongly depends on the solvent. Passing from water to methanol the equilibrium is displaced in favor of the cyclic form (27B).

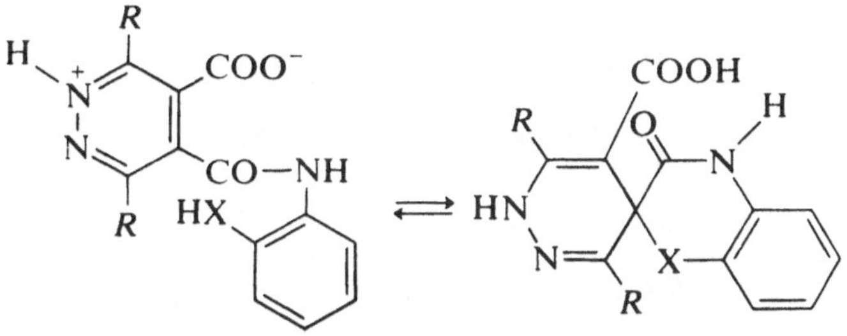

| (27A) | R = H, Me | (27B) |

Replacement of the NH group by a less nucleophilic OH group prevents the cyclization (27A, X = O) → (27B).

The examples mentioned above testify that normally ring-chain equilibrium brought about by intramolecular addition to a C=C bond is observed only if the C=C bond is polarized by strongly electron-donating or electron-withdrawing substituents.

Finally ring-chain equilibrations reflecting intramolecular addition of a C—MgX group to C=C bonds will be briefly discussed. They have been called "Grignard reagent cyclization cleavage rearrangements" (28A) ⇌ (28B) ⇌ (28A') and are thoroughly treated in Hill's reviews [73, 74]. We shall therefore not discuss them in detail.

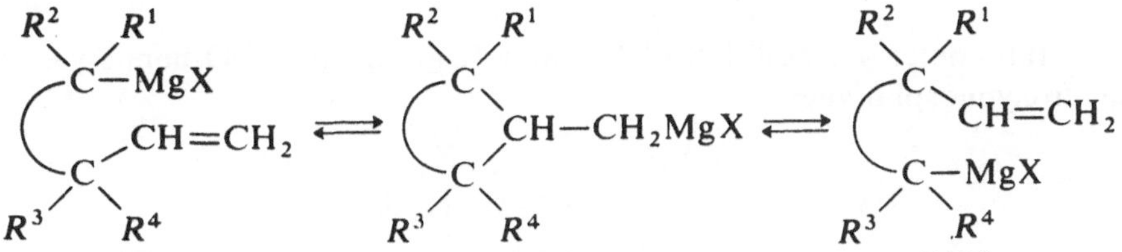

| (28A) | (28B) | (28A') |

Generally, these rearrangements require high activation energies, i.e., equilibria are attained after many hours at 100°C or even at higher temperatures. They are usually strongly shifted toward the cyclic form (28B) if five or six-membered rings are formed, and are displaced in favor of the open form (28A) or (28A′), however, if the cyclic isomer (28B) contains a three or four-membered ring. A concerted four-center mechanism has been assumed [73–75] for the rearrangements.

Rate and equilibrium constants of the reversible rearrangement (29A) ⇌ (29B) may be considered as an example (see Table 41) [75–77]. If $R^1 \neq R^3$, then ring opening can proceed in two different directions yielding

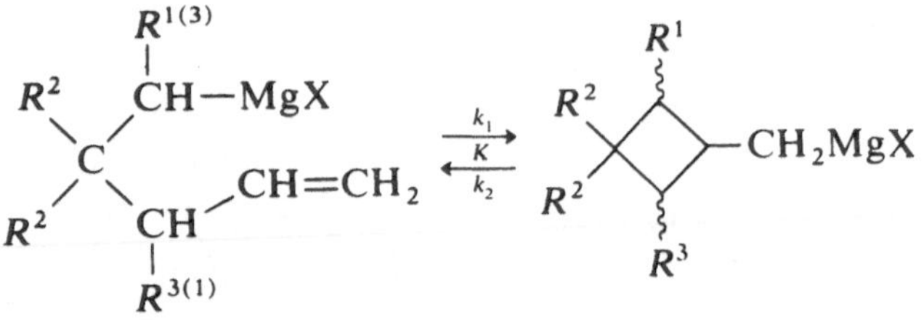

(29A)   X = Cl, Br   (29B)

two open isomers. The influence of the introduction of methyl substituents (R^1, R^2, R^3) on the ring-chain equilibrium constants, the rate constants, and the activation parameters is shown in Table 41 (some of the reported data were obtained [75] after extrapolation of determinations carried out under different conditions). Introduction of methyl substituents stabilizes the cyclic form as a result of the Thorpe-Ingold effect (see p. 27 for a discussion), especially if R^2 = Me (see [76]). However, if R^1 = Me the cyclization which is accompanied by a rearrangement of the less stable secondary Grignard reagent into the more stable primary one is strongly favored. This difference in stability between primary and secondary Grignard reagents was estimated [75] to be $-\Delta G° \sim$ 2.8–5 kcal/mol.

4.3. ADDITION TO THE P=O GROUP

Intramolecular addition of OH or NH groups to P=O bonds yields hydroxyphosphoranes:

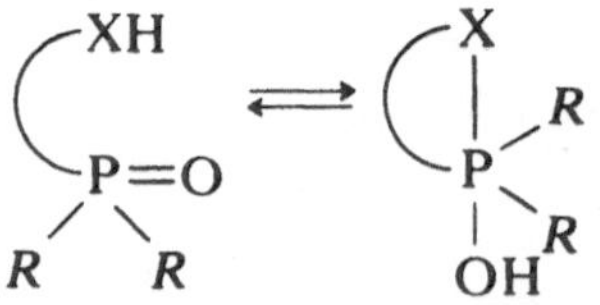

(30A)   (30B)
X = O, NR

They are, together with their conjugated bases, i.e., phosphorane oxide anions, usually considered as intermediates or transition states in substitu-

Table 41. Ring-Chain Equilibria, Rate Constants, and Activation Parameters for Cyclobutylmethyl Grignard Reagent and Its Methyl Derivatives (29A) $\rightleftarrows$ (29B) at 100° in Tetrahydrofuran[a]

In formula (29)				Cyclization			Ring opening			
R^1	R^2	R^3	$K = [(29B)]/[(29A)]$	k_1 (s^{-1})	$\Delta H^{\neq}$ (kcal/mol)	$\Delta S^{\neq}$ (e.u.)	k_2 (s^{-1})	$\Delta H^{\neq}$ (kcal/mol)	$\Delta S^{\neq}$ (e.u.)	Refs.
H	H	H	9×10^{-5}	2×10^{-8}	28.5	-15	2.2×10^{-4}	26.55	-4.6	75,77
H	H	Me	7.5×10^{-3}	5.2×10^{-7}			7×10^{-5}			75
H	Me	H	4×10^{-3}	8.8×10^{-7}			2.2×10^{-4}			76
Me	H	H	1.33	1.6×10^{-6}	23	-24	1.2×10^{-6}			75
Me	H	Me	3.44[b]	6.78×10^{-6}	19.6	-28	1.97×10^{-6}	25.0	-16	75

[a] Determined indirectly: after reaching equilibrium mixtures were hydrolyzed and the resultant hydrocarbons were analyzed by GLC.
[b] Ether solvent.

tion reactions at a tetracoordinated phosphorus in compounds bearing a P=O bond [78–80].

Thus, for instance, the formation of the cyclic product (32) in a reaction of hydroxyneopentyl hypophosphite with diazomethane has been taken as evidence for the equilibrium (31A) ⇌ (31B) [81].

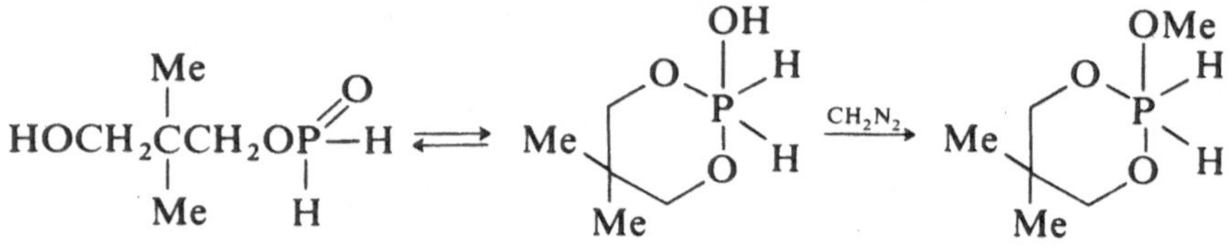

Recently the chemistry of phosphorus has developed considerably [82]. The presence of hydroxyphosphoranes of the type (33B) as intermediates in reactions has been repeatedly discussed [83–85].

Ramirez and coauthors [86, 87] have first described the stable hydroxy-phosphorane (33B) obtained by the action of hydrogen chloride on trimethylsilyloxyspirophosphorane (34, $R = Me_3SiO$).

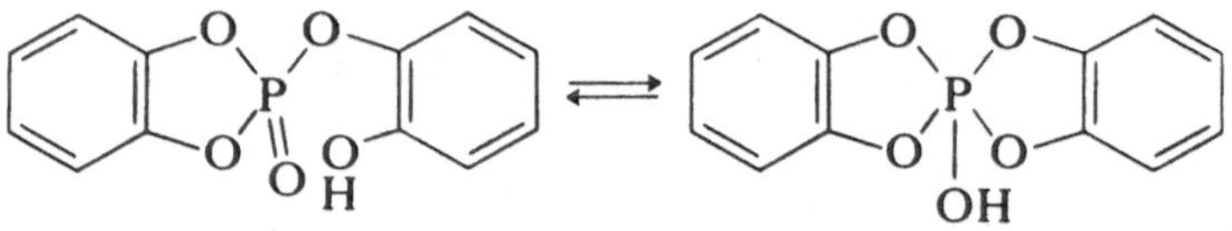

Later (33B) was obtained [88] by hydrolysis of the chlorospirophos-phorane (34, $R = Cl$).

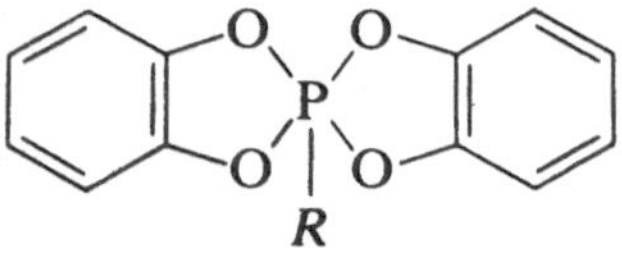

A characteristic feature of the ring-chain equilibrium between o-hydroxyphenyl-o-phenylene phosphate and hydroxy-bis-o-phenyl-enedioxyphosphorane (33A) ⇌ (33B), which was observed [86, 88] in aprotic solvents, is a reversible nucleophilic addition of a phenolic hydroxyl to a P=O bond. Both isomers were characterized by their ^{31}P-NMR spectra [86]: (33A) $\delta^{31}P = +6.8$ ppm, (33B) $\delta^{31}P = -26.6$ ppm (positive values downfield of $\delta H_3PO_4 = 0$; −48°C; CD₃CN)†.

†In different papers the chemical shifts $\delta^{31}P$ are denoted as positive values in different directions: downfield or upfield from 85% H_3PO_4.

The open isomer (33A) is stabilized by an intermolecular hydrogen bond OH···O=P (2.68 Å O···O distance) which was established [87] for the crystalline compound by an x-ray determination.

On passing from less to more polar and basic solvents the equilibrium is shifted toward the open form (33A) which indicates that the POH group of (33B) is less acidic than the phenolic hydroxyl group of (33A). Examples: $K_T = 3$ for acetone-d_6; $K_T = 1.5$ for CD_3CN ($-48°C$). Coalescence of both ^{31}P-signals were observed: in CD_3CN at $+10°C$, but in acetone-d_6 at $+55°C$.

Methylation of a solution of (33) with diazomethane yields only the cyclic methylderivative (34, R = OMe) while acetylation with acetylchloride provides a 4:1 mixture of open and cyclic acetylderivatives.

Ring-chain tautomerism of a number of hydroxyphosphoranes (35A) $\rightleftarrows$ (35B) and some analogues (36A) $\rightleftarrows$ (36B) has also been investigated by ^{31}P-NMR spectroscopy [89-91].

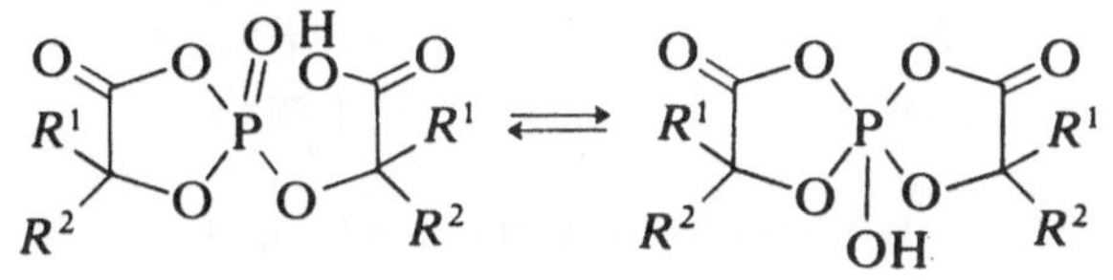

(35A)   (35B)

R^1 = H; R^2 = H, Me, i-Pr, Ph
$R^1 = R^2$ = Me, Ph

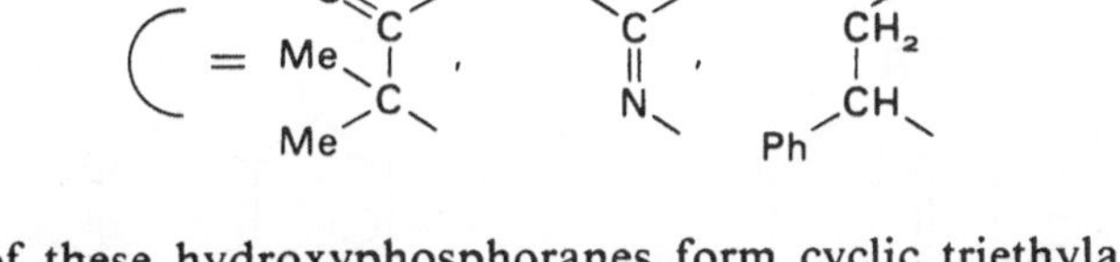

(36A)   (36B)

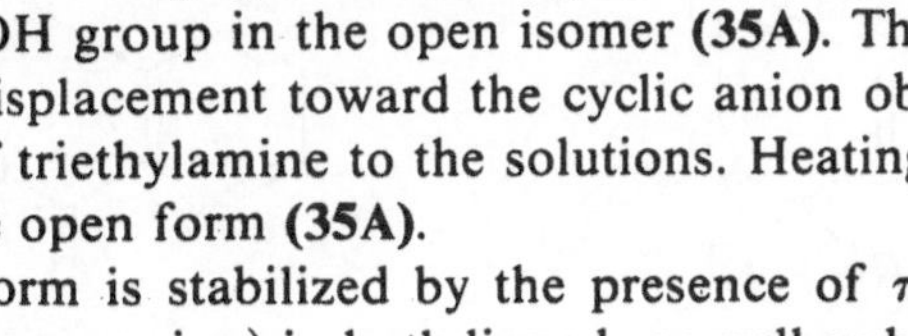

Some of these hydroxyphosphoranes form cyclic triethylammonium salts that led the authors [90] to assume a greater acidity for the POH group than for the COOH group in the open isomer (35A). This is supported by an equilibrium displacement toward the cyclic anion observed as a result of an addition of triethylamine to the solutions. Heating shifts the equilibrium toward the open form (35A).

The cyclic form is stabilized by the presence of π-electron systems (C=O, C=N, benzene ring) in both ligands as well as by the introduction of one or, more effectively, two methyl groups (**35**, $R^1 = R^2 = $ Me) (Thorpe-Ingold effect).

Martin and Granoth [92] have investigated the equilibrium (**37A**, $R^1 = $ Me, $R^2 = $ H) $\rightleftarrows$ (**37B**) using ^{1}H and ^{31}P-NMR spectroscopy. The acidity of hydroxyphosphorane was estimated ($pK_a \sim$ 10–11).

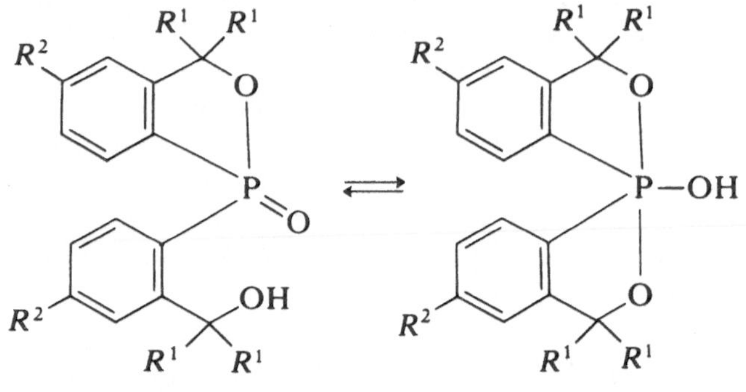

(37A)   (37B)

Only one broad signal is observed at room temperature in the ^{31}P-NMR spectrum of (**37**). Decreasing the temperature gave two ^{31}P signals at 52 ppm (**37A**) and -27 ppm (**37B**) which were detected in CH_3OD. The proportion of the cyclic isomer (**37B**) increases with decreasing temperature ($+5$ to $-50°C$). Basic solvents, such as pyridine, shift the equilibrium toward the hydroxyphorphorane (**37B**); the cyclic anion (**38**) is formed on addition of sodium methoxide to the methanolic solution.

Surprisingly, the much more acidic ($pK_a = 5.3 \pm 0.2$) trifluoromethyl derivative (**37**, $R^1 = CF_3$, $R^2 = $ Me) exists as a stable cyclic isomer (**37B**) and does not show ring-chain equilibria in solution [93]. The enhanced electronegativity of the apical oxygens in the fluoralkoxy ligands increases the electronegativity difference between the central phosphorus and the

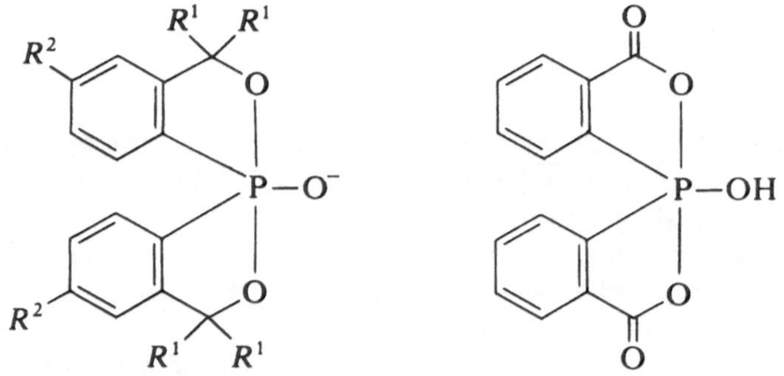

(38)   (39)

apical oxygens, thus stabilizing the cyclic isomer (**37B**, $R^1 = CF_3$, $R^2 = Me$). Evidently, the same effect is responsible for the stability of the cyclic carbonyl analogue **(39)** which according to indirect data [94] possesses a greater acidity than (**37B**, $R^1 = CF_3$, $R^2 = Me$).

Besides spirobicyclic hydroxyphosphoranes, monocyclic analogues form the ring-chain equilibria.

Recently ^{31}P-NMR spectral evidence was given [95] for the equilibrium **(40A)** ⇌ **(40B)** strongly displaced toward **(40A)**. An equilibrium between the corresponding anions **(41A)** ⇌ **(41B)** has also been discussed.

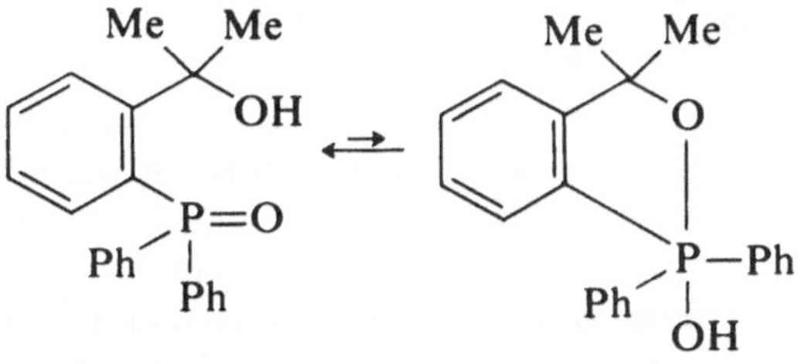

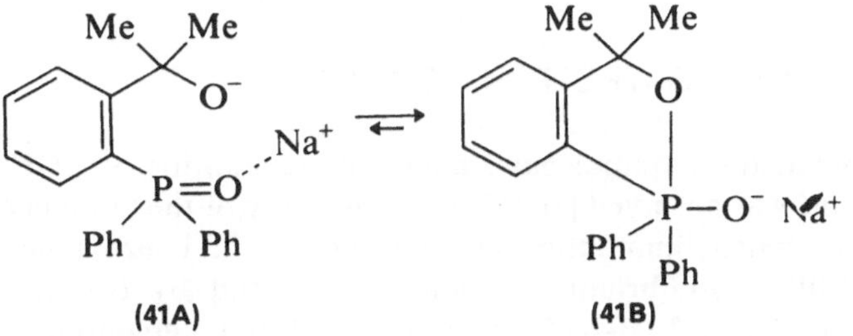

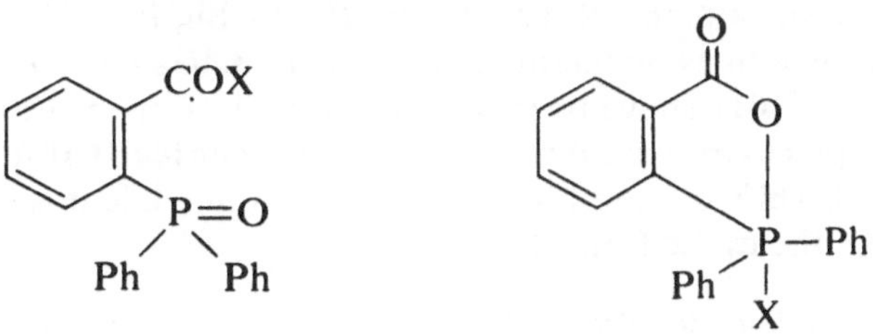

From a reaction of 2-diphenylphosphinylbenzoic acid with thionylchloride a stable cyclic chlorophosphorane (**42B**, X = Cl) was obtained [96], being formed by an intramolecular addition of a COCl group to a P=O bond. The open isomer **(42A)** could not be detected by spectroscopic methods. 2-Diphenylphosphinyl benzoic acid on the other hand exists in the open form (**42A**, X = OH) and no equilibrium with the cyclic hydroxyphosphorane (**42B**, X = OH) has been observed [96].

In summary, the known hydroxy and chlorophosphoranes have a five-membered ring which follows the rule [78, 79, 97, 98] that trigonal bipyramidal phosphoranes are strongly stabilized by bridging of an apical and an equatorial position by a five-membered ring.

Another type of spirophosphorane formed by ring-chain tautomerism (**43A**) ⇌ (**43B**) has been observed. Here the cyclic isomer is the result of an intramolecular addition of an X—H group to a phosphorus atom, the latter undergoing a coordination change.

(43A)   **(43B)**

X = O, NR

The influence of structural and external factors on the equilibrium (**43A**) ⇌ (**43B**) position has been treated in a review [99] where this phenomenon is discussed in terms of Pearson's concept† of hard and soft acids and bases. The regularities of this kind of ring-chain tautomerism will not be discussed here in detail particularly since no addition to a multiply-bonded phosphorus atom is involved in ring formation.

4.4. ADDITION TO THE P=N GROUP

The reversible intramolecular nucleophilic addition of a OH group to a P=N bond was observed [100, 101] in the study of the tautomeric system of 2-hydroxyphenyliminophosphoranes and 1,2,3-benzoxaphospholines (**44A**) ⇌ (**44B**). Equilibrium constants determined by two independent methods, i.e., ^{31}P and ^{1}H-NMR spectroscopy [101], agree well.

Influences of substituents R^3 and R^4 (by R^1 = t-Bu, R^2 = H) on equilibrium constants as well as on thermodynamic parameters ΔH and ΔS have been investigated [101] showing that increasing electron density on the oxygen atom displaces the equilibrium in favor of cyclic (**44B**). If R^3 = Ph$_3$C the equilibrium is shifted more toward the open form while it is displaced more in favor of the cyclic form for R^3 = t-Bu.

† R. G. Pearson, Ed., *Hard and Soft Acids and Bases.* Dowden, Hutchinson and Ross, Stroudsberg, Pa. 1973; T.-L. Ho, *Chem. Rev.* **1975**, *75*, 1.

Decreasing electron density at the phosphorus atom also shifts the equilibrium toward the cyclic form. For instance, cyclic isomers (**44B**, R^4 = F) which contain electron-withdrawing fluorine atoms at phosphorus are stable [103]. On the basis of such a substituent influence on the equilibrium position the cyclization is believed to proceed as an intramolecular addition of the lone pair of the oxygen atom to the electron deficient phosphorus atom and a subsequent proton migration [101].

Increasing encumbrance of substituents at the phosphorus atom (**44**, R^4 = Me, Et, Pr) displaces the equilibrium toward the open form; after introducing a *sec*-alkyl substituent at the phosphorus atom (**44**, R^1 = *t*-Bu, R^2 = H, R^3 = Ph$_3$C, *t*-Bu, R^4 = *i*-Pr) the cyclization does not proceed at all [101]. Sterically demanding substituents in a position ortho to the hydroxy group (**44**, R^1 = H, *t*-Bu, Ph$_3$C) strongly stabilize the cyclic structure [102].

(**44A**)   (**44B**)

R^1 = *t*-Bu; R^2 = H; R^3 = Ph$_3$C or *t*-Bu; R^4 = Me, Et, Pr, *i*-Pr, Ph [101].
R^1 = H, Me, *t*-Bu; R^2 = H, Me, *t*-Bu; R^3 = H, Me, *t*-Bu, Ph, Ph$_2$CH, (2-MeOC$_6$H$_4$)$_2$CH, (4-MeOC$_6$H$_4$)$_2$CH, Ph$_3$C; R^4 = Me, Et, Pr, Bu, Ph [102].

Rate constants for transformations (**44A**) $\rightleftarrows$ (**44B**) and activation parameters $\Delta H^{\neq}$ and $\Delta S^{\neq}$ have been determined [102] using dynamic ^{1}H-NMR spectroscopy (see Table 42; lifetimes are given in [102]). Equilibrium constants K_T and thermodynamic parameters ΔH and ΔS were obtained from ^{31}P-NMR spectra. However, since equilibrium constants K_T and rate constants k_1 and k_2 have been established by different methods the relationship $K_T = k_1/k_2$ is not entirely satisfactory.

Raising the temperature of solutions shifts the equilibrium (**44A**) $\rightleftarrows$ (**44B**) toward the open form (ΔH and ΔS are negative).

Solvents exert a large influence on the position of the equilibrium (**44A**) $\rightleftarrows$ (**44B**). Proton-accepting solvents (pyridine, tripropylamine) displace the equilibrium in favor of the ring form while chloroform and methylene chloride act in the opposite direction. On the basis of the thermodynamic equilibrium parameters (ΔH, ΔS) for different solvents [101]

Table 42. Ring-Chain Equilibria (44A) $\rightleftarrows$ (44B), Rate Constants, Thermodynamic and Activation Parameters as Determined by
^{31}P-NMR and Dynamic ^{1}H-NMR Methods in Pyridine-d$_5$ [102]

| R^1 | R^2 | R^3 | R^4 | ^{31}P-NMR | | | Dynamic ^{1}H-NMR | | | | | |
| | | | | K at 27°C | $-\Delta H$ (kcal/mol) | $-\Delta S$ (e.u.) | (A) → (B) | | | (B) → (A) | | |
							k_1 (s^{-1})	$\Delta H^{\neq}$ (kcal/mol)	$\Delta S^{\neq}$ (e.u.)	k_2 (s^{-1})	$\Delta H^{\neq}$ (kcal/mol)	$\Delta S^{\neq}$ (e.u.)
H	H	Me	Ph	0.33	4.61	17.6						
H	H	t-Bu	Ph	0.31	3.06	12.5						
H	Me	H	Ph	0.29	4.34	16.9						
H	t-Bu	H	Ph	0.26	3.84	15.5						
t-Bu	H	Ph$_2$CH	Et	2.09	4.79	14.5						
t-Bu	H	Ph$_2$CH	Pr	1.63	3.80	11.7	3.6	17.4	2.3	2.0	21.6	15.3
t-Bu	H	Ph$_2$CH	Bu	1.60	3.52	10.8	4.8	16.5	0.1	2.7	20.1	11.0
t-Bu	H	Ph$_2$CH	Ph	2.88	3.51	9.6	0.524	15.9	-6.6	0.252	19.1	2.7
t-Bu	H	(2-MeOC$_6$H$_4$)$_2$CH	Ph	2.93	4.60	13.2	0.287	16.4	-5.9	0.113	20.6	6.3
t-Bu	H	(4-MeOC$_6$H$_4$)$_2$CH	Ph	2.60	4.65	13.6	0.286	16.5	-5.6	0.106	21.1	7.7
t-Bu	H	Ph$_3$C	Ph	1.47	3.59	11.2	0.488	17.4	-1.7	0.426	20.2	7.6

chloroform is believed to specifically interact with the basic iminophosphorane group, thus stabilizing the open structure.

A similar ring-chain equilibrium was detected [104] in solutions of 4-alkyl-2,6-bis(phosphoranylideneamino)phenole **(45A)** ⇄ **(45B)**.

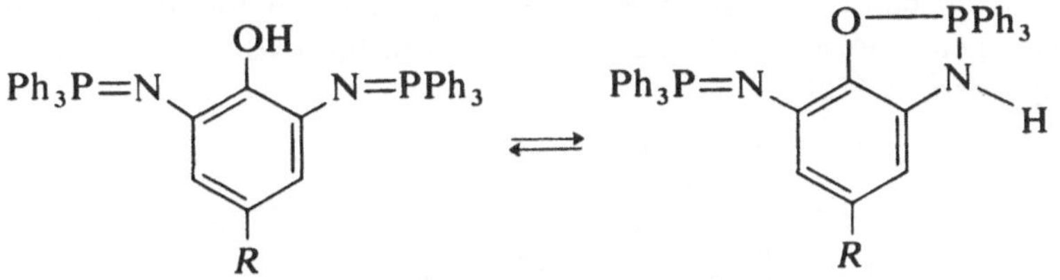

(45A)   R = Me, t-Bu, cyclo-C_6H_{11} (45B)

4.5. ADDITION TO THE S=O GROUP

Relatively few data are available regarding intramolecular additions to the S=O bond yielding cyclic hydroxysulfuranes **(46A)** ⇄ **(46B)**. However, the chemistry in this field has rapidly developed (see review [98]).

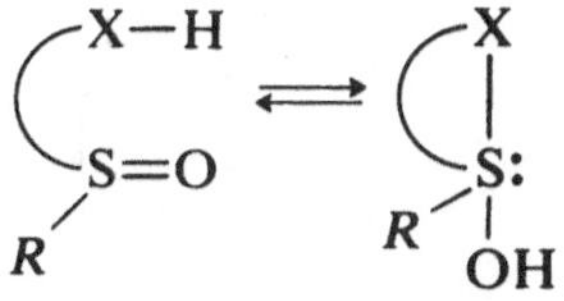

(46A)   (46B)

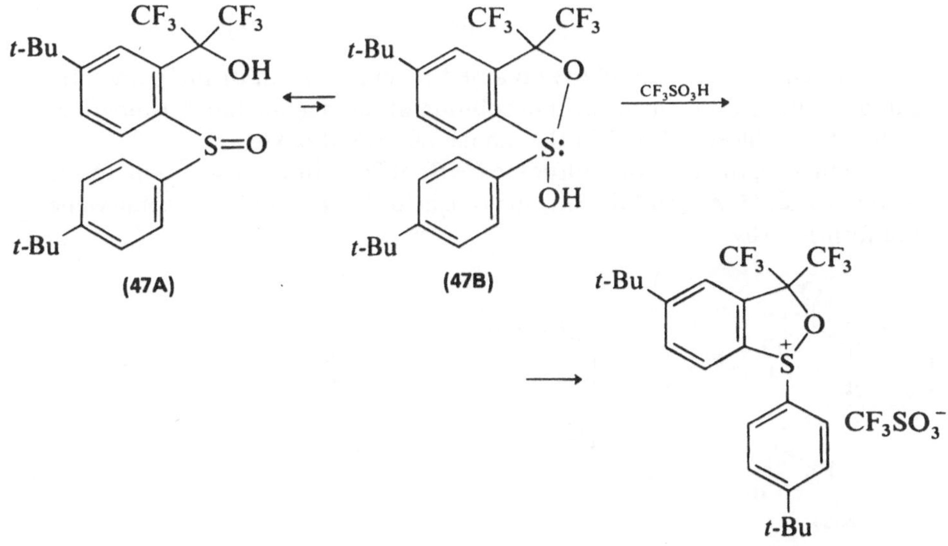

(47A)   (47B)   (48)

Solutions of the sulfoxide (47) in $CDCl_3$ have been investigated by ^{19}F-NMR spectroscopy showing [105] that the open structure (47A) predominates. Addition of triflic acid (CF_3SO_3H) to the solution yielded the cyclic sulfonium triflate (48).

A cyclic chlorosulfurane (50) was formed in a reaction of acetylchloride or hydrogen chloride with the sulfoxide (49), the hydrolysis of which led to ring opening [106].

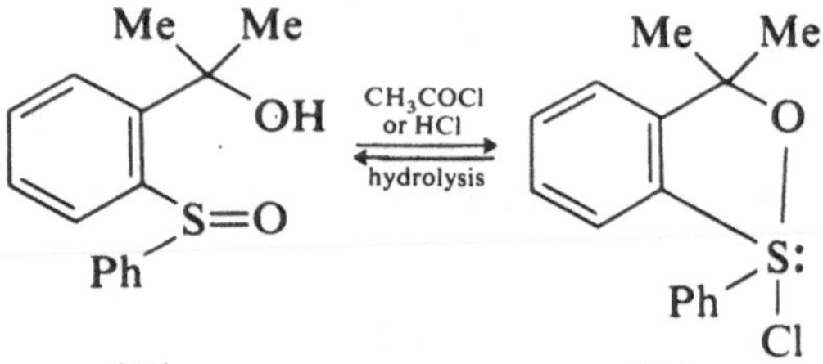

(49)   (50)

o-Carboxyphenyl-methyl and -phenyl sulfoxides exist in the open form (51, R = Me, Ph) [107–110].

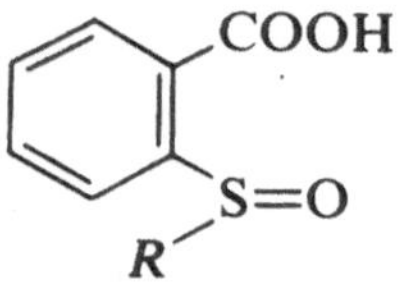

(51) 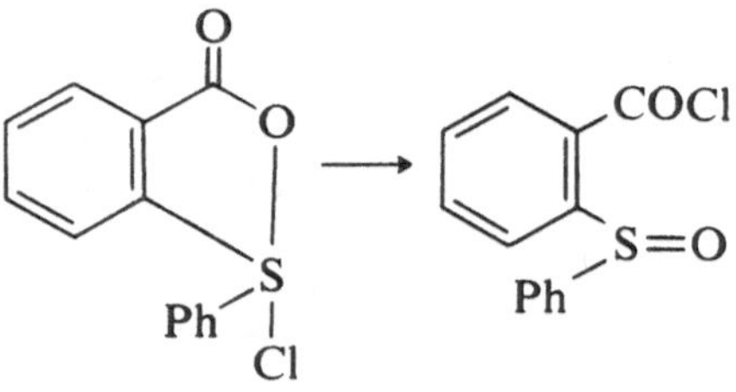   (52B)   (52A)

The chlorosulfurane (52B) obtained by chlorination of the potassium salt of 2-phenylthiobenzoic acid in solution at room temperature isomerizes to the open chloride (52A) in the course of several days.

Similarly, the t-butoxysulfurane (53B) at 70°C in chlorobenzene forms an open ester (53A). Additionally the free acid (51, R = Ph) and isobutylene are formed [109].

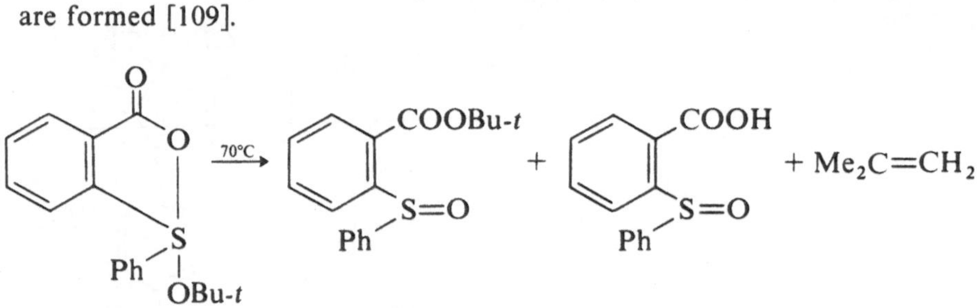

(53B)   (53A)

4.6. ADDITION TO THE Se=O GROUP

Unlike *o*-carboxyphenylmethyl and phenylsulfoxides (**51**, R = Me, Ph) the analogous selenoxides are cyclic (**54,** R = Me, Ph) in the solid state [108] as well as in solution [110–112].

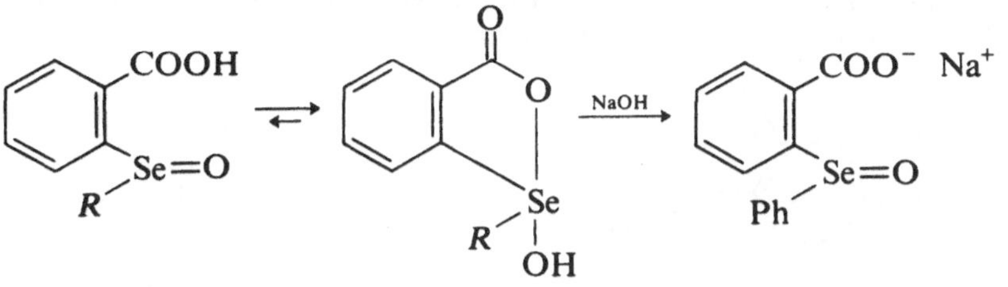

(54A)	**(54B)**	**(55)**

Following ^{1}H, ^{13}C, and ^{77}Se-NMR spectroscopy [112] the acid (**54,** R = Me) in methanol-d_4 appears to be a cyclic selenurane (**54B**). However, the sodium salt of this acid has open structure (**55**).

4.7. ADDITION TO THE I=O GROUP

Intramolecular nucleophilic additions to the I=O bond proceed easily and irreversibly if five-, but more rarely if six-, membered iodinane rings are formed.

2-Iodosobenzoic acid (**56**) [113–116], its amides (**57**) [115, 117], esters (**58**) [114, 116], and a mixed acetic acid anhydride (**59**) [116, 118] only appear

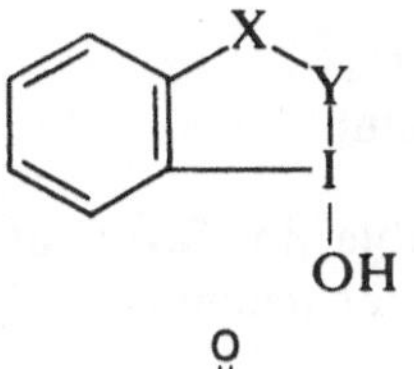

$$(56, X-Y = \overset{\overset{O}{\|}}{C}-O)$$

$$(57, X-Y = \overset{\overset{O}{\|}}{C}-NR)$$

$$(60, X-Y = SO_2-NH)$$

$$(61, X-Y = \overset{\overset{O}{\|}}{\underset{\underset{Me}{|}}{P}}-O)$$

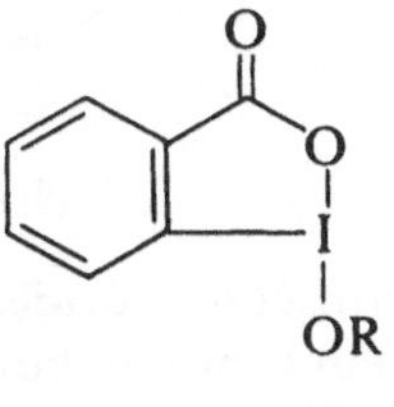

(**58**, R = alkyl)
(**59**, R = COMe)

as cyclic isomers as do 2-iodosobenzenesulfonamide **(60)** [119], 2-iodoso-phenylmethylphosphinic acid **(61)** [120], 2-iodosophenylacetic acid **(62)** [121], 2-iodosophenyl phosphate **(63)** [122] and 2-iodoxybenzoic acid **(64)** [116].

Stable cyclic hydroxyiodinanes **(65,** R = Me, CF_3) bearing a gem-dimethyl or gem-trifluoromethyl group have been synthesized [123] as well as the exclusively cyclic chloroiodinane **(66)** [123, 124].

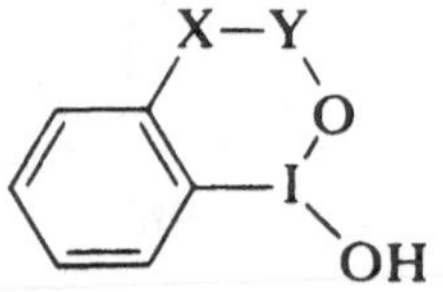

(62, X—Y = CH_2—C)

(63, X—Y = O—P)

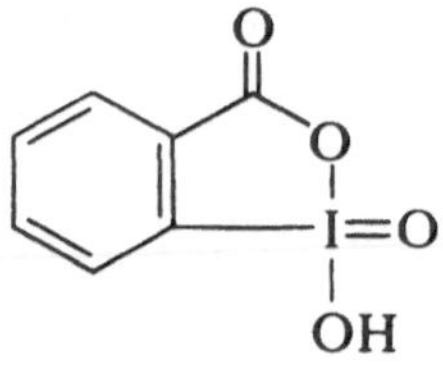

(64)

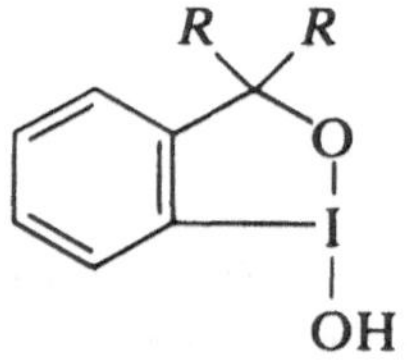

(65)

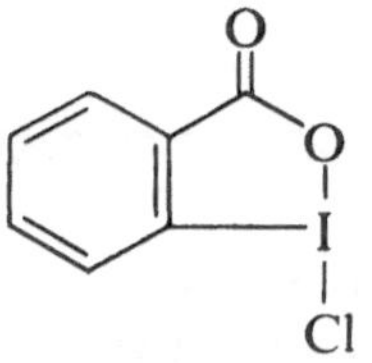

(66)

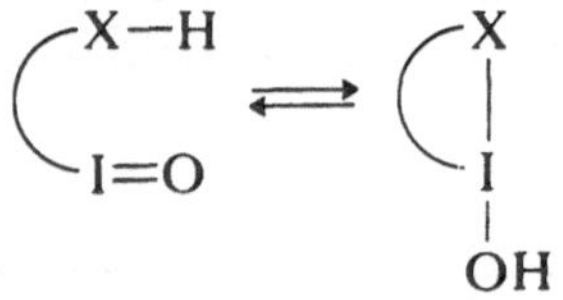

(67A) X = O, NR **(67B)**

No spectroscopic evidence is available in favor of equilibrium **(67A)** ⇌ **(67B)** in solution, however, a more far-reaching study of the compounds seems advisable.

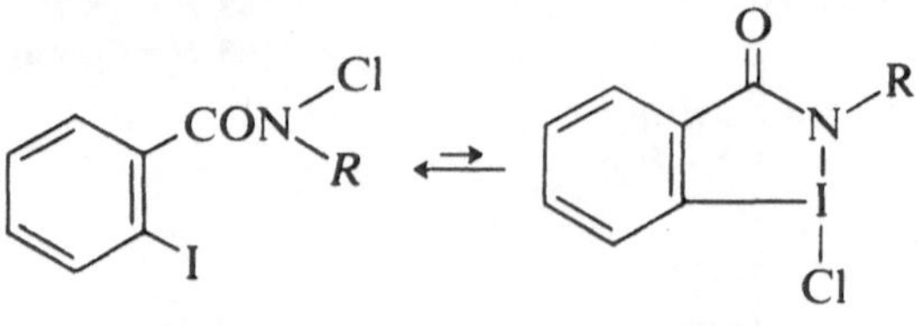

(68A)

(68B)

A ring-chain equilibrium **(68A)** ⇌ **(68B)** has been shown [125] to exist for *N*-chloro-2-iodobenzamides. In this case, however, the cyclic structure **(68B)** is formed as a result of an intramolecular addition of the N—Cl group to an iodine atom, the latter changing its coordination.

REFERENCES

1. A. I. Meyers and J. C. Sircar, In: *The Chemistry of the Cyano Group*, Ed. Z. Rappoport, Interscience Publishers, London, 1970, p. 341.
2. E. N. Zilberman, *The Reactions of Nitriles*, Khimiya, Moscow, 1972 (In Russian).
3. E. C. Taylor and R. V. Ravindranathan, *J. Org. Chem.* **1962**, *27*, 2622.
4. H. Junek, H. Fischer-Colbrie, H. Aigner, and A. M. Braun, *Helv. Chim. Acta* **1972**, *55*, 1459.
5. R. G. Dubenko and E. F. Gorbenko, *Khim. Geterotsikl. Soedin.* **1967**, 923.
6. H. Nohira, Y. Nishikava, Y. Furuya, and T. Mukayama, *Bull. Chem. Soc. Jpn.* **1965**, *38*, 897.
7. R. Kwok and P. Pranc, *J. Org. Chem.* **1967**, *32*, 738.
8. M. Regitz and D. Stadler, *Chem. Ber.* **1968**, *101*, 2351.
9. G. Kille and J.-P. Fleury, *Bull. Soc. Chim. Fr.* **1967**, 4619.
10. J.-P. Fleury and A. Baysang, *Bull. Soc. Chim. Fr.* **1969**, 4102.
11. J.-P. Fleury, A. Baysang, and D. Clerin, *Bull. Soc. Chim. Fr.* **1969**, 4108.
12. S. I. Suminov and A. N. Kost, *Zh. Obshch. Khim.* **1963**, *33*, 2208.
13. A. Foucoud, *Bull. Soc. Chim. Fr.* **1964**, 123.
14. E. C. Taylor and R. W. Hendess, *J. Am. Chem. Soc.* **1965**, *87*, 1995.
15. J. Le Ludec, D. Danion, and R. Carrie, *Bull. Soc. Chim. Fr.* **1966**, 3895.
16. T. L. Patton, *J. Org. Chem.* **1967**, *32*, 383.
17. E. C. Taylor, A. McKillop, Y. Shvo, and G. H. Havks, *Tetrahedron* **1967**, *23*, 2081.
18. Z. Höhn, *Z. Chem.* **1970**, *10*, 386.
19. W. Klötzner, R. Franzmair, and H. Bretschneider, *Monatsh. Chem.* **1970**, *101*, 1263.
20. J. N. Wells, W. J. Wheeler, and L. M. Davisson, *J. Org. Chem.* **1971**, *36*, 1503.
21. T. S. Safonova, M. P. Nemeryuck, L. A. Mishkina, and N. I. Traven', *Khim. Geterotsikl. Soedin.* **1972**, 944.
22. R. E. Valters, A. E. Bace, and S. P. Valtere, *Latv. PSR Zinat. Akad. Vestis, Kim. Ser.* **1972**, 726.
23. V. Szabo, J. Borda, and E. Theisz, *Acta Chim. Acad. Sci. Hung.* **1980**, *103*, 271.
24. G. Cignarella and U. Teotino, *J. Am. Chem. Soc.* **1960**, *82*, 1594.
25. D. E. Balode, R. E. Valters, and S. P. Valtere, *Khim. Geterotsikl. Soedin.* **1978**, 1632; *see also* R. E. Valters, R. B. Kampare, S. P. Valtere, D. E. Balode, and A. E. Bace, *Khim. Geterotsikl. Soedin.* **1983**, 1635.
26. R. E. Valters and S. P. Valtere, *Khim. Geterotsikl. Soedin.* **1972**, 281.
27. A. Braun and J. Tcherniac, *Ber. Dtsch. Chem. Ges.* **1907**, *40*, 2709.
28. H. Matsui and S. Ishimoto, *Tetrahedron Lett.* **1966**, 1827.
29. M. Renson and R. Colienne, *Bull. Soc. Chim. Belg.* **1964**, *73*, 491.
30. H. des Abbayes, *Bull. Soc. Chim. Fr.* **1970**, 3661.
31. G. Pangdon, G. Thuillier, and P. Rumpf, *Bull. Soc. Chim. Fr.* **1970**, 1991.
32. E. C. Taylor and P. K. Loeffler, *J. Am. Chem. Soc.* **1960**, *82*, 3147.
33. J. A. Settepani and A. B. Borkovec, *J. Heterocycl. Chem.* **1966**, *3*, 188.
34. F. S. Babichev and A. K. Tiltin, *Ukr. Khim. Zh.* **1970**, *36*, 62.

35. J. T. Shaw, D. M. Taylor, F. J. Corbett, and J. D. Ballentine, *J. Heterocycl. Chem.* **1972**, *9*, 125.

36. G. Coispeau and J. Elguero, *Bull. Soc. Chim. Fr.* **1970**, 2717.

37. P. E. Gagnon, J. L. Boiwin, and R. N. Jones, *Can. J. Res.* **1949**, *27B*, 190.

38. P. E. Gagnon, J. L. Boiwin, P. A. Boiwin, and R. N. Jones, *Can. J. Chem.* **1951**, *29*, 182.

39. P. E. Gagnon, J. L. Boiwin, and A. Chisholm, *Can. J. Chem.* **1952**, *30*, 904.

40. R. E. Valters, E. A. Baumanis, L. K. Stradina, and E. E. Liepins, *Khim. Geterotsikl. Soedin.* **1981**, 516.

41. J. Eby and J. A. Moore, *J. Org. Chem.* **1967**, *32*, 1346.

42. S. G. Tadevosyan, E. N. Teleshov, I. V. Vasil'eva, and A. N. Pravednikov, *Zh. Org. Khim.* **1980**, *16*, 353.

43. A. R. Butler, *J. Chem. Soc., Perkin Trans. 2* **1974**, 1239.

44. J. Zabicky, *Chem. Ind.* **1964**, 236.

45. I. V. Vasil'eva, S. G. Tadevosyan, E. N. Teleshov, and A. N. Pravednikov, *Dokl. Akad. Nauk SSSR* **1981**, *256*, 398.

46. M. St. C. Flett, *Spectrochim. Acta* **1962**, *18*, 1537.

47. L. J. Bellamy, *Advances in Infrared Group Frequencies*, Methuen and Co, Ltd., Bungay, 1968, Chapter 5 and 8.

48. M. E. Baguley and J. A. Elvidge, *J. Chem. Soc.* **1957**, 709.

49. K. W. Breukink and P. E. Verkade, *Rec. Trav. Chim.* **1960**, *79*, 443.

50. G. W. Stacy, A. J. Papa, F. W. Villaescusa, and S. C. Ray, *J. Org. Chem.* **1964**, *29*, 607.

51. G. W. Stacy, F. W. Willaescusa, and T. E. Wollner, *J. Org. Chem.* **1965**, *30*, 4074.

52. G. W. Stacy and T. E. Wollner, *J. Org. Chem.* **1967**, *32*, 3028.

53. D. L. Eck and G. W. Stacy, *J. Heterocycl. Chem.* **1969**, *6*, 147.

54. G. Simchen and G. Entenmann, *Angew. Chem.* **1973**, *85*, 155.

55. G. Simchen and M. Häfner, *Justus Liebigs Ann. Chem.* **1974**, 1802.

56. G. Simchen and G. Entenmann, *Justus Liebigs Ann. Chem.* **1977**, 1249.

57. P. R. Jones, *Chem. Rev.* **1963**, *63*, 461.

58. T. I. Temnikova, *The Course of Theoretical Principles of Organic Chemistry*, Khimiya, Leningrad, 1968, Chapter 26 (In Russian); I. N. Nazarov, *Uspekhi Khimii*, **1951** 20, 71 and 309.

59. V. I. Staninets and E. A. Shilov, *Uspekhi Khimii* **1971**, *40*, 491; Yu. I. Gevaza and V. I. Staninets, *Khim. Geterotsikl. Soedin.* **1982**, 1443.

60. A. Lattes, *Khim. Geterotsikl. Soedin.* **1975**, 7.

61. M. D. Dowle and D. I. Dawies, *Chem. Soc. Rev.* **1979**, *8*, 171.

62. T. Teitei, *Aust. J. Chem.* **1982**, *35*, 1231.

63. E. Napolitano, R. Fiaschi, and A. Marsili, *Tetrahedron Lett.* **1983**, *24*, 1319.

64. L. R. Letsinger, E. N. Oftedahl, and I. R. Nazy, *J. Am. Chem. Soc.* **1965**, *87*, 742.

65. R. P. Linstead and H. N. Rydon, *J. Chem. Soc.* **1933**, 580.

66. B. D. Whelton and A. C. Huitric, *J. Org. Chem.* **1970**, *35*, 3143.

67. T. Hashimoto and S. Nagase, *J. Pharm. Soc. Jpn.* **1960**, *80*, 1637.

68. H. H. Baer and L. Urbas, In: *The Chemistry of the Nitro and Nitroso Groups*, Ed. H. Feuer, Interscience Publishers, New York, 1970, Vol. 2, Chapter 3.

69. H. H. Baer and F. Kienzle, *Can. J. Chem.* **1965**, *43*, 190.

70. R. T. Borchardt and L. A. Cohen, *J. Am. Chem. Soc.* **1972**, *94*, 9175.

71. G. Pitacco and V. Ennio, *Tetrahedron Lett.* **1978**, 2339.

72. S. Chimichi, R. Nesi, M. Scotton, C. Mannucci, and G. Adembri, *J. Chem. Soc., Perkin Trans. 2*, **1980**, 1339; See also S. Chimichi, R. Nesi, F. De Sio, R. Pepino, and A. Degl'Innocenti, *Gazz. Chim. Ital.* **1982**, *112*, 249; R. Nesi, S. Chimichi, F. De Sio, and M. Scotton, *Org. Magn. Reson.* **1983**, *21*, 42.

73. E. A. Hill, *J. Organomet. Chem.* **1975**, *91*, 123.
74. E. A. Hill, *Adv. Organomet. Chem.* **1977**, *16*, 131.
75. E. A. Hill and M. M. Myers, *J. Organomet. Chem.* **1979**, *173*, 1.
76. E. A. Hill, D. C. Link, and P. Donndelinger, *J. Org. Chem.* **1981**, *46*, 1177.
77. E. A. Hill and H. R. Ni, *J. Org. Chem.* **1971**, *36*, 4133.
78. F. H. Westheimer, *Acc. Chem. Res.* **1969**, *1*, 70.
79. R. F. Hudson and C. Brown, *Acc. Chem. Res.* **1972**, *5*, 204.
80. S. J. Benkovic and K. J. Schray, In: *Transition States in Biochemical Processes*, Ed. R. D. Gandour and R. L. Schowen, Plenum, New York, 1978, p. 493.
81. E. E. Nifant'ev and L. M. Matveeva, *Zh. Obshch. Khim.* **1969**, *39*, 1555.
82. B. A. Arbuzov and N. A. Polezhaeva, *Uspekhi Khimii* **1974**, *43*, 933.
83. F. Ramirez, M. Nowakovski, and J. F. Marecek, *J. Am. Chem. Soc.* **1976**, *98*, 4330.
84. A. Munoz, M. Galagher, R. Klaebe, and R. Wolf, *Tetrahedron Lett.* **1976**, 673.
85. G. Kemp and S. Trippett, *Tetrahedron Lett.* **1976**, 4381.
86. F. Ramirez, M. Nowakovski, and J. F. Marecek, *J. Am. Chem. Soc.* **1977**, *99*, 4515.
87. R. Sarma, F. Ramirez, B. McKeever, M. Nowakovski, and J. F. Marecek, *J. Am. Chem. Soc.* **1978**, *100*, 5391.
88. G. Kemp and S. Trippett, *J. Chem. Soc., Perkin Trans 1* **1979**, 879.
89. Ch. Bui-Cong, A. Munoz, M. Sanchez, and A. Klaebe, *Tetrahedron Lett.* **1977**, 1587.
90. A. Munoz, B. Garrigues, and M. Koenig, *J. Chem. Soc., Chem. Commun.* **1978**, 219.
91. A. Munoz, B. Garrigues, and M. Koenig, *Tetrahedron* **1980**, *36*, 2647.
92. I. Granoth and J. C. Martin, *J. Am. Chem. Soc.* **1978**, *100*, 5229.
93. I. Granoth and J. C. Martin, *J. Am. Chem. Soc.* **1979**, *101*, 4618.
94. Y. Segall and I. Granoth, *J. Am. Chem. Soc.* **1979**, *101*, 3687.
95. I. Granoth, R. Alkabets, and E. Shirin, *J. Chem. Soc., Chem. Commun.* **1981**, 981.
96. I. Granoth, R. Alkabets, and Y. Segall, *J. Chem. Soc., Chem. Commun.* **1981**, 622.
97. S. J. Benkovic, *Compr. Chem. Kinetics* **1969**, *10*, 1.
98. J. C. Martin and E. F. Perozzi, *Science* **1976**, *191*, 154.
99. A. Munoz, *Bull. Soc. Chim. Fr.* **1977**, 728.
100. H. B. Stegmann, G. Bauer, E. Breitmaier, E. Herrmann, and K. Scheffler, *Phosphorus* **1975**, *5*, 207.
101. H. B. Stegmann, R. Haller, and K. Scheffler, *Chem. Ber.* **1977**, *110*, 3817.
102. H. B. Stegmann, R. Haller, A. Burmester, and K. Scheffler, *Chem. Ber.* **1981**, 114, 14.
103. H. B. Stegmann, H. V. Dumm, and K. B. Ulmschneider, *Tetrahedron Lett.* **1976**, 2007.
104. H. B. Stegmann, H. Müller, K. B. Ulmschneider, and K. Scheffler, *Chem. Ber.* **1979**, *112*, 2444.
105. J. C. Martin and E. F. Perozzi, *J. Am. Chem. Soc.* **1974**, *96*, 3155.
106. T. M. Balthazor and J. C. Martin, *J. Am. Chem. Soc.* **1975**, *97*, 5634.
107. P. Livant and J. C. Martin, *J. Am. Chem. Soc.* **1977**, *99*, 5761.
108. B. Dahlen, *Acta Crystallograph.* **1973**, *B29*, 595.
109. P. Livant and J. C. Martin, *J. Am. Chem. Soc.* **1976**, *98*, 7851.
110. W. Nakanishi, S. Murata, Y. Ikeda, T. Sugawara, Y. Kawada, and H. Iwamura, *Tetrahedron Lett.* **1981**, *22*, 4241.
111. W. Nakanishi, Y. Ikeda, and H. Iwamura, *J. Org. Chem.* **1982**, *47*, 2275.
112. W. Nakanishi, S. Matsumoto, Y. Ikeda, T. Sugawara, Y. Kawada, and H. Iwamura, *Chem. Letters* **1981**, 1353.
113. R. Bell and K. J. Morgan, *J. Chem. Soc.* **1960**, 1209.
114. G. P. Baker, F. G. Mann, N. Sheppard, and A. J. Tetlow, *J. Chem. Soc.* **1965**, 3721.
115. W. Wolf and L. Steinberger, *J. Chem. Soc., Chem. Commun.* **1965**, 449.
116. H. Siebert and M. Handrich, *Z. anorg. und allgem. Chem.* **1976**, *426*, 173.

117. H. J. Barber and M. A. Henderson, *J. Chem. Soc., Sect. C* **1970**, 862.
118. J. Z. Gougoutas and J. C. Clardy, *J. Solid State Chem.* **1972**, *4*, 226.
119. H. Jaffe and J. E. Leffler, *J. Org. Chem.* **1975**, *40*, 797.
120. T. M. Balthazor, J. A. Miles, and B. R. Stults, *J. Org. Chem.* **1978**, *43*, 4538.
121. J. E. Leffler, L. K. Dyall, and P. W. Inward, *J. Am. Chem. Soc.* **1963**, *85*, 3443.
122. J. E. Leffler and H. Jaffe, *J. Org. Chem.* **1973**, *38*, 2719.
123. R. L. Amey and J. C. Martin, *J. Org. Chem.* **1979**, *44*, 1779.
124. L. J. Andrews and R. M. Keefer, *J. Am. Chem. Soc.* **1959**, *81*, 2374.
125. T. M. Balthazor, D. E. Godar, and B. R. Stults, *J. Org. Chem.* **1979**, *44*, 1447.

Generalizations Concerning the Influence of Structural and External Factors on the Relative Stability of Ring and Chain Isomers

Experimental data discussed in the previous chapters mainly apply to reversible intramolecular nucleophilic addition reactions of hydroxy, amino, and mercapto groups to polar multiple bonds, the $C=O$ and $C=N$ bonds being represented most broadly. Special attention was paid to the influence of structure and external factors on the stability of open and cyclic isomers and on the mode of their interconversions.

The aim of the present chapter is to acquaint the reader with general rules concerning these influences.

Equilibrium constants for a large number of compounds determined under fixed experimental conditions are required to formulate general rules. The structure of the connecting link should be varied widely; substituents with different electronic and steric effects should be bonded at significant places and, finally, the results of these investigations should be compared with those of systems possessing related pairs of interacting groups.

Unfortunately, there are only a few systematic investigations of this kind, e.g., in the field of acylcarboxylic acids (Bowden), their derivatives (Valters, Flitsch, Chiron, and Graff), and hydroxy and mercaptoderivatives of imines, oximes, and hydrazones (Potekhin). Most available results stem from studies of ring-chain tautomerism in a narrow range of compounds which have found interest for the chemical properties of the compounds and their synthesis.

Certainly, equilibrium constants are most important for the estimation of the relative stabilities of isomers. The use of the equilibrium constants in a broader field is often limited by the fact that quantitative investigations were carried out using different methods and various conditions (see Table 3). Besides, particularly important areas have not been studied adequately at all.

For many equilibria of the type $(1A) \rightleftarrows (1B)$, $(2A) \rightleftarrows (2B)$, and $(3A) \rightleftarrows (3B)$ the stability of the cyclic form has been characterized by the ring-chain equilibrium constant K or by the free energy difference ΔG, which is proportional to the logarithm of K. It would be desirable to discuss the influence of structural and external factors using these parameters. There are, however, objections which should be taken into account. First of all,

$$
\begin{array}{c}
\text{(1A)} \\
\overset{X-H}{\underset{C=O}{\bigg\rvert}}\ R^1
\end{array}
\quad
\underset{k_2}{\overset{k_1}{\underset{\xrightleftharpoons{\ K\ }}{}}}
\quad
\begin{array}{c}
\text{(1B)} \\
\overset{X}{\underset{C-OH}{\bigg\rvert}}\ R^1
\end{array}
$$

$$
\begin{array}{c}
\text{(2A)} \\
\overset{X-H}{\underset{C=N}{\bigg\rvert}}\ R^1 \quad R^2
\end{array}
\quad
\underset{k_2}{\overset{k_1}{\underset{\xrightleftharpoons{\ K\ }}{}}}
\quad
\begin{array}{c}
\text{(2B)} \\
\overset{X}{\underset{C-NHR^2}{\bigg\rvert}}\ R^1
\end{array}
$$

$$
\begin{array}{c}
\text{(3A)} \\
\overset{X-H}{\underset{N=C}{\bigg\rvert}}\ R^1,\ R^2
\end{array}
\quad
\underset{k_2}{\overset{k_1}{\underset{\xrightleftharpoons{\ K\ }}{}}}
\quad
\begin{array}{c}
\text{(3B)} \\
X,\ C,\ R^1,\ R^2,\ N,\ H
\end{array}
$$

(3A) X = O, NR³, S (3B)

even small structural modifications very often shift the system out of the balanced equilibrium state [1, 2], i.e., one isomer can be detected only and thus no equilibrium constant is available. Moreover, the free energy difference enthalpy and entropy components ($\Delta H, \Delta S$), are rarely available. Furthermore, as pointed out by Jencks [3], the use of these parameters ($\Delta H, \Delta S$) is connected with uncertainties which mainly originate from the need to introduce corrections for solvation effects.

Energetic aspects of intramolecular reactions which separate enthalpic and entropic influences and their comparison with those of intermolecular reactions are discussed in Page's review [4]. It is difficult to estimate potential energy differences between open and cyclic structures utilizing the changes in bond lengths, valence angle deformations, nonbonded and electrostatic interactions, steric strain, solvation, and formation of hydrogen bonds.

Many data are available in the literature on the structure of tautomeric compounds in the solid state. However, these data may rightfully be used for the estimation of structural factors that influence isomer stability only in the case when the absence of tautomeric equilibria in solutions is proven by spectroscopic methods. Moreover, the influence of isolation conditions on the structure of isomer obtained in solid state should also be estimated by taking into account that under some conditions a thermodynamically less stable isomer can be isolated. Undoubtedly, crystal-field forces may also play an important role. The situation improves if the structures of a large series of crystalline compounds are investigated and the conditions for isolation are essentially the same.

Little attention has been paid to chemical reactions of pure isomers. The reactivity of individual isomers or isomeric mixtures cannot be used generally to estimate their relative stability. The reasons for this can be: 1) different rates of reactions of open or cyclic isomers; 2) the influence of the reagent on the position of the ring-chain equilibrium; 3) the influence of reaction conditions (heat, presence of catalysts) on the equilibrium position; 4) ring-chain equilibria of reaction products formed during the reaction or isolation. Finally, and most importantly 5) there are at least two reactive centers in the open and/or cyclic isomers and an attack of the reagent on both centers must be taken into account. Moreover, the preferential direction of this attack is determined not only by the structure of the substrate but also by external factors as well as the structure of the reagent. Esterification of acylcarboxylic acids provides a good illustration of these difficulties (see Sec. 2.1.4).

Certainly, the reactivity of tautomeric bifunctional compounds is of great theoretical and practical importance. However, the scarcity of experimental data does not allow the proper analysis of this complicated problem.

5.1. STRUCTURAL INFLUENCES

5.1.1. Electronic Effects of Substituents at Interacting Groups

If the connecting link allows the interacting functional groups to approach sufficiently close for an intramolecular addition (1A) → (1B), (2A) → (2B) or (3A) → (3B), the ring-chain equilibrium will be found more to the side of the cyclic form the more electrophilic the carbon atom of the multiple bond and/or the more nucleophilic the heteroatom X.

Nucleophilicity of X — H Group:

$$X = O < NR^3 < S$$

† Here and elsewhere the arrow depicts an equilibrium shift toward ring **B**.

Leaving the remaining structural features unaltered the stability of the cyclic isomer obtained as a result of an intramolecular nucleophilic addition of an X—H group to polar multiple bonds increases in the sequence X=O < NR < S, which can be explained by an enhancement of the nucleophilicity of the atom X [5] as well as by a diminished steric strain [4] of sulfur containing heterocycles as compared with oxygen or nitrogen analogue.

Substituent R^3 in the $X = NR^3$ Group:

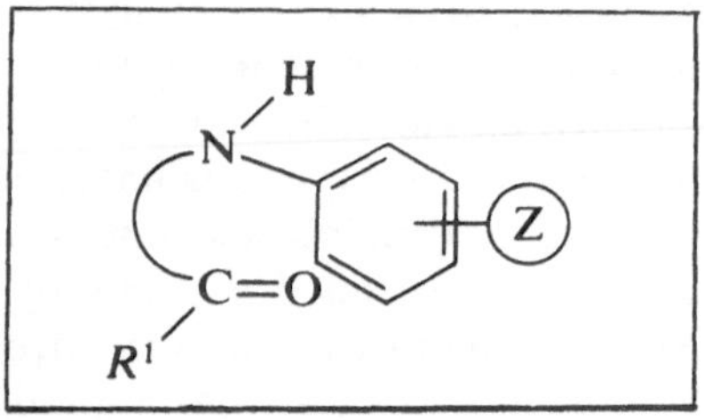

The influence of the nucleophilicity of the amino group on the stability of the cyclic isomer can be observed in a series of anilino derivatives bearing different substituents Z in the aromatic ring. The introduction of electron-donating substituents Z shifts the equilibrium toward the cyclic form while electron-withdrawing substituents act in the opposite direction. This is confirmed by the negative value of the ρ-constant on correlating the ring-chain equilibrium constants of 3,3-dimethyllevulinic acid N-arylamides (see Sec. 2.1.3: **(47A)** ⇌ **(47B)** and Table 10) with substituent coefficients σ^+.

A similar destabilization of cyclic isomers is brought about by decreasing the nucleophilicity of the nitrogen atom, i.e., on introduction of aryl (R^3 = Ar in **(1–3)**) and acyl (R^3 = COR) substituents, respectively, or dialkylammonium groups to the nitrogen atom (see Sec. 2.1.3: **(71)** → **(72)**).

N-Methylated derivatives (R^3 = Me) almost always show a greater stability of the cyclic form than their N-unsubstituted analogues. Exceptions to this rule (see, for instance, Sec. 2.3.2: **(166A)** ⇌ **(166B)**) have been explained by a steric effect of the methyl group destabilizing the ring structure. By increasing the steric bulk of N-alkyl substituents the estimation of the influence of their electronic effects on the stability of cyclic isomers is impossible since steric effects act considerably stronger.

C=O or C=NR² Group:

Equilibria involving azomethines **(2A)**⇌**(2B)** in solution are more in favor of the cyclic form that those of carbonyl derivatives **(1A)** ⇌ **(1B)** of similar structure (see Table 43). However, the substituents R^2 at the nitrogen atom of the azomethine group are also of great importance.

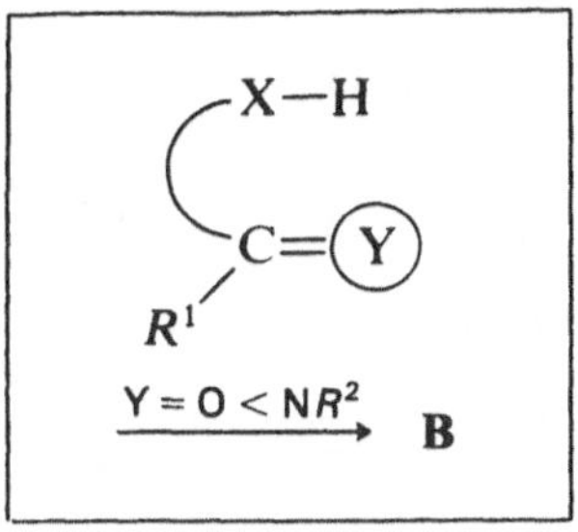

Substituents R^1 and R^2 at $C=O$ or $C=N$ Bonds. Substituents at carbonyl and azomethine groups affect the electrophilicity of the carbon atom. Electron-withdrawing substituents increase the electrophilicity and stabilize the cyclic isomer while electron-donating substituents act in the reverse direction.

Tautomeric equilibrium constants of acylcarboxylic acids increase in the following order for substituents at the $C=O$ group: aryl < alkyl < H (see Table 3). The stability of the open-structured aroylcarboxylic acids was

Table 43. Ring-Chain Equilibrium Constants for Hydroxyketones, Ketocarboxylic Acids, and Their N-Substituted Imines

Open isomer structure			Solvent	$Y = O$		$Y = NR^3$		
				$K_T = [B]/[A]$	Refs.	R^3	$K_T = [B]/[A]$	Refs.
	R	n						
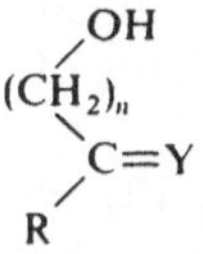	H	3	75% aqueous dioxan (Y = O)	7.8	6	Me	6.25	7
	H	4	CCl$_4$ (Y = NMe)	15.4	6	Me	>100	7
	Me	3	CCl$_4$	0.73	8	Me	4.0	7
	Me	4	CCl$_4$	0.95	8	Me	7.2	7
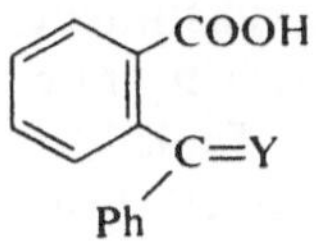			Dioxan	0.07	9	Ph	4.78	10
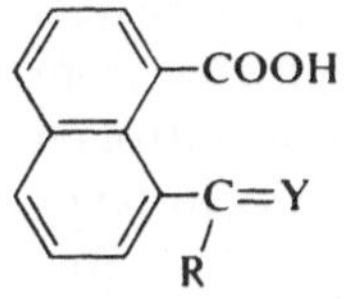			Dioxan	1.41	11	Ph	>100	12
				2.43	13			

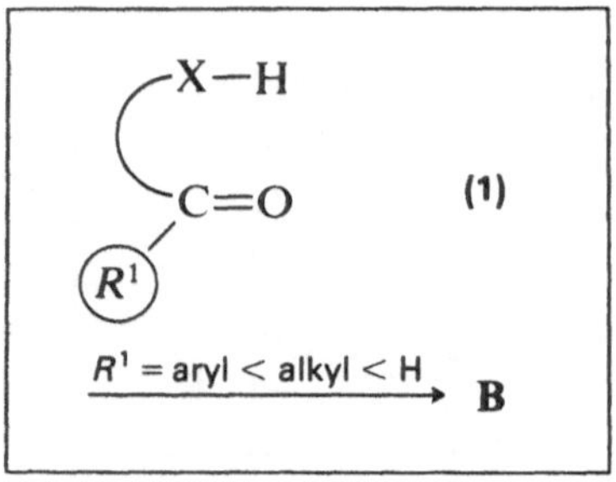

explained not only by the low electrophilicity due to aryl groups but also by a conjugation occurring in the open isomer.

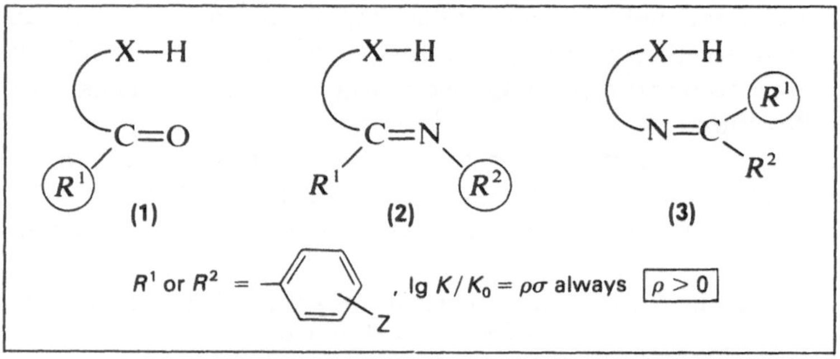

Introduction of electron-withdrawing substituents Z into o- or m-positions of phenyl groups placed at the $C=O$ or $C=N$ bonds (R^1 and R^2 in formulas (1), (2) and (3)) always displaces the ring-chain equilibrium in favor of a cyclic form. The reaction constant ρ is positive wherever a linear correlation of ring-chain equilibrium constants with σ or σ^+ coefficients of substituent Z (see Table 44) was observed.

Using N-($tert$-alkyl)-2-aroylbenzamides (see Sec. 2.1.3, Table 12), it has been shown that electron-withdrawing substituents in the aryl group bonded to the keto group favors the cyclization reaction more than steric hindrance can hinder it, i.e., stable cyclic isomers with a bulky substituent at the nitrogen atom may be obtained. At the same time increasing rate constants of isomeric interconversions are observed (see Sec. 2.1.3, Table 13).

The stabilization of protonated cyclic isomers of N-($tert$-alkyl)-2-(4-dimethylaminobenzoyl)- and 2-(2-imidazolylcarbonyl)benzamides (see Sec. 2.1.3: (66) → (67) and (69A) → (70)) is worth noting; this originates from an $-I$-effect of the substituent at the keto group. Protonation of the nitrogen atom of this substituent changes the ring-chain equilibrium totally in favor of the cyclic isomer, while a deprotonation gives rise to the reverse transformation.

Table 44. Examples of Linear Correlations of Ring-Chain Equilibrium Constants ($K_T = [B]/[A]$) with Polar Coefficients of Substituents Z in an Aryl Group Placed at C=O or C=N Bonds

(A) ⇌ (B)	Solvent	Method	$\lg K_T/(K_T)_0 =$	n^a	r	Refs.
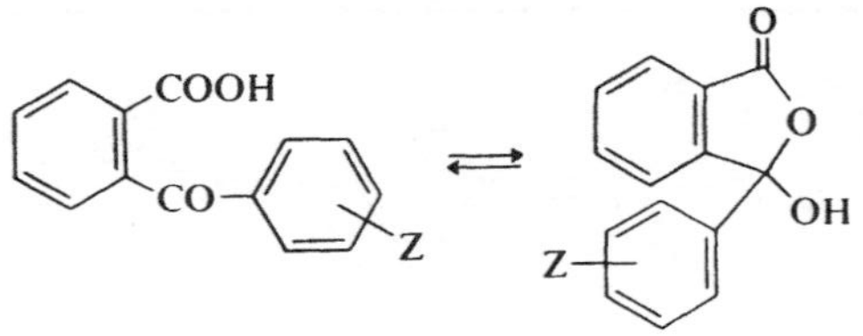	MeNO$_2$ MeCN Dioxan CHCl$_3$	IR	0.87σ 0.60σ 0.54σ 1.15σ	9		14
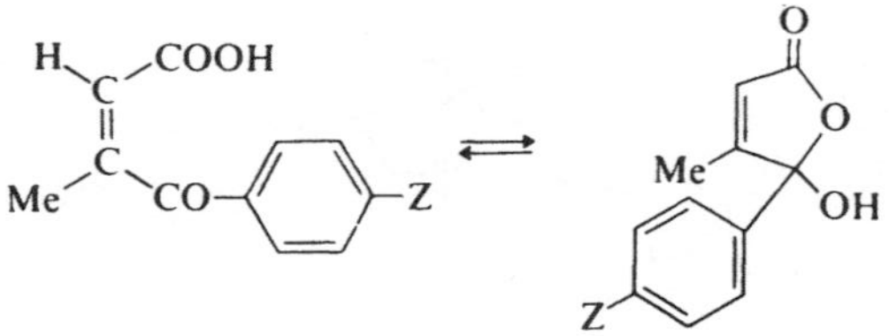	Dioxan	IR	$0.374\sigma_I + 1.885\sigma_R^0$	6	0.990	15
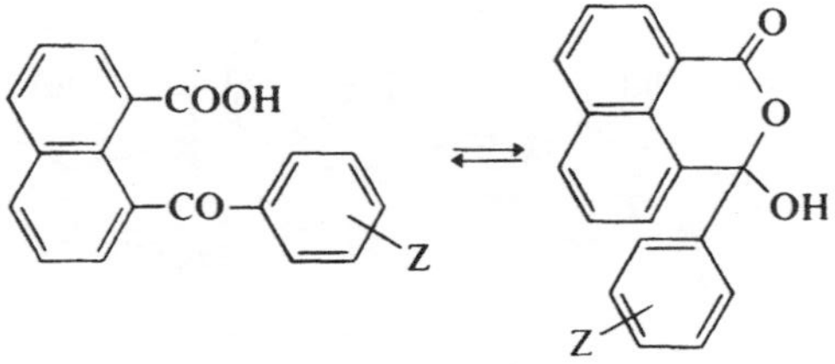	Dioxan	IR	$1.286\sigma_I + 2.918\sigma_R^0$	8	0.970	11
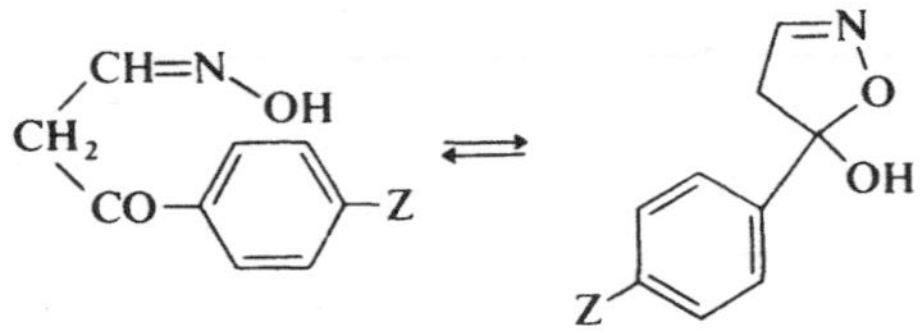	Pyridine-d$_5$	^{1}H-NMR	1.26σ $1.41\sigma_I + 1.26\sigma_R^+$	9	0.993	16

continued

Table **44**. (*cont.*)

(A) ⇄ (B)	Solvent	Method	$\lg K_T/(K_T)_0 =$	n^a	r	Refs.
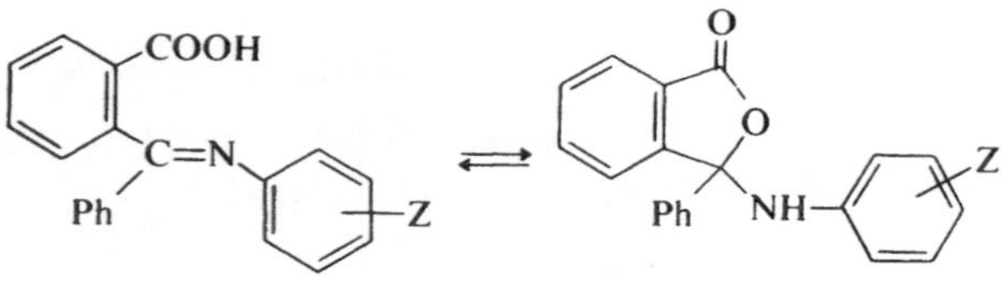 *(structure: isobenzofuranone–Ph–NH–C₆H₄Z ⇄ 2-(PhC=N–C₆H₄Z)benzoic acid)*	Dioxan	UV	0.58σ $0.35\sigma^+$	7	0.94 0.97	10
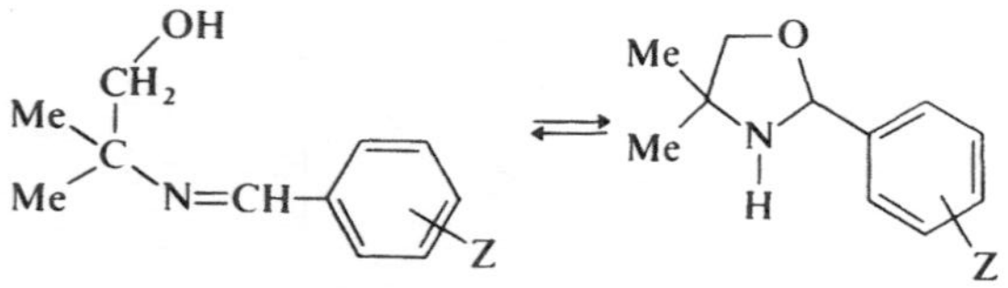 *(structure: oxazolidine with Me, Me, C₆H₄Z ⇄ Me₂C(CH₂OH)–N=CH–C₆H₄Z)*	CCl₄	¹H-NMR	0.856σ $0.543\sigma^+$	8	0.973 0.999	17
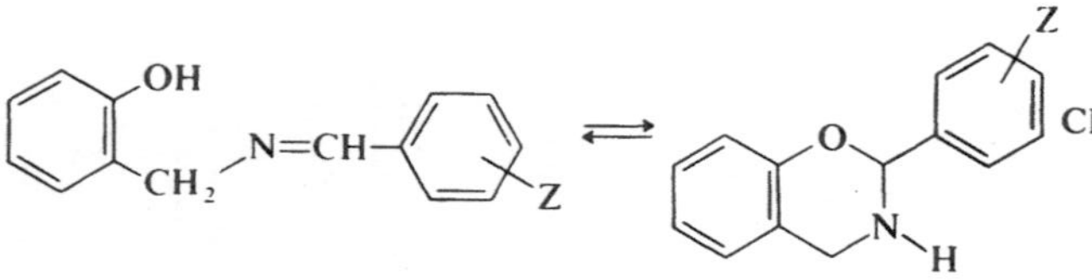 *(structure: benzoxazine with C₆H₄Z ⇄ 2-hydroxybenzyl–CH₂–N=CH–C₆H₄Z)*	CDCl₃	¹H-NMR	0.738σ $0.680\sigma^+$	5	0.993 0.999	17 18

^{a} Number of studied compounds with different substituents Z.

(2)

R^1 = alkyl < H ⟶ **B**

(3)

R^1 = alkyl, R^2 = alkyl < H ⟶ **B** but R^1 = Ph, R^2 = H < Me ⟶ **B**

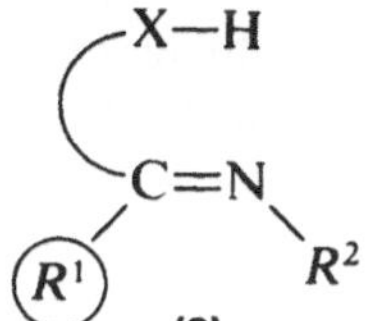

The influences of substituents R^1 and R^2 at the C=N group of tautomeric systems **(2A)** ⇄ **(2B)** and **(3A)** ⇄ **(3B)** are more complicated. Ring-chain equilibria of imines of hydroxyaldehydes (**2**, R^1 = H) are shifted considerably more toward the cyclic form than those of imines of hydroxyketones (**2**, R^1 = alkyl) (see Sec. 3.1.2: **(19A)** ⇄ **(19B)**, Table 35).

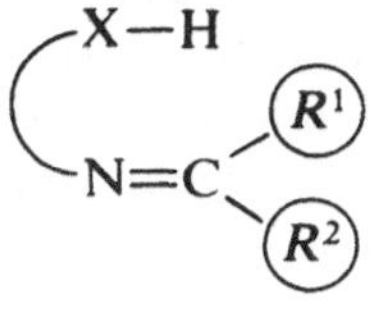

Similarly, the stability of cyclic isomers of *N*-(2-hydroxyalkyl)hydrazones of aliphatic aldehydes is superior to that of analogous ketone derivatives (see Sec. 3.1.2: **(11A)** ⇄ **(11B)**, Table 31).

In the series of *N*-(2-hydroxy and 2-mercaptoalkyl)hydrazones of aromatic aldehydes and acetophenone a reverse effect is detected (see Sec. 3.1.2: **(13A)** ⇄ **(13B)** and Sec. 3.3: **(56A)** ⇄ **(56B)**): passing from a benzaldehyde derivative to an acetophenone derivative the equilibrium is displaced in favor of the cyclic form. This has been explained by a steric hindrance of conjugation in the N—N=C—Ar fragment of the open isomer caused by the methyl group.

(2)

R^2 = alkyl < Ph ⟶ **B**

H < COR ⟶ **B**

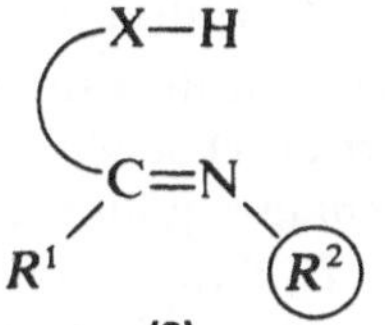

Substitution of an alkyl group by an aryl group at the imine nitrogen atom (R^2 in formula **(2)**) (see Sec. 3.1.4: **(19A)** ⇄ **(19B)**) or introduction of an acyl group at the same position (see Sec. 3.2.3: **(45A)** ⇌ **(45B)**) strongly stabilizes the cyclic form. This curious effect is explained by an energy gain as a result of conjugation of the nitrogen lone pair with the aryl or acyl C=O group in the cyclic isomer.

5.1.2. Steric Effects of Substituents at Interacting Groups

An estimation of steric influences on ring-chain equilibria is frequently based on a discussion of the optimal route on which interacting centers can approach. A calculation or an empirical evaluation of energy differences due to nonbonded interactions in both molecules, i.e., the open and the cyclic isomer, would be desirable.

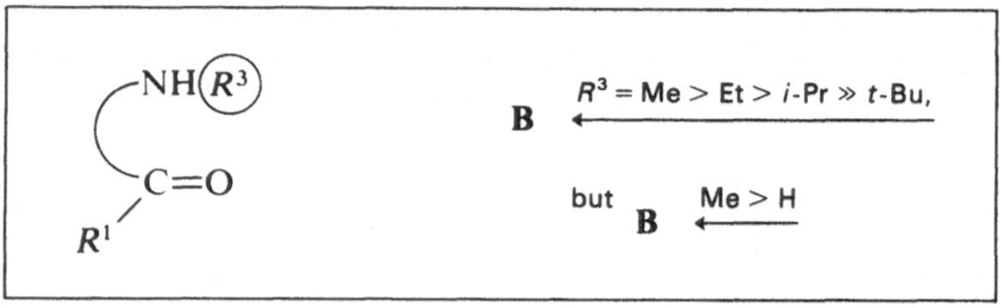

Sterically demanding substituents R^3 at the nitrogen atom of aminocarbonyl compounds shift the equilibrium **(1A**, $X = NR^3$**)** ⇌ **(1B)** toward the open form. *N*-Monosubstituted acylcarboxylic acid amides (see Sec. 2.1.3: **(3A)** ⇌ **(3B)**), *N*-(3-oxoalkyl)-*N'*-alkylthioureas (see Sec. 2.3.3: **(167A)** ⇌ **(167B)**), *S*-(2-oxoalkyl)dithiocarbamates (see Sec. 2.3.4: **(170A)** ⇌ **(170B)**) may be taken for illustration. Many instances are known where the introduction of *tert*-alkyl substituents at the nitrogen atom prevents the formation of ring-chain equilibria. However, it should be noted that for *N*-methylamides of keto carboxylic acids (see Sec. 2.1.3: **(3A)** ⇌ **(3B)**) the stability of the cyclic form is almost always increased compared with that of *N*-unsubstituted compounds which can be explained by the action of the +I-effect of a methyl group.

A similar influence of steric encumbrance of the substituent R^3 placed at the nitrogen atom holds for imines **(2A**, $X = NR^3$**)** ⇌ **(2B)**, and **(3A**, $X = NR^3$**)** ⇌ **(3B)**. Examples are *N*-(α-alkylaminoacyl)hydrazones (see Sec. 3.2.2: **(38A)** ⇌ **(38B)**), *N*-alkyl-2-benzoylbenzamide aniles (see Sec. 3.2.3: **(44A)** ⇌ **(44B)**) a.o.

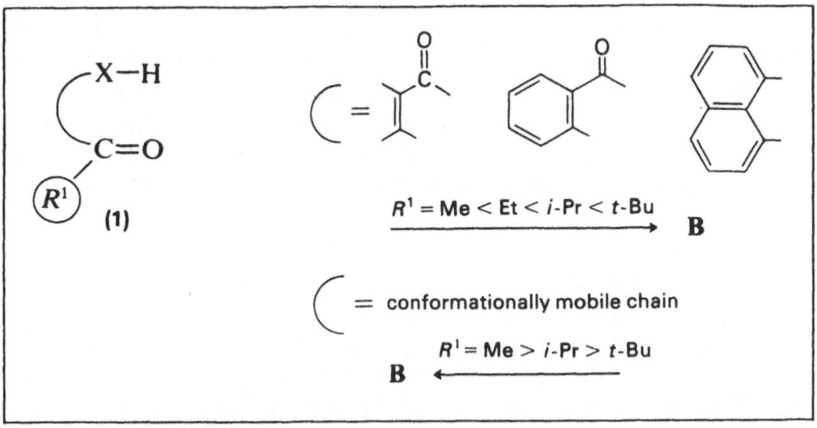

In discussing the influence of branched alkyl groups at the ketone group a distinction between two cases should be taken into account regarding the structure of the connecting link.

1. The connecting link contains double bonds or aromatic rings in conjugation with the keto group:

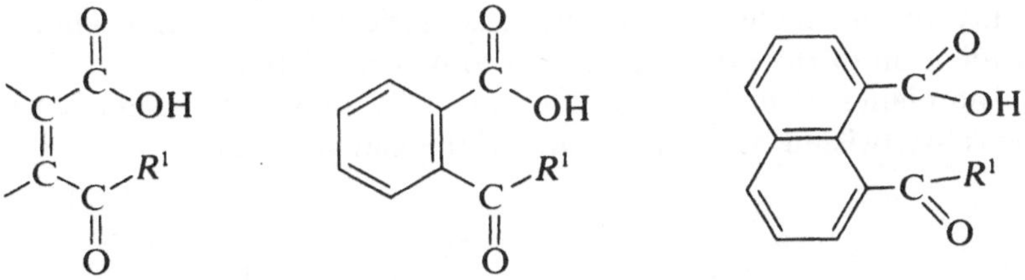

An increase in the steric bulk of the alkyl substituent R^1 in the sequence Me < Et < i-Pr < t-Bu twists the keto group out of the plane of the ethylene bond or the aromatic ring and this conformation of the C=O group is most favorable for an intramolecular addition to it. Furthermore, a diminished conjugation increases the electrophilicity of the carbonyl group stabilizing the cyclic isomer.

Thus, equilibrium constants of Z-3-acylacrylic [19], 2-acylbenzoic [9], and 8-acyl-1-naphthoic [11] acids increase in the sequence of substituents at the keto group: Me < Et < i-Pr < t-Bu and the ring-chain equilibrium constants of 8-acyl-1-naphthoic acids are correlated with Taft's steric coefficients [11]:

$$\lg K_T/(K_T)_0 = -0.60\ E_s, \qquad r = 0.977, \qquad s = 0.12, \qquad n = 4.$$

This effect is observed also for the ring-chain equilibrium of 2-acylben-zoic acid methyl esters which takes place under conditions of acidic or basic catalysis (see Sec. 2.1.4: **(100A)** $\rightleftarrows$ **(100B)** and Table 18). On increasing the steric bulk of the alkyl substituent both rate constants of the equilibrium (k_1 and k_2) decrease. However, equilibrium constants ($K_T = k_1/k_2$) increase at the same time and correlate with Taft's steric coefficients [20]:

$$\lg K_T/(K_T)_0 = -0.35 E_s.$$

2. The connecting link is conformationally mobile and an sp^3-carbon is placed at the α-position of the keto group. Here, an increasing volume of the substituent R^1 at the keto group destabilizes the cyclic structure. This has been shown in the series of S-(2-oxoalkyl)dithiocarbamates (see Sec. 2.3.4: **(170A)** $\rightleftarrows$ **(170B)**), 2-acylmethylthio-benzimidazoles and -perimidines (see Sec. 2.3.7: **(183A)** $\rightleftarrows$ **(183B)** and **(187A)** $\rightleftarrows$ **(187B)**, Table 26) and 1-(2-oxoalkyl)-3-phenyltriazenes (see Sec. 2.3.8: **(195A)** $\rightleftarrows$ **(195B)**, Table 27).

There are, however, some exceptions to this rule. 1-Hydroxyimino-3-alkyl-propan-3-ones (see Sec. 2.2.2: **(141A)** $\rightleftarrows$ **(141B)**, Table 24) may be taken as an example where the influence of alkyl substituents has a rather complicated character. The absence of a correlation of equilibrium constants with Taft's steric constants E_s has been explained by the conformational mobility of the molecule, i.e., besides steric and electronic effects the conformation of the open isomer has to be considered.

The planes of o,o'-disubstituted phenyls, 9-anthryl and related fused systems are twisted out of the plane of the ketone group:

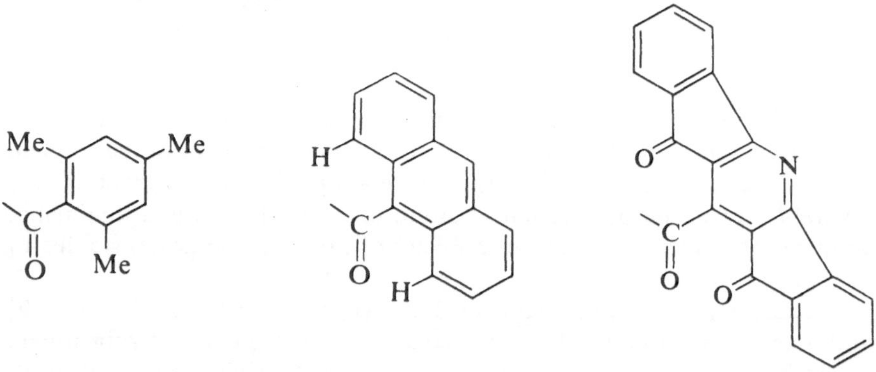

As a consequence an attack of the nucleophilic part of the molecule to the shielded C=O group is made impossible. Therefore amides of 2-mesityloyl, 2-(9-anthroyl)benzoic acids and related derivatives (see Sec. 2.1.3: Table 12) exist independently of the structure of the substituent at the nitrogen atom only as the open isomers, and they are not capable of cyclization even under conditions of basic catalysis. The same has been reported for 2-mesityloylbenzoic and 8-mesityloyl-1-naphthoic acids (see Sec. 2.1.1), 2-mesityloyl and 2-(9-anthroyl)phenyldimethylcarbinols (see Sec. 2.2.1: **(118A) ⇌ (118B)**) and dimesityloylformoine (see Sec. 2.2.1: **(135A) ⇌ (135B)**).

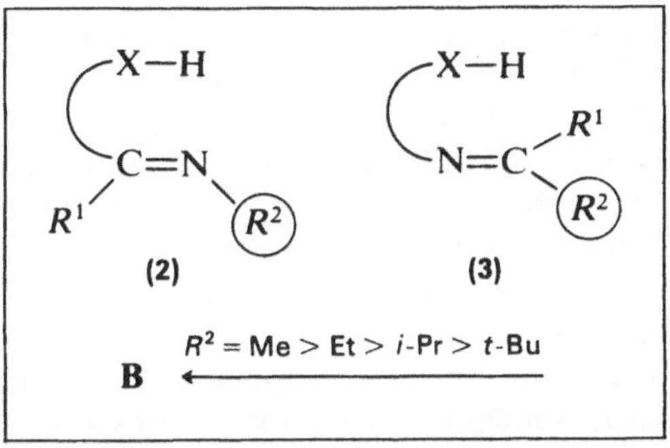

$$R^2 = \text{Me} > \text{Et} > i\text{-Pr} > t\text{-Bu}$$

B

(2)   (3)

  Increasing the steric bulk of substituents at the C=N group (R^1 and R^2 in formulas **(2)** and **(3)**) in compounds containing conformationally mobile connecting links destabilizes the cyclic isomers and shifts the equilibrium in favor of the open form.

  Thus, for instance, equilibrium constants of N-methyl-N-(1-hydroxy-1-phenyl-2-propyl)hydrazones of aliphatic aldehydes (see Sec. 3.1.2: **(12A) ⇌ (12B)**, Table 32), N-2-hydroxy- and N-2-mercaptoalkylhydrazones of aldehydes and ketones (see Sec. 3.1.2: **(11A) ⇌ (11B)**, Table 31 and Sec. 3.3: **(56A) ⇌ (56B)**), N-alkylimines of hydroxyketones and hydroxyaldehydes (see Sec. 3.1.4: **(19A) ⇌ (19B)**, Table 35) decrease in the sequence of alkyl substituents at the C=N group: Me > Et > i-Pr > t-Bu.

5.1.3. Structure of the Connecting Link

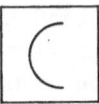

  A study of the influence of the connecting links should be based on a comparison of equilibria and rates of tautomeric compounds by varying the structure of the connecting links and leaving the interacting functions unaltered. Unfortunately, investigations of this kind have only rarely been reported. Moreover, equilibrium constants have frequently been determined under different conditions and by different methods which diminished the value of the information.

Thus, the estimation of the influence of the structure of the connecting link on the facility of ring formation and on ring stability is a rather complicated problem. Two theoretical approaches are possible here: 1) estimation of the fulfillment of the structural requirements for a forced approach of the interacting groups in the orientation optimal for an intramolecular reaction; 2) estimation of the energy of ring formation from the thermodynamic point of view, i.e., deriving the enthalpic participation in the free energy difference from the changes in bond distances, angle deformations, nonbonded and electrostatic interactions, steric strain etc. [4].

Many investigations in this direction have been carried out with the purpose of finding a model system explaining the enormous rates of enzymatic reactions as compared to normal bimolecular reactions. As a result, new concepts such as "stereopopulation control" [21-25] and "orbital steering" [26-28] appeared.

According to the concept of stereopopulation control the enormous acceleration of the lactonization reactions of o-hydroxydihydrocinnamic acids observed after introduction of methyl groups into the chain between interacting functions may be explained by an increase in the content of the reactive conformer, both interacting centers of which are optimally orientated. Later it was shown [29, 30] that this effect cannot make so large an acceleration and the increase in the strain energy of the open structured compound by the introduction of methyl groups plays the most important role. In addition, a reexamination showed [31] that this rate acceleration is smaller than first observed [26-28].

Koshland's "orbital steering concept" aims to explain the acceleration of the lactonization reactions of hydroxycarboxylic acids, which was observed in the course of a systematic modification of the connecting link. Following this concept the highest possible acceleration may be achieved, if the structural preconditions for an optimal angular orientation of intramolecularly interacting groups are realized [26-28]. The concept has been criticized [32] for being too simplified for complicated enzymatic reactions and for overestimating the influence of an optimal angular orientation. An experimental disproof has been presented [33], showing no rate dependence on angular orientation within the confines of 10° changes for a lactonization reaction (for a discussion see [34]).

Another experimental approach to this problem has been suggested by Bürgi and Dunitz [35-38]. Based on X-ray structure data of suitable bifunctional compounds they have found the optimal geometry of the reaction path for interacting functions.

On the basis of simple considerations of stereochemical requirements for an intramolecular reaction Baldwin has proposed [39] simple empirical rules concerning the favored routes of cyclization reactions.

If the intramolecular addition of a nulceophilic atom occurs at a double or triple bond with the formation of the smallest of two possible rings, the cyclization is named exo-cyclic, while the formation of the bigger ring is called endo-cyclic:

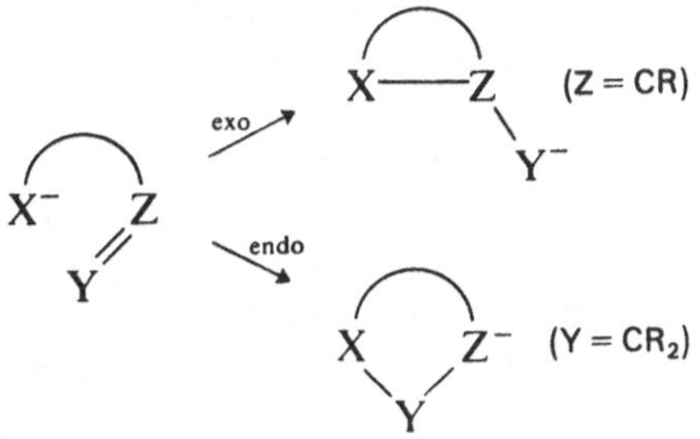

Tetrahedral, trigonal, or digonal geometry of the carbon atom at which intramolecular substitution (tet) or addition occurs is designated by prefixes "tet," "trig," and "dig." The number of members in the ring is indicated.

Some examples may be given:

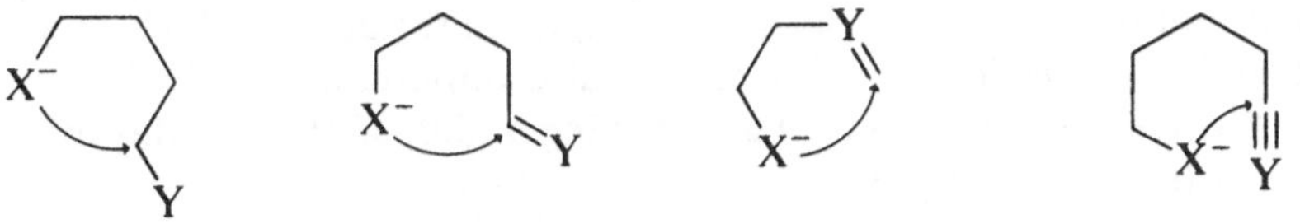

| 5-exo-tet | 5-exo-trig | 5-endo-trig | 6-exo-dig |

If X is a first-row element cyclizations of type (3-7)-exo-tet, (3-7)-exo-trig, (6,7)-endo-trig, (5-7)-exo-dig, and (3-7)-endo-dig should be favorable processes. Cyclizations of type (3-5)-endo-trig and (3,4)-exo-dig should be disfavorable [39].

Baldwin has suggested a very convenient classification system for ring-forming reactions. The proposed rules are of general character and very easy to apply. However, their practical use is bound up with some limitations, since many structural aspects, e.g., the structure of the connecting link, the presence of substituents having different electronic and steric effects, and the nature of interacting groups have been neglected. Some exceptions have been observed in the field treated in this book. Thus, a "disfavored" 5-endo-trig reaction has been observed for N-2-hydroxyalkyl-imines of aromatic aldehydes as is shown in Sec. 3.1.1: **(3A) ⇌ (3B)**. A more rapid "disfavored" 5-endo-trig cyclization as compared with a "favored" 6-endo-trig cyclization was also reported [40] (see Sec. 3.2.1: **(31A) ⇌ (31B)**).

The majority of the cyclic isomers discussed in this monograph contain five- or six-membered heterocycles. The preference in stability of the five- or six-membered ring system depends on many factors. Sometimes the formation of five-membered rings is sterically more favored while in other cases a six-membered ring forms more easily and/or is more stable.

3-Acylpropionamides, for instance, being substituted in a definite pattern at the amide nitrogen atom and the keto group, form stable five-membered cyclic isomers (see Sec. 2.1.3: **(45B)**, Table 9) while six-membered hydroxylactams totally failed to be obtained from 4-acylbutyramides having the same substitution pattern (see Sec. 2.1.3: **(89A)**).

The greater stability of five-membered hydroxylactams compared with six-membered analogues is supported by the isomerization with ring contraction for benzile-o-carboxamides (see Sec. 2.1.3: **(78C)** → **(78B)**).

3-Acylpropanols and 4-acylbutanols (see Sec. 2.2.1: **(109A)** ⇌ **(109B)**, Tables 20 and 21) are examples of an opposite effect: ring-chain equilibrium constants for the butanols forming six-membered rings are greater than those of the propanol derivatives. This effect is even more pronounced for 3-acylpropanol and 4-acylbutanol N-alkylimines (see Sec. 3.1.4: **(19A)** ⇌ **(19B)**, Table 35).

`A general rule can be stated: the stability of the cyclic isomer increases with increasing rigidity of the connecting link, i.e., decreasing conformational mobility. A possibility to assume a conformation which favors the interaction of the functional groups is also important.

Bowden [11] proposes the following order of the stability of cyclic isomers (see also Table 3):

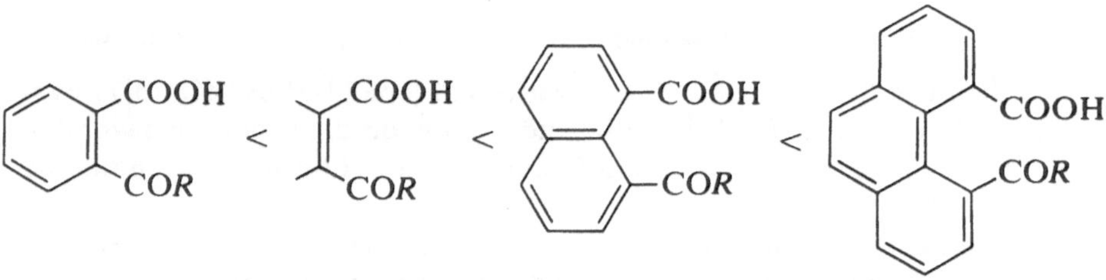

This order corresponds to a steric approach of the C=O and COOH groups.

An analogous sequence was observed [41] in the case of 3- and 4-benzoylcarboxylic acid amides (see Table 15):

$$PhCO(CH_2)_3CONHR < PhCO(CH_2)_2CONHR <$$

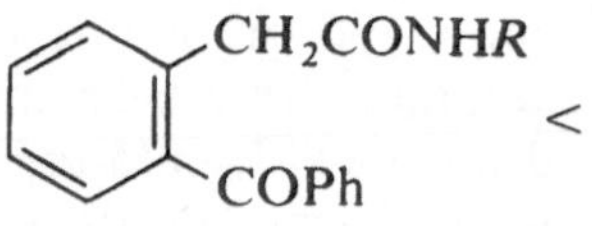

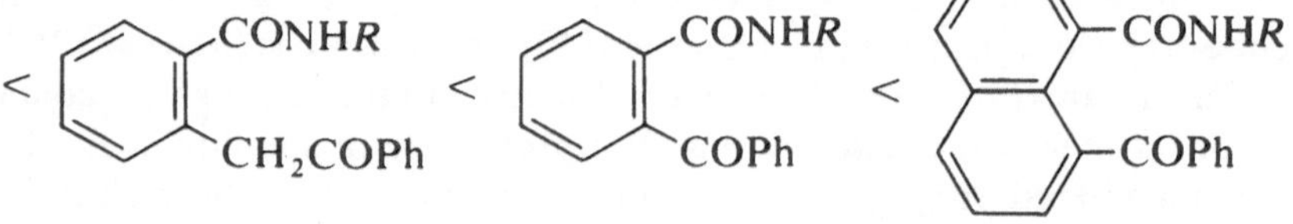

Here keto groups can assume a conformation allowing the nucleophilic part of the molecule to approach nearly perpendicularly to its plane. Noncompliance with this requirement hinders cyclization. Anthraquinone-1-carboxamides, for instance, are not capable of forming hydroxylactams because the keto group is rigidly fixed in the plane of the aromatic ring (see Sec. 2.1.3: **(81A)**).

Different pairs of functional groups have different steric (geometrical) requirements for an optimal path of approach in the course of an intramolecular reaction. Therefore it is impossible from different systems to find direct correlations between ring-chain equilibrium constants or cyclization rate constants and the structure of the connecting link. A critical outlook at trajectory considerations of cyclization reactions is presented in a recent report [34].

5.1.4. Steric Effects of Substituents at the Connecting Link

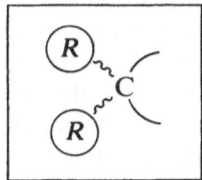

Thorpe-Ingold effect. The introduction of alkyl groups at the carbon atoms placed in the chain between the interacting groups and, especially, the introduction of two geminal groups increases the cyclization rate and displaces the equilibrium toward the ring isomer. This effect is commonly called "gem-dimethyl," "gem-dialkyl" or the "Thorpe-Ingold effect" and its influence has been repeatedly discussed in previous chapters.

The most satisfactory theoretical explanation of this effect has been given by Allinger and Zalkov [42] (see Sec. 2.1.1).

The influence of alkyl substituents at the connecting link on the ring-chain equilibrium constants, as well as on the cyclization rate constants, depends on the steric encumbrance of these substituents, their position at the connecting link, and the conformational mobility of the connecting link itself. If the chain between the interacting groups consists of three sp^3-carbon atoms, the greatest influence is exerted by two alkyl groups at the central carbon atom.

The introduction of methyl groups to the α-carbon of a functional group, in particular a C=O group, can lead to the opposite effect due to the steric shielding of the reactive center. Thus, destabilization of the cyclic isomer was observed on introducing a methyl group at the α-carbon of the

C=O group of N-(3-oxoalkyl)-N'-substituted thioureas (see Sec. 2.3.3: **(167A)** ⇌ **(167B)**) and 1-(2-oxoalkyl)-3-aryltriazenes (see Sec. 2.3.8: **(195A)** ⇌ **(195B)**, Table 27).

The influence of alkyl substituents is greatly enhanced if the conformational mobility of the connecting link is decreased. Thus, a strong stabilization of the cyclic isomer was observed on introducing two methyl groups in a connecting link of type -o-$C_6H_4CR_2$- in 2-acylmethylbenzoic and o-acylphenylacetic acids (see Sec. 2.1.1: **(20A)** ⇌ **(20B)**, **(21A)** ⇌ **(21B)**, Table 5).

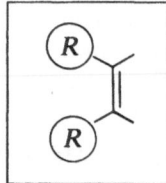

Alkyl substituents at sp^2-carbon atoms of the connecting link stabilize the cyclic isomer as well, evidently as a result of increasing steric strain of the open isomer [43]. This has been observed particularly in the case of Z-3-acetylacrylic acids (see Sec. 2.1.1: **(18A)** ⇌ **(18B)**, Table 4) where the introduction of methyl substituents at both ethylene carbons increases the ring-chain equilibrium constant by four orders of magnitude!

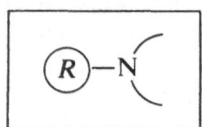

A large stabilization of the cyclic isomer also has been observed by an introduction of alkyl substituents at a nitrogen atom that is part of the connecting link. Examples have been reported of compounds being unsubstituted at the nitrogen atom (R = H) and existing in the stable open form exclusively. Ring-chain equilibria were observed only for derivatives containing N-alkyl substituents. Increasing the steric bulk of this substituent shifts the equilibrium further toward the ring form.

This effect is well documented by investigations of the tautomerism of N-alkyl-N-(2-hydroxyalkyl)-N-(2-oxoalkyl)amines (see Sec. 2.2.1: **(114A)** ⇌ **(114B)**), 2-[N-alkyl-N-(2-hydroxyethyl)amino]-1,4-benzoquinones (see Sec. 2.2.1: **(116A)** ⇌ **(116B)**, Table 22), N-alkyl-N-(2-hydroxy- or 2-mercaptoalkyl)-hydrazones of aldehydes and ketones (see Sec. 3.1.2: **(11A)** ⇌ **(11B)**, Table 31 and Sec. 3.3.1: **(56A)** ⇌ **(56B)**, Table 38), as well as other systems.

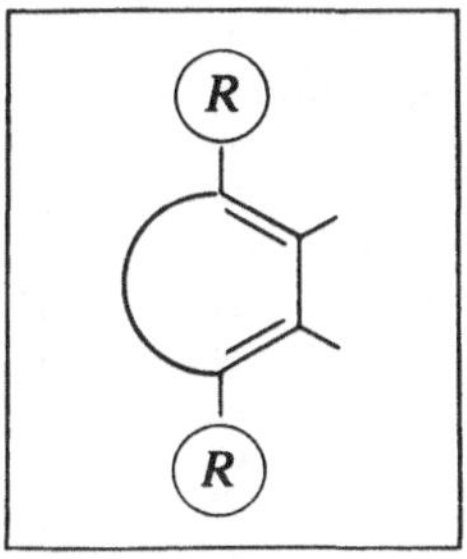

"Support" Effect or Steric Assistance Effect. If the connecting link contains an aromatic ring, the introduction of methyl or other groups in the *ortho*-positions to the interacting groups creates steric strain which favors intramolecular cyclization [44].

Quantitative evidence using ring-chain tautomerism of 2-acetyl [45] and 2-benzoylbenzoic acids [46] is presented in Table 45.

This effect also governs the acylo- or arylo-tropic tautomerism of tropolone derivatives (see Sec. 2.4: **(211A) ⇌ (211B)**). Introduction of methyl or benzyl substituents in positions 3 and 7 of the tropolone ring sterically favors the formation of a cyclic bipolar transition state or intermediate and accelerates the migration of acyl or aryl groups [47].

From the examples discussed above (see also [48]) it appears that the influences of substitution at the connecting link depend on many factors

Table 45. Steric Assistance Effect of
ortho-Substituents on Ring-Chain
Equilibrium of 2-Acetyl [45] and
2-Benzoylbenzoic Acids [46]

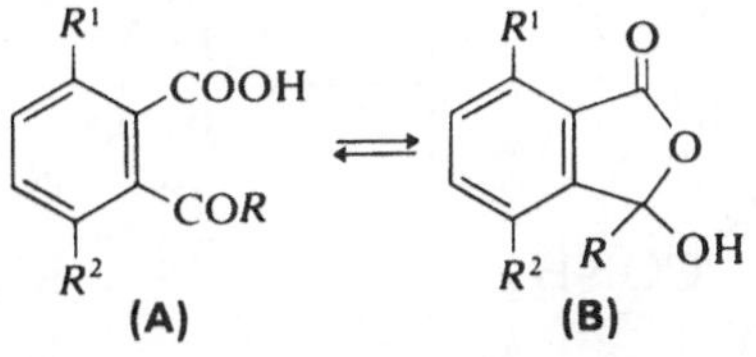

		$K_T = [B]/[A]$	
R^1	R^2	$R = Me$	$R = Ph$
H	H	3.2	<0.1
Me	H	16	2.85
H	Me	0.52	0.22
Me	Me		13.3

Table 46. Influence of Methyl Substituents on Cyclization Rate Constants
of 3-Ureidopropionic Acids (4) → (5) [49]

	R^1	R^2	R^3	R^4	R^5	$\lg k_{rel}$
	H	H	H	H	H	0.00
	Me	H	H	H	H	0.07
	H	H	Me	H	H	0.45
Erithro-	Me	H	Me	H	H	−0.08
Threo-	Me	H	H	Me	H	0.47
	Me	Me	H	H	H	−0.46
	H	H	Me	Me	H	1.25
	H	H	H	H	Me	1.66

and a quantitative prediction is difficult. Progress is expected from force-field
calculations.

The introduction of alkyl substituents not only shifts the equilibrium
toward the cyclic form, it also accelerates cyclization reactions if a transition
from open-structured ground state into cyclic transition state takes place
in the rate-limiting step.

The rate constants of cyclization of the methyl substituted 3-
ureidopropionic acids (4) → (5) (see Table 46) [49] are taken as an example
where the introduction of two methyl groups at the β-carbon atom (regarding the carboxylic group) (4, $R^3 = R^4 = Me$) as well as the introduction of
a methyl group at nitrogen ($R^5 = Me$) strongly accelerates the cyclization
reaction while two methyl groups in the α-position retard it, obviously, due
to the steric shielding of the reactive center.

Blagoeva, Kurtev, and Pojarlieff [49] found a good linear correlation
($r = 0.932$–0.993) of free activation energies for the cyclization of methyl
substituted 3-ureidopropionic acids (4) → (5) as well as for previously
investigated anhydrizations of succinanilic [50] and succinic [51] acids,
hydrolysis of p-bromophenylglutarates [52], and lactonization of 4-hydroxy-
butyric and 5-hydroxyvaleric acids [53] with steric strain energies estimated
on the basis of the formation enthalpies of homomorphic hydrocarbons [54].

Empirical force-field calculations [44, 55–60] were successfully used for a quantitative prediction of an acceleration of hydroxycarboxylic acid lactonizations by modification of the structure of the connecting link and/or introduction of methyl groups [30, 61].

The potential of these calculations has been well demonstrated by De Los De Tar [62, 63] using cyclization reactions of bromoalkylamines and their in-chain alkylsubstituted derivatives. A good coincidence of calculated and experimentally determined cyclization rates has been observed. Activation enthalpies were calculated by the force-field method and activation entropies obtained from analogous formal cyclization reactions of homomorphic alkanes.

The results of calculations show that the driving force of the Thorpe-Ingold effect is mainly an enthalpic effect (and not an entropic effect as was supposed earlier [21, 22]) caused by steric strain in the ground state of the open structured starting material.

5.1.5. Intramolecular Hydrogen Bonds

If the structure of the open isomer has the steric requirements for the formation of an intramolecular hydrogen bond with a proton-donating group participating (X—H···) [64], then stabilization of this isomer may ensue. The formation of such a bond blocks the proton-donating group X—H. Moreover, the open isomer may be fixed in a conformation unfavorable for intramolecular ring formation.

3-Acetyl-2-(2-hydroxy-3-acetylphenylamino)-5-methoxy-1,4-benzoquinone (6) [65], bis-(1,3-cyclohexanedione-2-yl)methanes (7) [66–68], Z-1-amino-2-(2-oxoalkyl)ethylenes (8) [69], and N-(2-hydroxybenzyl)imines of aromatic aldehydes (9) [70] serve as illustrations.

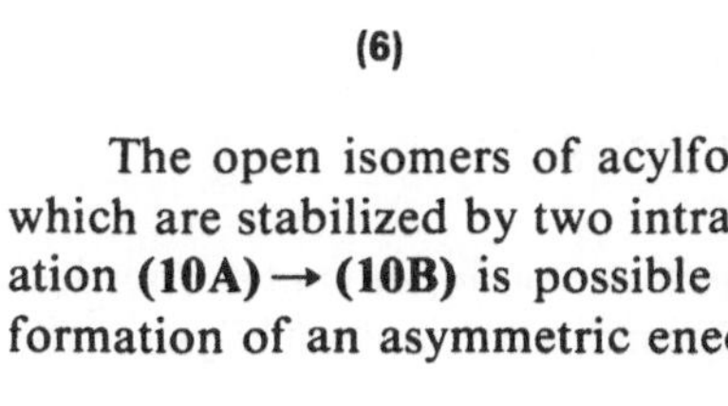

(6) 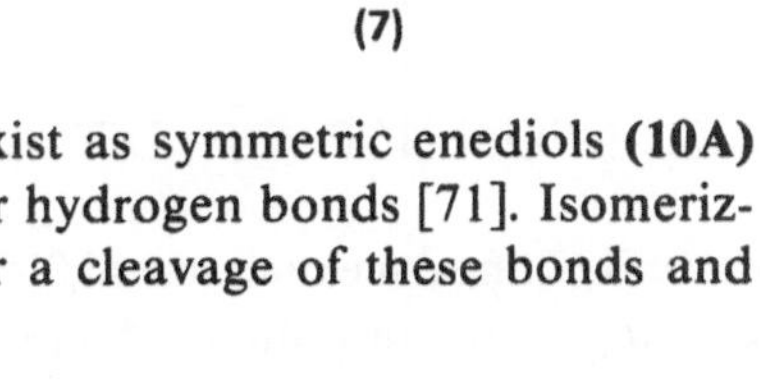   (7)

The open isomers of acylformoins exist as symmetric enediols (10A) which are stabilized by two intramolecular hydrogen bonds [71]. Isomerization (10A) → (10B) is possible only after a cleavage of these bonds and formation of an asymmetric enediol.

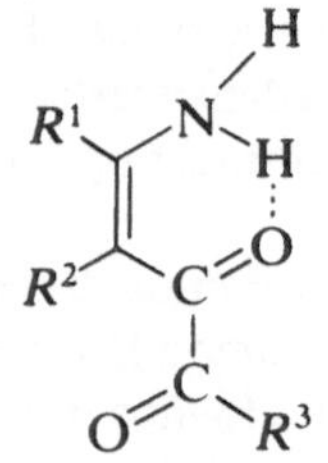

(8)

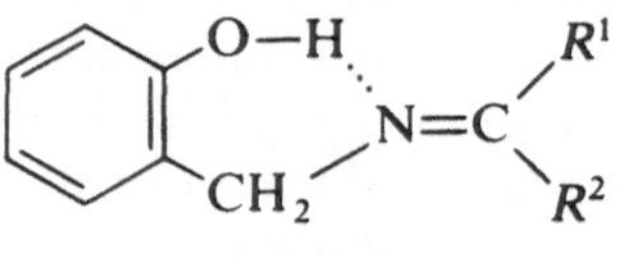

(9)

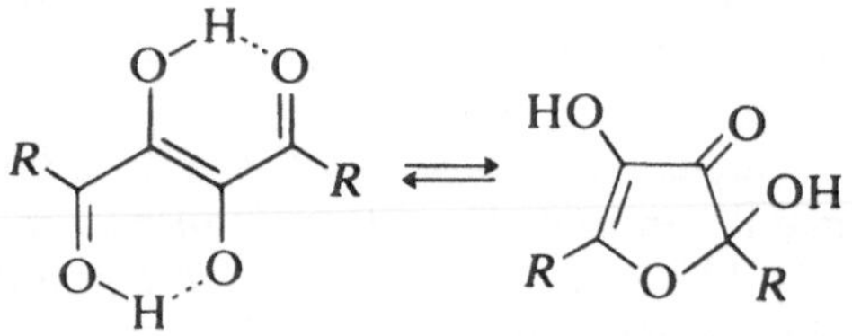

(10A)   **(10B)**

Intramolecular hydrogen bonds are cleaved in polar proton-accepting solvents. This cleavage leads to a destabilization of open-structured hydroxy- or amino-carbonyl derivatives and a displacement of the equilibrium in favor of the cyclic form. This is not observed for open azomethine analogues which, however, are stabilized by proton-accepting solvents.

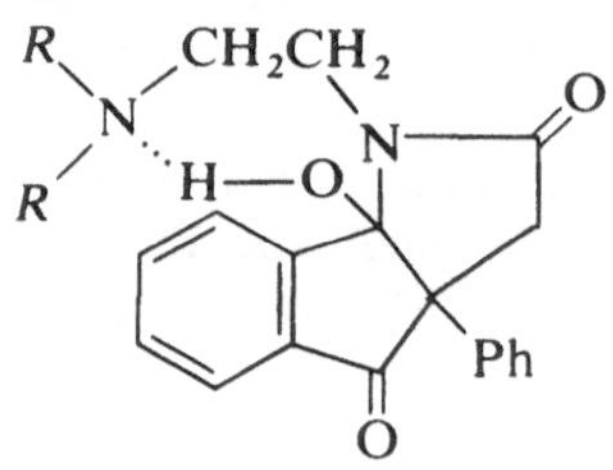

(11) 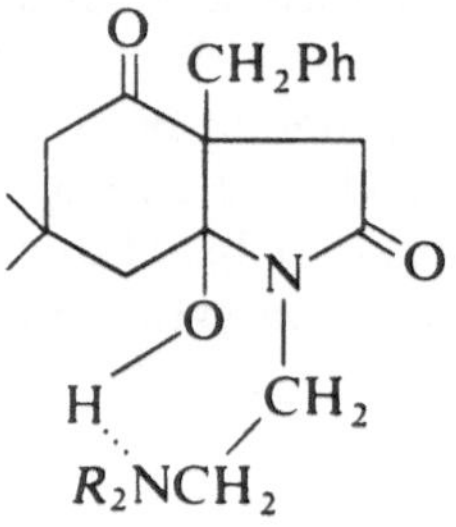   **(12)**

A stabilization of cyclic isomers by means of intramolecular hydrogen bonds occurs only rarely. Examples are N-(β-dialkylaminoethyl)amides of (2-phenyl-1,3-indanedione-2-yl)acetic **(11)** [72] and (2-benzyldimedone-2-yl)acetic acids **(12)** [73].

5.2. INFLUENCE OF EXTERNAL FACTORS

The influence of external factors (temperature, state, solvent properties) on ring-chain tautomerism has only rarely met with interest. Systematic physical investigations have not been reported and the available experimental data do not allow any resonable generalizations.

Equilibria should be studied in solution and in the gas phase to differentiate the influences of structure and solvent. Unfortunately, only a few mass-spectrometric investigations have been carried out which, moreover, could not yield exact quantitative equilibrium constants in the gas phase.

We first consider some general statements. The tautomeric equilibrium constant ($K_T = [\mathbf{B}]/[\mathbf{A}]$) is a quantitative measure for the investigated systems. A control of this constant by external factors is commonly discussed.

Rate constants k_1 and k_2 of tautomerization

$$\mathbf{A} \underset{k_2}{\overset{k_1}{\rightleftharpoons}} \mathbf{B}$$

have been but rarely determined because of experimental difficulties. However, a change of external conditions can influence rate constants k_1 and k_2 of the interconversion while leaving equilibrium constants ($K_T = k_1/k_2$) almost unaffected. 3,3-Dimethyllevulinic acid amides may serve as an example (see Sec. 2.1.3, (47A)). Solvent properties influence the equilibrium constants only weakly while the equilibration rates vary to a large extent.

Protolytic ring-chain tautomeric interconversions, being considered here, differ from other protolytic tautomerizations in one significant respect.

The thoroughly investigated keto-enol tautomerism proceeds through one common anion (13A):

(13K)   (13A)   (13E)

In the protolytic ring-chain tautomerism, however, two anions (14C) and (14D) can participate. Usually one anion is less stable than the other, the relative stability being determined by the acidity of the X—H and Y—H groups: if the X—H group is more acidic then the open structured anion (14C) is the more stable, which was observed for acylcarboxylic acids (see Sec. 2.1.1). A more acidic Y—H group stabilizing the cyclic anion (14D) has been noticed for the amides of acylcarboxylic acids (see Sec. 2.1.3).

Protolytic transformations of the keto-enol system (13K) $\rightleftharpoons$ (13A) $\rightleftharpoons$ (13E) proceed rapidly, ring-chain tautomeric interconversions generally take place considerably slower. Obviously, the formation and/or the cleavage of a C—X bond is responsible for this difference. Apart from the anions **(14C)** and **(14D)**, bipolar **(14E)** and **(14F)** or even protonated **(14G)** and **(14H)** intermediates are conceivable. The conversion mechanism **(14A)** ⇌ **(14B)** is determined by acid-base properties of X—H and Y—H groups and by protolytic characteristics of the medium.

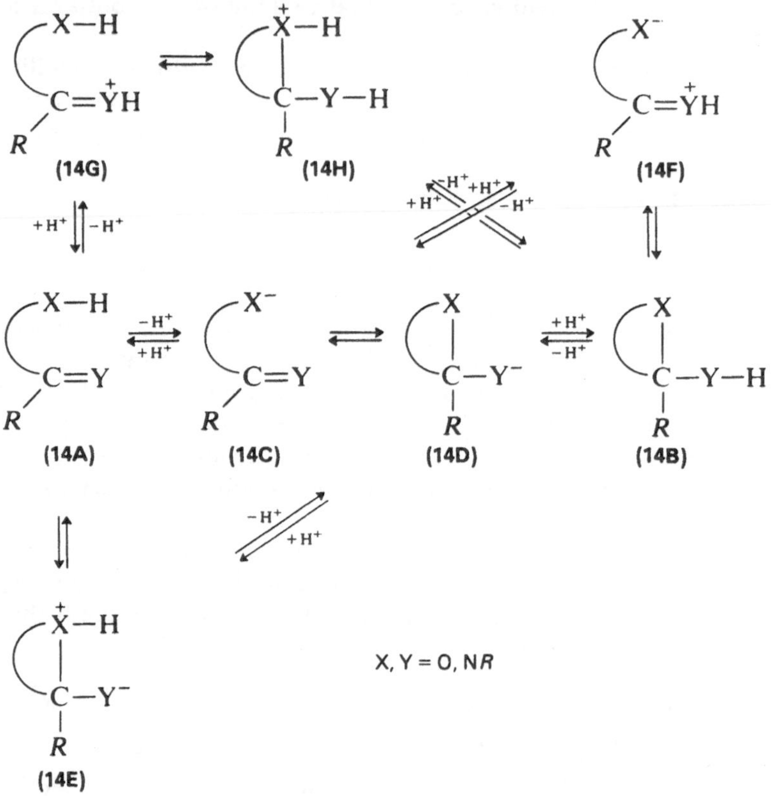

Many ring-chain tautomeric systems are known which require 10 days or even more for equilibration in solution. Water, acids, or bases strongly accelerate this process. Therefore, sometimes equilibrium constants have to be established in the presence of these catalysts.

Proton-donating or proton-accepting properties of the medium affect directly the proton migration and therefore also determine the rate of tautomerizations.

Solvents control the position of equilibria by solvation. Open and cyclic isomers usually have different polarities. As a consequence, in polar solvents the equilibrium is shifted toward a more polar form while in solvents of low polarity the opposite is observed.

In addition, the formation of intermolecular hydrogen bonds with solvent molecules is of great significance for stabilization of either tautomer. The solvent may act as a proton donor or proton acceptor, the latter being observed more often. Tautomeric systems which are exclusively controlled by intermolecular hydrogen bonds have been observed [74]. In various systems solvent-proton-accepting properties differently influence the stability of the cyclic form.

Hydroxycarbonyl Compounds (**14**, $X = Y = O$). Solvent polarity and proton-accepting ability show comparatively little influence on the ring-chain equilibria of hydroxyketones (see Sec. 2.2.1: (**109A**) $\rightleftarrows$ (**109B**), Table 21). The insensitivity of the proton-acceptor properties of the solvent is easily understood since both types (open and ring) of isomers have similar proton-donating hydroxyl groups. The equilibrium position is strongly affected by the addition of water to the solutions (see Sec. 2.2.1: (**109A**) $\rightleftarrows$ (**109B**)). The considerable decrease in enthalpy and entropy taking place is explained by hydration of the open isomer.

In the series of 2-aroylbenzoic acids polar solvents shift the equilibrium toward the more polar lactol form (see Table 47) [14]. The highest equilibrium constants was detected in dioxan though its dielectric constant is lower than those of acetonitrile and nitromethane. This can be explained by a participation of the dioxan molecules in the formation of intermolecular hydrogen bonds with the lactolic hydroxyl group.

The sensitivity of the ring-chain equilibrium of 2-(3- or 4-X-benzoyl)benzoic acids to electronic influences of substituents X depends on the solvent being used. As a rule increasing stabilization of the cyclic isomers by solvent diminishes the sensitivity, i.e., decreases the reaction constant ρ of the equilibria (see Sec. 2.1.1: (**10A**) $\rightleftarrows$ (**10B**), $R = C_6H_4X$, $X = H$, 4-Cl, 4-Br, 4-CN, 4-NO$_2$, 3-NO$_2$-4-Me, 3-NO$_2$-4-Cl, and Table 44 for detailed information) [14]. Diluting solutions of acylcarboxylic acids shifts equilibria toward the open forms. In acidic media the equilibria are displaced in favor of the less acidic ring forms.

Table 47. Solvent Influence on the
Ring-Chain Equilibrium Constant of
2-Benzoylbenzoic Acid [14]

Solvent	$K_T = [B]/[A]$
Chloroform	0.006
Nitromethane	0.025
Acetonitrile	0.030
Dioxan	0.033

Aminocarbonyl Compounds (14, *X* = *NR*, *Y* = *O*). An increase of proton-acceptor ability and polarity of the solvents displaces the equilibria toward the ring form. Z-1-Amino-2-(2-oxoacyl)ethylenes (see Sec. 2.3.2: **(164A)** ⇌ **(164B)**, Table 25), 6-acylmethylmercapto-4-chloro-5-methyl-aminopyrimidines (see Sect. 2.3.2: **(166Ab)** ⇌ **(164Bb)**, 3-aryl-1-(2-oxoalkyl)triazenes (see Sec. 2.3.8: **(195A)** ⇌ **(195B)**, Table 27) are examples.

The influence of proton-accepting solvents may be explained by a greater energy of formation of intermolecular hydrogen bonds of the type **(15B)** as compared to that of type **(15A)**. Besides, the formation of an intermolecular hydrogen bond of type **(15A)** increases the nucleophilicity of the nitrogen atom favoring an intramolecular addition of N—H groups to the C=O groups.

(15A)   (15B)

There are, however, exceptions, as for example the 5-(2-acetylamino-phenyl)tetrazole equilibrium which is shifted toward the open form in pyridine (see Sec. 2.3.1: **(152A)** → **(152B)**).

The stabilization of cyclic isomers of aminocarbonyl compounds repeatedly observed is caused by self-association. Sometimes dilution of these solutions with nonpolar solvents leads to displacement of the equilibrium in favor of the open tautomer (see, for instance, Sec. 2.3.7: **(183A)** ⇌ **(183B)**).

Hydroxyderivatives of Azomethines. Unlike aminocarbonyl compounds hydroxy derivatives of Schiff's bases and their analogues bear a hydroxyl group in the open isomers. This explains the opposite effect of solvent properties on equilibrium positions as compared to the carbonyl analogues considered above. Increasing proton-acceptor ability and polarity of solvents shift equilibria toward the open form, obviously, due to a preferable formation of intermolecular hydrogen bonds O—H···solvent. Such solvent influences were observed for benzaldehyde *N*-(1-hydroxy-2-methyl-2-propyl)imines (see Sec. 3.1.1: **(3A)** ⇌ **(3B)**, Table 30), 5-hydroxy-2-pentanone *N*-methylimines (see Sec. 3.1.4: **(19A)** ⇌ **(19B)**), 2-benzoyl-benzoic acid anils (see Sec. 3.1.5: **(21A)** ⇌ **(21B)**, Table 37), and other azomethines.

Paukstelis and Hammaker [74] have detected a most interesting linear correlation between hydroxyl group frequency displacements ($\Delta \nu_{O-H}$) in the IR spectra of di-*tert*-butyl-carbinol solutions and enthalpies of benzaldehyde *N*-(1-hydroxy-2-methyl-2-propyl)imine ring-chain equilibria in the same solvents. As a consequence the displacement of the hydroxyl band frequency of di-*tert*-butylcarbinol ($\Delta\nu_{O-H}$) was used as a measure of the proton-acceptor ability of solvents involved in intermolecular hydrogen bonds with the hydroxyl group of the open isomer ($Ph-CH=N-CMe_2CH_2O-H\cdots$solvent).

For *N*-alkylimines of hydroxycarbonyl compounds (see Sec. 3.1.4: **(19A)** ⇌ **(19B)**, Table 35), on passing from the liquid state to solutions in solvents of low polarity there is a shift in the equilibrium toward the cyclic form. Hence, it appears that self-association stabilizes the open form.

Temperature Influences. In the majority of the equilibria discussed in this monograph an increasing temperature was shown to displace the ring-chain equilibrium in favor of the open isomer. This often has been used preparatively for a thermal isomerization of cyclic isomers to open ones.

+   +   +

In conclusion it should be noted that our knowledge of the influence of solvation on ring-chain equilibria is rather limited and investigations in this field would be welcome.

A deeper understanding of these influences could be achieved in at least two ways:

1) From temperature dependencies of ring-chain equilibria in different solvents (see [74, 75]) information concerning thermodynamic parameters (ΔH and ΔS) of the equilibria is to be expected; this could be correlated with solvent parameters.

2) Ring-chain equilibria in the gaseous phase should be investigated. A comparison of these results with those obtained for solution would lead to a better understanding of solvent effects.

It is not yet clear whether experimental observations which have been attributed to structure are due to solvent effects and vice versa.

REFERENCES

1. V. I. Minkin, L. P. Olekhnovich, and Yu. A. Zhdanov, *Molecular Design of Tautomeric Systems*, Rostov on Don University Publishing House, Rostov on Don, 1977 (in Russian); V. I. Minkin, L. P. Olekhnovich, and Yu. A. Zhdanov, *Acc. Chem. Res.* **1981**, *14*, 210.

2. R. E. Valters, *Uspekhi Khimii* **1982**, *51*, 1374.

3. W. P. Jencks, *Catalysis in Chemistry and Enzymology*, McGraw-Hill Book Company, New York, 1969, Chapter 5.

4. M. I. Page, *Chem. Soc. Rev.* **1973**, *2*, 295.

5. W. P. Jencks, In: *Progress in Physical Organic Chemistry*, Ed. S. G. Cohen, A. Streitwieser, Jr., and R. W. Taft, Interscience Publishers, New York, 1964, Vol. 2, p. 63.
6. Ch. D. Hurd and W. H. Saunders, Jr., *J. Am. Chem. Soc.* **1952**, *74*, 5324.
7. A. A. Potekhin and S. L. Zhdanov, *Khim. Geterotsikl. Soedin.* **1979**, 1317.
8. J. E. Whitting and J. T. Edward, *Can. J. Chem.* **1971**, *49*, 3799.
9. K. Bowden and G. R. Taylor, *J. Chem. Soc., Sect. B* **1971**, 1390.
10. R. E. Valters and V. P. Ciekure, *Khim. Geterotsikl. Soedin.* **1975**, 1476.
11. K. Bowden and A. M. Last, *J. Chem. Soc., Perkin Trans. 2* **1973**, 1144.
12. R. E. Valters and V. R. Zin'kovska, *Khim. Geterotsikl. Soedin.* **1973**, 1127.
13. R. E. Valters, V. R. Zin'kovska, A. V. Burkevica, and S. P. Valtere, *Latv. PSR Zinat. Akad. Vestis, Kim. Ser.* **1974**, 118.
14. M. V. Bhatt and K. M. Kamath, *J. Chem. Soc., Sect. B* **1968**, 1036.
15. K. Bowden and M. P. Henry, *J. Chem. Soc., Perkin Trans 2* **1972**, 201.
16. R. Escale, R. Jaquier, B. Ly. F. Petrus, and J. Verducci, *Tetrahedron* **1976**, *32*, 1369.
17. J. V. Paukstelis and L. L. Lambing, *Tetrahedron Lett.* **1970**, 299.
18. A. F. McDonagh and H. E. Smith, *J. Org. Chem.* **1968**, *33*, 1.
19. K. Bowden and M. P. Henry, *J. Chem. Soc., Perkin Trans 2*, **1972**, 206.
20. K. Bowden and G. R. Taylor, *J. Chem. Soc., Sect. B* **1971**, 1395.
21. S. Milstien and L. A. Cohen, *Proc. Nat. Acad. Sci. USA* **1970**, *67*, 1143.
22. S. Milstien and L. A. Cohen, *J. Am. Chem. Soc.* **1972**, *94*, 9158.
23. R. T. Borchardt and L. A. Cohen, *J. Am. Chem. Soc.* **1972**, *94*, 9166.
24. R. T. Borchardt and L. A. Cohen, *J. Am. Chem. Soc.* **1972**, *94*, 9176.
25. J. M. Karle and I. L. Karle, *J. Am. Chem. Soc.* **1972**, *94*, 9182.
26. A. Dafforn and D. E. Koshland, Jr., *Proc. Nat. Acad. Sci. USA* **1971**, *68*, 2463.
27. D. R. Storm and D. E. Koshland, Jr., *J. Am. Chem. Soc.* **1972**, *94*, 5805.
28. D. R. Storm and D. E. Koshland, Jr., *J. Am. Chem. Soc.* **1972**, *94*, 5815.
29. C. Danforth, A. W. Nicholson, J. C. James, and G. M. Loudon, *J. Am. Chem. Soc.* **1976**, *98*, 4275.
30. R. E. Winans and Ch. F. Wilcox, Jr., *J. Am. Chem. Soc.* **1976**, *98*, 4281.
31. M. Caswell and G. L. Schmir, *J. Am. Chem. Soc.* **1980**, *102*, 4815; *see also* De Los F. De Tar, *J. Am. Chem. Soc.* **1982**, *104*, 7205.
32. B. Capon, *J. Chem. Soc., Sect. B* **1971**, 1207; Th. C. Bruice, A. Brown, and D. O. Harris, *Proc. Nat. Acad. Sci. USA* **1971**, *68*, 658; M. I. Page and W. P. Jencks, *Proc. Nat. Acad. Sci. USA* **1971**, *68*, 1678.
33. F. M. Menger and L. E. Glass, *J. Am. Chem. Soc.* **1980**, *102*, 5404.
34. F. M. Menger, *Tetrahedron* **1983**, *39*, 1013.
35. D. J. Chadwick and J. D. Dunitz, *J. Chem. Soc., Perkin Trans 2* **1979**, 276.
36. H. B. Bürgi, J. D. Dunitz, and E. Shefter, *J. Am. Chem. Soc.* **1973**, *95*, 5065.
37. H. B. Bürgi, J. D. Dunitz, J. M. Lehn, and G. Wipf, *Tetrahedron* **1974**, *30*, 1563; H. B. Bürgi, *Angew. Chem.* **1975**, *87*, 461.
38. H. B. Bürgi and J. D. Dunitz, *Acc. Chem. Res.* **1983**, *16*, 153.
39. J. E. Baldwin, *J. Chem. Soc., Chem. Commun.* **1976**, 734; J. E. Baldwin and M. J. Lusch, *Tetrahedron* **1982**, *38*, 2939.
40. J. B. Lambert and M. V. Majchrzak, *J. Am. Chem. Soc.* **1980**, *102*, 3588.
41. R. E. Valters, *Ring-Chain Isomerism of Keto, Imino, and Cyano Carboxamides*, Synopsis of Thesis of Doctoral Dissertation, Riga, 1975 (in Russian).
42. N. L. Allinger and V. Zalkov, *J. Org. Chem.* **1960**, *25*, 701.
43. A. I. Kirby and P. W. Lancaster, *J. Chem. Soc., Perkin Trans. 2*, **1972**, 1206.
44. F. H. Westheimer, In: *Steric Effects in Organic Chemistry*, Ed. M. S. Newman, John Wiley and Sons, Inc., New York, 1956, Chapter 12.
45. R. P. Bell, D. W. Earls, and J. B. Henshall, *J. Chem. Soc., Perkin Trans 2*, **1976**, 39.

46. M. S. Newman and C. Courduvelis, *J. Org. Chem.* **1965**, *30*, 1795.
47. L. P. Olekhnovich, N. I. Borisenko, Z. N. Budarina, V. P. Metlushenko, Yu. A. Zhdanov, and V. I. Minkin, *Zh. Org. Khim.* **1982**, *18*, 1785.
48. B. Capon and S. T. McManus, *Neighboring Group Participation*, Plenum Press, New York, 1976, Vol. 1.
49. I. B. Blagoeva, B. I. Kurtev, and I. G. Pojarlieff, *J. Chem. Soc., Perkin Trans.* 2, **1979**, 1115.
50. T. Higuchi, L. Eberson, and A. K. Herd, *J. Am. Chem. Soc.* **1966**, *88*, 3805.
51. L. Eberson and H. Welinder, *J. Am. Chem. Soc.* **1971**, *93*, 5821.
52. Th. C. Bruice and W. C. Bradbury, *J. Am. Chem. Soc.* **1965**, *87*, 4846.
53. O. H. Wheeler and E. J. De Rodriguez, *J. Org. Chem.* **1964**, *29*, 1227.
54. B. I. Istomin and V. A. Palm, *Reakts. Spos. Org. Soedin.* **1972**, *9*, 433.
55. E. M. Engler, J. D. Andose, and P. v. R. Schleyer, *J. Am. Chem. Soc.* **1973**, *95*, 8005.
56. N. L. Allinger, M. T. Tribble, M. A. Miller, and D. H. Wertz, *J. Am. Chem. Soc.* **1971**, *93*, 1637.
57. R. H. Boyd, *J. Chem. Phys.* **1968**, *49*, 2574.
58. De Los F. De Tar, *J. Am. Chem. Soc.* **1974**, *96*, 1254.
59. De Los F. De Tar and C. J. Tenpas, *J. Org. Chem.* **1976**, *41*, 2009.
60. De Los F. De Tar, D. F. McMullen, and N. P. Luthra, *J. Am. Chem. Soc.* **1978**, *100*, 2484.
61. De Los F. De Tar, *J. Am. Chem. Soc.* **1974**, *96*, 1255.
62. De Los F. De Tar and N. P. Luthra, *J. Am. Chem. Soc.* **1980**, *102*, 4505.
63. De Los F. De Tar and W. Brooks, Jr., *J. Org. Chem.* **1978**, *43*, 2245.
64. M. Tichy, In: *Advances in Organic Chemistry, Methods and Results*, Ed. R. A. Raphael, E. C. Taylor, and H. Wynberg, Interscience Publishers, New York, 1965, Vol. 5, p. 115.
65. W. Schäfer and H. Schlude, *Tetrahedron* **1971**, *27*, 4721.
66. D. F. Martin, M. Shamma, and W. C. Fernelius, *J. Am. Chem. Soc.* **1958**, *80*, 5851.
67. S. N. Ananchenko, I. V. Berezin, and I. V. Torgov, *Izv. Akad. Nauk SSSR, Otd. Khim. Nauk* **1960**, 1644.
68. G. V. Kondrat'eva, G. A. Kogan, and S. I. Zav'yalov, *Izv. Akad. Nauk SSSR, Otd. Khim. Nauk* **1962**, 1441.
69. G. Bianchi, A. Gamba-Invernizzi, and R. Gandolfi, *J. Chem. Soc., Perkin Trans.* 1, **1974**, 1757.
70. A. F. McDonagh and H. E. Smith, *J. Org. Chem.* **1968**, *33*, 1.
71. Y. Miyagi, S. Kimura, and R. Goto, *Bull. Chem. Soc. Jpn.* **1968**, *41*, 2927.
72. R. E. Valters, S. P. Valtere, and A. E. Kipina, *Zh. Org. Khim.* **1968**, *4*, 445.
73. R. E. Valters and S. P. Valtere, In: *Biological Active Compounds*, Nauka, Leningrad, 1968, p. 218 (in Russian).
74. J. V. Paukstelis and R. M. Hammaker, *Tetrahedron Lett.* **1968**, 3557.
75. G. A. Dafforn and D. E. Koshland, Jr., *J. Am. Chem. Soc.* **1977**, *99*, 7246.

Index